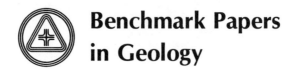

Benchmark Papers
in Geology

Series Editor: Rhodes W. Fairbridge
Columbia University

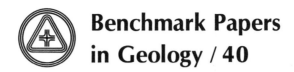

**Benchmark Papers
in Geology / 40**

A BENCHMARK® Books Series

DIAGENESIS OF DEEP-SEA
BIOGENIC SEDIMENTS

Edited by

Gerrit J. van der Lingen
Sedimentation Laboratory,
New Zealand Geological Survey

**Dowden, Hutchinson
& Ross, Inc.**

STROUDSBURG, PENNSYLVANIA

Copyright © 1977 by **Dowden, Hutchinson & Ross, Inc.**
Benchmark Papers in Geology, Volume 40
Library of Congress Catalog Card Number: 77-7496
ISBN: 0-87933-283-2

79   78   77   1   2   3   4   5
Manufactured in the United States of America.

LIBRARY OF CONGRESS CATALOGING IN PUBLICATION DATA
Main entry under title:
Diagenesis of deep-sea biogenic sediments.
   (Benchmark papers in geology ; 40)
   Includes index.
   CONTENTS: Carbonate diagenesis: Bathurst, R. G. C.
Problems of lithification in carbonate muds. Schlanger,
S. O. and Douglas, R. G. The pelagic ooze-chalk-limestone
transition and its implications for marine stratigraphy.
[etc.]
   1. Marine sediments—Addresses, essays, lectures.
2. Diagenesis—Addresses, essays, lectures.
I. Van der Lingen, Gerrit J.
GC380.15.D5   551.4′6083   77-7496
ISBN 0-87933-283-2

Exclusive Distributor: **Halsted Press**
A Division of John Wiley & Sons, Inc.
ISBN: 0-470-99229-8

SERIES EDITOR'S FOREWORD

The philosophy behind the "Benchmark Papers in Geology" is one of collection, sifting, and rediffusion. Scientific literature today is so vast, so dispersed, and, in the case of old papers, so inaccessible for readers not in the immediate neighborhood of major libraries that much valuable information has been ignored by default. It has become just so difficult, or so time consuming, to search out the key papers in any basic area of research that one can hardly blame a busy man for skimping on some of his "homework."

This series of volumes has been devised, therefore, to make a practical contribution to this critical problem. The geologist, perhaps even more than any other scientist, often suffers from twin difficulties—isolation from central library resources and immensely diffused sources of material. New colleges and industrial libraries simply cannot afford to purchase complete runs of all the world's earth science literature. Specialists simply cannot locate reprints or copies of all their principal reference materials. So it is that we are now making a concerted effort to gather into single volumes the critical material needed to reconstruct the background of any and every major topic of our discipline.

We are interpreting "geology" in its broadest sense: the fundamental science of the planet Earth, its materials, its history, and its dynamics. Because of training and experience in "earthy" materials, we also take in astrogeology, the corresponding aspect of the planetary sciences. Besides the classical core disciplines such as mineralogy, petrology, structure, geomorphology, paleontology, and stratigraphy, we embrace the new fields of geophysics and geochemistry, applied also to oceanography, geochronology, and paleoecology. We recognize the work of the mining geologists, the petroleum geologists, the hydrologists, the engineering and environmental geologists. Each specialist needs his working library. We are endeavoring to make his task a little easier.

Each volume in the series contains in Introduction prepared by a specialist (the volume editor)—a "state of the art" opening or a summary of the object and content of the volume. The articles, usually some thirty to fifty reproduced either in their entirety or in significant extracts, are selected in an attempt to cover the field, from the key papers of the last century to fairly recent work. Where the original works are in foreign languages, we have endeavored to locate or commission translations. Geologists, because of their global subject, are often acutely aware of the oneness of our world. The selections cannot, therefore, be restricted to any one country, and whenever possible an attempt is made to scan the world literature.

To each article, or group of kindred articles, some sort of "highlight commentary" is usually supplied by the volume editor. This commentary should serve to bring that article into historical perspective and to emphasize its particular role in the growth of the field. References, or citations, wherever possible, will be reproduced in their entirety—for by this means the observant reader can assess the background material available to that particular author, or, if he wishes, he, too, can double check the earlier sources.

A "benchmark," in surveyor's terminology, is an established point on the ground, recorded on our maps. It is usually anything that is a vantage point, from a modest hill to a mountain peak. From the historical viewpoint, these benchmarks are the bricks of our scientific edifice.

RHODES W. FAIRBRIDGE

PREFACE

Diagenesis is such a vast subject that many Benchmark volumes would be necessary to cover all aspects. This volume restricts itself to the diagenesis of deep-sea biogenic sediments, which can be divided conveniently into two main fields: carbonate diagenesis and silica diagenesis. The former concerns itself with the transition ooze-chalk-limestone, the latter with the transition siliceous sediment-porcelanite-chert.

Before the start of the Deep Sea Drilling Project in 1968, deep-sea sediments could be studied directly only from samples collected from the uppermost few meters of the sediment column. Sampling devices like piston corers could not penetrate much deeper than about 30 m. Lithified biogenic sediments were hardly ever sampled. In the few exceptional cases where these were encountered, lithification was probably due to exceptional conditions. It was therefore not possible to study "normal" progressive diagenetic changes. Direct study of diagenetically altered biogenic sediments was possible only from rocks exposed on land, for which a deep-sea origin was inferred. But again, it was not possible to observe progressive changes.

The Deep Sea Drilling Project has changed all that. Complete sediment columns, overlying igneous oceanic basement, have now been drilled through and have been sampled in detail. For the first time in history, progressive diagenesis could be studied.

By fortunate coincidence, the right instruments for studying deep-sea sediments became available only a few years before the start of deep-sea drilling. The most important of these were the scanning electron microscope and associated analytical instruments. Diagenetic processes like dissolution, reprecipitation (neoformation), and pseudomorphism could now be studied in minute detail.

Deep-sea drilling and the new laboratory instruments made the study of deep-sea biogenic sediments an almost totally new field of research in the earth sciences. To do these new developments justice, it was necessary to devote all the available space in this Benchmark volume to the literature of the last ten years. Only by doing this could the reader be given a sufficiently complete and up-to-date picture of the "state of the art." Fortunately, many of the papers selected refer back to publications before the deep-sea drilling era, thus providing adequate historical background and continuity. It is mainly in the fields of nomenclature and mineralogy that earlier papers are relevant to present-day studies. Moreover, other Benchmark volumes have treated, and will treat, aspects of diagenesis we cannot cover in this volume. For instance, Carozzi (1975), in his Benchmark volume (No. 15), has included four historic papers on chert formation.

But even so, the modern literature from which a selection had to be made is already quite voluminous. Over a hundred papers were considered for inclusion. The final choice of twenty was made to cover as wide a field as possible, including related topics such as stable isotope composition, pore liquid geochemistry, acoustistratigraphy, heatflow, the silica cycle in the oceans, paleoclimatology, thermodynamics, and experimental studies.

Evaluation of the influence of diagenesis on the characteristics of oil-reservoir rocks is of great economic importance. The final paper in this volume has therefore been chosen to illustrate the application of modern diagenesis studies of biogenic pelagic sediments to an actual situation, viz. the Danian chalk of the Ekofisk field in the North Sea.

The volume editor is grateful to Professor Rhodes W. Fairbridge, the series editor, for his encouragement and constructive suggestions during the preparation of this volume. Dr. Malcolm G. Laird, New Zealand Geological Survey, Christchurch, kindly read the new manuscript parts and made many suggestions for their improvement. Logistic support was provided by Miss Christina Johnstone (typing) and Mr. Ernie Annear (drafting), both from the New Zealand Geological Survey, Christchurch, and Ms. Lee Leonard (drafting) and Mr. Albert Downing (photography), both of the Geology Department, University of Canterbury, Christchurch.

This book could not have been compiled without the generous assistance of the authors of the papers, who not only gave their permission for reproduction and provided the necessary reprints but wherever possible also made available original negatives or prints of their illustrations.

GERRIT J. van der LINGEN

CONTENTS

Contents

CONTENTS BY AUTHOR

INTRODUCTION

The term *diagenesis,* introduced in 1868 by von Gümbel, pertains to the lithification of sedimentary rocks after deposition. Though diagenesis has been a common concept in geology textbooks ever since, substantial progress in understanding it has been made only in recent decades. Like most other terms in geology, its precise meaning has "fluctuated," suffering from many national and personal "adaptations." An excellent review of the history of the diagenesis concept can be found in a paper by Dunoyer de Segonzac (1968). In this paper the author traces the development of the concept in the German, Russian, French, and Anglo-Saxon literature, respectively.

The Russian concept of diagenesis differs most from the others in that it is restricted to what others would call "early diagenesis." The opinions of workers in the other language areas differ mainly in the definitions of the limits of the diagenetic part of the rock cycle. On the one side the limit is the boundary with halmyrolysis ["Geochemical modification of sediments during deposition, due to reactions with sea water (ionic transfer), originally called 'submarine weathering' by Hummel (1922), but applies also to ionic rearrangement and replacement (Pettijohn 1957)" (Fairbridge 1967, p. 88)], and on the other side the limit is the boundary with the metamorphic realm.

It is outside the scope of this volume to discuss these terminology problems in detail. This book does not concern itself with the concept of diagenesis in general but restricts itself to the diagenesis of a specific sediment type. Thus the volume editor does not feel the need to add yet another definition to the multitude already in existence. For the readers of this book the modern definition by Fairbridge (1967, p. 88) suffices:

1

"Physical and chemical changes which the sediment undergoes after deposition and during lithification, without introduction of heat (over ca. 300°C) or great pressure (ca. 1,000 bars)." Fairbridge's definition adequately covers the meaning of diagenesis as implied by the authors of the reprints in this volume.

As stated in the Preface, the study of the diagenesis of deep-sea biogenic sediments is an almost totally new field of research. The Deep Sea Drilling Project (DSDP) began in 1968 and ended its first three phases in 1975 after drilling at 391 sites in all the oceans and major seas of the world. The DSDP made possible for the first time in history to sample complete sequences of deep-sea sediments in situ. At the same time new laboratory instruments made it possible to study the samples at the required level of detail. The scanning electron microscope (SEM) in particular and its associated analytical instruments such as the energy dispersive X-ray microanalyzer (EDAX) and the microprobe have been extremely important.

Continuous sampling is necessary to enable the study of progressive diagenetic changes, but in the early stages of the DSDP, two phenomena hampered the continuous sampling of sedimentary sequences overlying oceanic igneous basement. The first phenomenon was encountering extensive chert horizons of Eocene age and older (see Figure 1 and Table 1). These horizons formed almost impenetrable barriers, rapidly wearing out drill bits. Only after the introduction of improved drill bits did it become possible to penetrate thick chert sequences. The second phenomenon was the tendency to try to reach the second layer (igneous basement) as quickly as possible to verify the sea-floor spreading hypothesis. Because coring is time-consuming, it was kept to a minimum (so-called 'spot-coring'). One can understand this eagerness to reach the second layer. After all, the sea-floor spreading hypothesis was one of the stimulating forces behind the DSDP. But much useful information was thus not obtained. Moore (1972) was one of the scientists to express concern about this approach. He wrote, "Chief scientists who strive only for the superficial facts that make for good press releases have often flushed from the drill hole the material that holds the geological record of the oceans" (p. 30). Later in the project when the urge to prove sea-floor spreading had been satisfied, attention returned to a more integrated study of the "geological record of the oceans." Starting with Leg 17 in the Pacific, relatively complete sedimentary sequences were sampled from which progressive diagenesis could be studied (Schlanger et al. 1973; Matter 1974; Papers 2, 5, and 7).

The papers in this volume have been selected to give, within the space available, a representative account of the diagenesis studies based on data from the DSDP, and of associated research. The study of

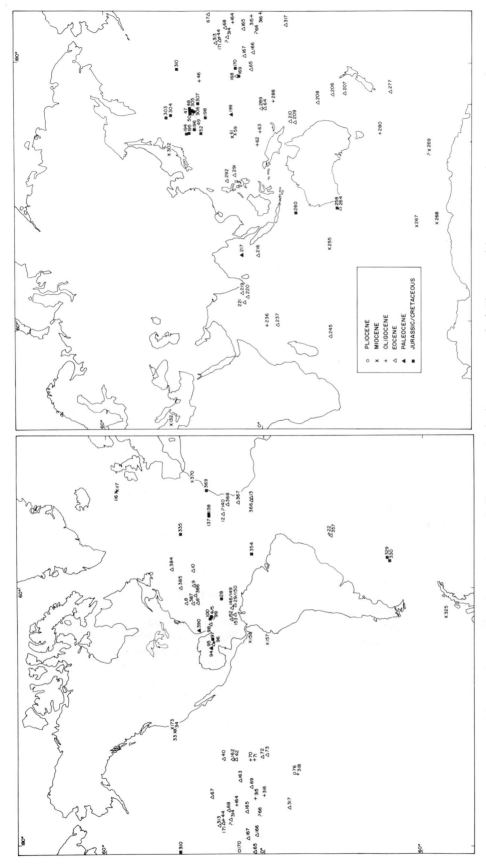

FIG. 1   World map, showing localities of DSDP drill sites at which chert was sampled. The age of the youngest chert encountered is indicated (see also Table 1). Base map courtesy of the Deep Sea Drilling Project.

3

Table 1 List of DSDP drill sites at which chert was sampled, together with the ages of the youngest chert encountered (see also Figure 1). Compiled by the volume editor.

Leg 1 Gulf of Mexico and Western Atlantic
Site 4 Early Cretaceous
      5 Late Cretaceous
      6 Middle Eocene

Leg 2 North Atlantic
Site 8 Eocene
      9 Eocene
     10 Early Eocene
     12 Eocene

Leg 3 South Atlantic
Site 13 Middle Eocene
     22 Upper Oligocene

Leg 4 Central Atlantic
Site 28 ?Late Cretaceous
     29 Middle Eocene

Leg 5 Northeast Pacific
Site 33 Middle Miocene
     34 Middle Miocene
     40 Early Eocene
     42 Middle Eocene

Leg 6 Central Pacific
Site 44 Early Oligocene
     46 Early Oligocene
     47 Late Cretaceous
     48 Late Cretaceous
     49 Plio-Pleistocene
     50 Jurassic-Early Cretaceous
     52 Cretaceous
     59 Early Miocene

Leg 7 Central Pacific
Site 61 Early Miocene
     62 Late Oligocene
     63 Early Oligocene
     64 Eocene
     65 Late Eocene
     66 Age unknown
     67 Paleo-Eocene

Leg 8 Central Pacific
Site 68 Middle Eocene
     69 Middle Eocene
     70 Late Oligocene
     71 Late Oligocene
     72 Late Eocene
     73 Late Eocene

Leg 9 East Pacific
Site 76 Pliocene

Leg 10 Caribbean
Site 94 Late Paleocene
     95 Middle-Early Eocene
     96 Early Oligocene
     97 Late Cretaceous

Leg 11 West Atlantic
Site 98 Early-Middle Eocene
     99 Pliocene
    100 Late Jurassic–Early Cretaceous

Leg 12 North Atlantic
Site 116 Early Miocene
     117 Oligo-Miocene

Leg 13 Mediterranean
Site 132 Late Miocene

Leg 14 Central Atlantic
Site 135 Cretaceous
     137 Late Cretaceous
     138 Late Cretaceous
     140 Age unknown

Leg 15 Caribbean
Site 146 Early Eocene
     149 Early Eocene
     150 Early Eocene
     152 Paleocene–Early Eocene
     153 Early Eocene

Leg 16 Eastern Central Pacific
Site 157 Late Miocene
     158 Middle Miocene
     162 Middle Eocene
     163 Late Eocene

continued

4

Table 1, *continued*

Leg 17 Central Pacific	*Leg 27 East Indian Ocean*
Site 164 Early Oligocene	Site 260 Early Cretaceous
165 Middle Eocene	
166 Late Eocene	*Leg 28 Between Australia and Antarctica*
167 Middle Eocene	Site 264 Middle Eocene
168 Late Eocene	267 Miocene
169 Late Cretaceous	268 Miocene
170 Late Cretaceous	269 ?Miocene
171 Middle Eocene	274 Age unknown
Leg 18 Northeast Pacific	*Leg 29 Subantarctic (South Tasman Sea)*
Site 173 Early Miocene	Site 277 Late Eocene
	280 Late Eocene–Early Oligocene
Leg 20 Western Central Pacific	
Site 194 Cretaceous	*Leg 30 Southwest Pacific*
195 Late Cretaceous	Site 288 Late Oligocene
196 Early Cretaceous	289 Middle Eocene
198 Late Cretaceous	
199 Late Paleocene	*Leg 31 West Pacific–Japan Sea*
	Site 291 Late Eocene
Leg 21 Southwest Pacific	292 Late Eocene
Site 206 Late Eocene	302 Late Miocene
207 Middle Eocene	
208 Middle Eocene	*Leg 32 North Pacific*
209 Late Eocene	Site 303 Cretaceous
210 Late Eocene	304 Early Cretaceous
	305 Late Cretaceous
	306 Early Cretaceous
Leg 22 East Indian Ocean	307 Cretaceous
Site 216 Middle Eocene	310 Late Cretaceous
217 Paleocene	313 Late Eocene
Leg 23 Northwest Indian Ocean	*Leg 33 Central Pacific*
and Red Sea	Site 314 ?Eocene
Site 219 Middle Eocene	315 Late Oligocene
220 Early Eocene	316 Early Oligocene
221 Middle Eocene	317 Late Eocene
	318 Late Oligocene
Leg 24 West Indian Ocean	
Site 236 Early Oligocene	*Leg 35 Antarctic, near South America*
237 Middle Eocene	Site 325 Early-Middle Miocene
Leg 25 West Indian Ocean	*Leg 36 Antarctic, near South America*
Site 245 Early Eocene	Site 327 Late Cretaceous
	330 Early Cretaceous
Leg 26 East Indian Ocean	*Leg 39 Southwest Atlantic*
Site 255 Miocene	Site 354 Late Cretaceous
258 Late Cretaceous	357 Middle-Late Eocene

continued

Table 1, *continued*

Leg 41 Eastern North Atlantic	*Leg 43 North Atlantic*
Site 366 Middle Eocene	Site 384 Early Eocene
367 Eocene	385 Middle Eocene
368 Early Eocene	386 Middle Eocene
369 Late Cretaceous	387 Middle Eocene
370 Early Miocene	
	Leg 44 Western North Atlantic
	Site 390 Paleocene

diagenesis can be divided into several subtopics. In Table 2 we have plotted a list of these subtopics against the papers reproduced in this volume. The table will enable the reader to quickly find his way through the reprinted material.

Deep-sea biogenic sediments consist mainly of microfossils, forming carbonate oozes or siliceous oozes, or mixtures of both. Their present-day distribution is a function of ocean currents, climatic conditions, and water depths (Lisitzin 1971). Their distribution in the past, as revealed by deep-sea drilling, is explained by invoking similar parameters (actualistic principle), and by the inferred evolution of crustal plates (Heezen et al. 1973; Papers 6, 10, 11, 12, 17, and 18).

The carbonate and siliceous components of deep-sea sediments have different diagenetic responses and thus are generally studied and described separately. For ease of discussion, mixtures of the two components are also studied and described separately. It seems likely, however, that in mixed sediments, the diagenesis of the one can influence that of the other (Paper 5).

Most deep-sea carbonate sediments differ from shallow-water carbonate rocks in two major aspects: grain size and chemistry. Foraminifera and nannofossils are the main constituents, which puts the sediment in the textural silt to clay class (generally called *ooze*). The calcite of these organisms is of the low-magnesium variety.

The first step in the lithification of a carbonate ooze is its transformation to chalk. Fundamentally, deep-sea chalk does not differ from epicontinental pelagic chalk. One of the best-known epicontinental chalks is the Upper Cretaceous to Lower Tertiary chalk of Northwest Europe. Historic problems related to epicontinental chalk have been revived in studies of deep-sea chalk. One is the origin of chert-nodules (flints) in chalk (a historic review is given by Shepherd 1972); another is the lithification without apparent compaction under overburden (Papers 1 and 14). The origin of chert in chalk is treated in the papers on silica diagenesis (Papers 4 to 13). The lithification of chalk

Table 2

Legend:
- ● Major subject of paper
- ○ Secondary subject of paper
- X Subject mentioned in paper

Paper	Carbonate diagenesis	Silica diagenesis	Silica phases	Carbonate diagenesis – experimental	Silica diagenesis – experimental	Silica cycle in the oceans	Silica source (volcanic vs biogenic)	Geochemistry (incl. interstitial liquids)	Stable isotopes	Diagenetic potential	Acoustistratigraphy	Hardgrounds	Fecal pellets	Heatflow	Compaction	Thermodynamics	X-ray diffraction	EDAX	Other instrument analyses	Horizon A	Chalk of Northwest Europe	Oil–reservoir studies	Paleo – climatology, – oceanography
1. Bathurst	●											X			○								
2. Schlanger & Douglas	●																						
3. Adelseck et al.				●						○	○		X										
4. Wise & Hsü	●	○	○						X							X							
5. van der Lingen and Packham	●	●									X												X
6. Heath and Moberley		●	X			X	X							X		X	X						
7. Lancelot		●	○				X							X			X						
8. Froehlich		●	○														X		X		X		
9. Flörke et al.		X	●		X												X						
10. Gibson and Towe							●																X
11. Weaver and Wise		●			●															○			
12. Gibson and Towe/Weaver and Wise																							
13. Matter et al.	●							○	○	X			X			X							
14. Neugebauer	●							○							○	X							
15. Knauth and Epstein		○	X						●														
16. Kastner et al.			X			X	X	○		X			X	X			X	X					X
17. Heath		X				●																	X
18. Leclaire		○				●	X																
19. Schrader						○							●										
20. Dunn	○																				X	●	

7

without compaction has recently been explained as a consequence of the geochemistry of chalk, more specifically its low magnesium character (Paper 14). Apparently, a rigid framework is established early in diagenesis by spot-welding adjacent grains (Neugebauer 1973; Mapstone 1975; Papers 1 and 14).

In the early stages of the DSDP, scientists were mainly establishing an inventory of the diagenetic processes, as observed with the SEM. Everything was new so that a certain amount of stocktaking had to precede the formulation of any theories and hypotheses. From those early observations emerged a generalized picture of progressive carbonate diagenesis (Schlanger et al. 1973; Packham and van der Lingen 1973; Matter 1974; Paper 5).

It should be realized that some dissolution and disintegration of fossil tests can take place even before burial. This process can start in the viscera of plankton-feeding animals and during settling through the water column (Lisitzin 1971, Figure 127). It is important to realize that much deep-sea biogenic sediment settles in the form of fecal pellets (for references, see Table 2). The amount of dissolution during settling is a function of water depth and the position of the lysocline and carbonate compensation depth. Further dissolution and break-up can occur at or near the sediment-water interface, in which infaunal activity no doubt plays an important role.

Some precipitation of secondary calcite must start before or shortly after burial, especially on discoasters. Hardly ever have discoasters without some calcite-overgrowth been observed, even in surface sediments (Burns 1972). However, diagenesis proper does not start until after the initial expulsion of water and the establishment of a grain-supporting framework (Packham and van der Lingen 1973). Lithification of the carbonate sediment during diagenesis takes place by a process of selective dissolution and reprecipitation (overgrowth) of calcite. This process is described and illustrated in Papers 2, 5, and 13, and a short summary is given on page 112 of Paper 5. The selective dissolution and overgrowth of calcareous nannofossils has been reproduced experimentally (Paper 3). Neugebauer (1975) has described the diagenetic alteration of foraminifera in detail.

Attempts have been made to explain the diagenetic processes in carbonate sediments theoretically (Paper 14). Important parameters seem to be the size and crystallographic morphology of the fossil particles. Crystal size (free energy) determines the susceptibility to dissolution, while the morphology determines the amount of deposition of syntaxial cement.

From the preceding discussion, it is clear that sediments with different compositions will react differently, even under similar diagenetic conditions. In other words, they have different *diagenetic potentials* (Schlanger and Douglas, Paper 2), which means that some sediments can lithify faster than others. Harder layers may be underlain by softer ones (Wise and Kelts 1972; Paper 4), which in a progressive diagenetic context, may give the impression of diagenetic reversals.

Differences in diagenetic potential could well reflect changes in paleo-oceanographic conditions and consequently would show their effect over wide areas. As differences in lithification can show up as seismic reflectors, it should be possible to correlate some reflectors over large areas, resulting in a so-called *acoustistratigraphy* (Paper 2).

Assuming that no calcite is introduced from outside the system, the decrease in volume during the transformation from ooze to chalk to limestone can be calculated (Papers 2 and 5). An ooze (near the sediment-water interface) has porosities between 70 and 80 percent, a chalk between 40 and 65, while a limestone has porosities between 25 and 40 percent. This corresponds with a decrease in volume by about two thirds. However, limestone densities and porosities are often strongly influenced by silica diagenesis (Paper 5; see also Wise and Weaver 1974).

Historically, silica diagenesis has received a disproportionate amount of attention. This attention is not surprising, given the fact that silica in either bedded or nodular chert provides little information on its origins, thus lending itself to much theorizing (Shepherd 1972). In contrast, carbonate sediments tend to retain aspects of their original character during most of their diagenetic history. It is only in the metamorphic realm that they lose most of their identity. Silica diagenesis has also received a lion's share of attention in modern studies of deep-sea sediments.

One of the major discoveries the DSDP made was that deep-sea sediments contain extensive chert horizons, e.g., the so-called *Horizon A* in the Atlantic Ocean and *Horizon A'* in the Caribbean Sea (Ewing et al. 1970; Papers 10 to 12). Surprisingly, chert is especially abundant in Eocene and older sediments. It is far less abundant in Oligocene and Miocene sediments and is almost absent in younger sediments. In Figure 1 we have plotted all the DSDP drilling sites that contain chert. For each site the age of the youngest chert encountered has been indicated (see also Table 1). The so-called maturation theory is based on such observations (Paper 6). This theory states that there is a progressive change with time and depth of burial from biogenic opal to "disordered" cristobalite to quartz.

Broadly speaking, three major types of silicification are observed in deep-sea sediments: silicified carbonate rock ("porcelanite"), silicified noncarbonate rock (often "bedded cherts"), and chert nodules (in carbonate rocks). Chert nodules in turn can consist of cristobalite or quartz (Keene 1975). The immediate source of diagenetic silica has been the subject of much controversy.

Two sources, biogenic and volcanic, have been proposed (Calvert 1974; Papers 10 to 12). Most scientists who have studied the samples from deep-sea drilling now seem to accept a biogenic source for most of the diagenetic silica in the oceanic environment. This is not to say, of course, that in other environments diagenetic silica may not have been derived from altered volcanic ash (Henderson et al. 1971).

The first step in silica diagenesis is the dissolution of biogenic silica (biogenic opal, or opal-A, Jones and Segnit 1971). This silica is subsequently reprecipitated as a poorly-crystalline silica phase. Oehler (1975, 1976) is of the opinion that this transformation takes place via a colloidal phase, but this theory is not generally accepted (Weaver and Wise 1972). It is an old idea that chert nodules in chalk solidified from colloidal gel lumps (Shepherd 1972).

With sufficient space to grow freely, the poorly-crystalline silica crystallizes as tiny spherules, only a few microns in diameter. The spherules are made up of thin blades; hence the name *lepispheres* ("spheres of blades," Wise and Kelts 1972). Their detailed morphology can only be studied with an SEM. The first SEM photographs of lepispheres were published in 1971 (see Editor's Comments on Paper 4). Lepispheres have been given a variety of mineralogical names: "lussatite," "α-cristobalite," "disordered cristobalite," "poorly-crystalline tridymite," "opal-CT," etc. (Paper 5). Their exact mineralogy has been the subject of much discussion (Oehler 1973; Buurman and van der Plas 1971; Wilson et al. 1974; Jones and Segnit 1975; Klasik 1975; Paper 9). At present the name "opal-CT" seems to be generally accepted for the mineralogy and "lepisphere" for the morphology.

According to the maturation theory, opal-CT is eventually transformed into quartz, given enough time and/or burial depth. Whether this is a solid-solid or a solution-reprecipitation process is not known as yet (Paper 15). Diagenetic quartz can occur in various forms, as microcrystalline quartz, as chalcedony, or as so-called quartz whiskers (Paper 8).

However, Lancelot has proposed an alternative hypothesis (Paper 7) and states that the particular diagenetic silica phase being formed depends on the composition and porosity of the host lithology rather than on time and depth of burial. Opal-CT forms in noncarbonate, low-porosity sediments while quartz precipitates directly from solution in porous carbonate sediments.

The two theories have provoked much discussion (von Rad and Rösch 1975; Calvert 1974; Greenwood 1973; Paper 16). Recent experimental evidence, however, seems to suggest that both theories contain elements of the truth (Paper 16).

In recent times geochemical and thermodynamic aspects of diagenesis have received more attention. Changes in the geochemistry of interstitial liquids have been compared with changes in the geochemistry of the sediments, which has provided much new information. One of the most promising lines of research, however, seems to be the study of stable isotope composition of both diagenetic carbonate and diagenetic silica phases (Papers 13 and 15).

For the time being it is hardly necessary to collect more material from the deep seas to further our understanding of diagenetic processes. The material collected from the 391 drill sites of the DSDP has only been studied "initially." A wealth of material is in cool storage waiting for its secrets to be unraveled.

THE DEEP SEA DRILLING PROJECT

For those readers who are not familiar with the organization of the DSDP and the technical capabilities of its drilling vessel *Glomar Challenger,* a short synopsis is given.

The DSDP is part of the ocean-sediment coring program of the National Science Foundation (NSF) of the United States. It has an advisory scientific planning body, the so-called Joint Oceanographic Institutions for Deep Earth Sampling (JOIDES). Original members of JOIDES were: Lamont-Doherty Geological Observatory of Columbia University, Rosenstiel School of Marine and Atmospheric Science of the University of Miami, Department of Oceanography of the University of Washington, Woods Hole Oceanographic Institution, and Scripps Institution of Oceanography of the University of California.

The University of California (Scripps) is the prime contractor. The principal subcontractor is Global Marine, Inc., who designed, built, owns, and operates the *Glomar Challenger.* "All qualified scientists from the academic community, government agencies, and industrial organizations, worldwide" can participate in the project. Any qualified scientist, worldwide, can request core samples for study.

Each cruise lasts for about two months and is assigned its own leg number. For each cruise a "shipboard scientific party" is invited. Their responsibilities include publishing an "Initial Reports" volume on the results of their cruise.

The drilling vessel *Glomar Challenger* is 122 m long. It can handle a drill string of up to 6860 m long. The ship uses a dynamic positioning system that can keep the ship within a 24-m radius above the drill hole, without using any anchors. Hydrophones in the ship's hull pick up the signals from a sonic beacon dropped onto the sea floor upon arrival at the drill site. The signals are fed into a computer that calculates the ship's position and automatically sends corrective orders to the four tunnel thrusters and the main propellers. Navigation is by satellite. Weather forecasts are assisted by weather-satellite photographs, transmitted to the ship by shore-based stations.

Each drill site has its own, consecutive number. If more than one hole is drilled at a site, the extra holes are designated with a letter following the site number.

Coring is done with a 9-m long core barrel, 75 mm in diameter, in which a plastic liner is inserted. After retrieval the plastic liner containing the core is cut into six 1.5-m sections. These sections are numbered 1 to 6 from top to bottom. The core-catcher sample is labeled "CC." In the sedimentology laboratory the core sections are cut in half lengthwise. One half is designated the "working half" and is used for description and sampling by the shipboard scientists. The other half is called the "archive half" and is photographed and stored for future reference. Each sample is given a unique number. For example, "207A–3–5–90–92" means "Site 207, hole A (second hole), core 3, core-section 5, interval 90–92 cm (measured from the top of the section).

The DSDP has recently started a new series of deep-sea drilling, called the "International Phase of Ocean Drilling" (IPOD). For this new phase other countries have joined the United States in funding the project. The JOIDES planning body has been enlarged by the membership of oceanographic institutes from these countries. The first leg of IPOD started in November 1975. For the first few years the *Glomar Challenger* will still be used. As with the DSDP, first accounts of the results of IPOD cruises are being published in *Geotimes*.

REFERENCES

Burns, D. A. (1972) Discoasters in Holocene sediments, southwest Pacific Ocean. *Marine Geology* **12:** 301–306.

Buurman, P., and van der Plas, L. (1971) The genesis of Belgian and Dutch flints and cherts. *Geol. Mijnbouw* **50:** 9–27.

Calvert, S. E. (1974) Deposition and diagenesis of silica in marine sediments. In Hsü, K. J., and Jenkyns, H. C. (eds.), *Pelagic sediments: On land and under the sea. Internat. Assoc. Sedimentologists' Spec. Pub. No. 1,* 273–299.

Dunoyer de Segonzac, G. (1968) The birth and development of the concept of diagenesis. *Earth-Sci. Rev.* **4:** 153–201.

Ewing, J.; Windisch, C.; and Ewing, M. (1970) Correlation of Horizon A with JOIDES bore-hole results. *Jour. Geophys. Research* **29:** 5645–5653.

Fairbridge, R. W. (1967) Phases of diagenesis and authigenesis. In Larsen, G., and Chilingar, G. V. (eds.), Diagenesis in sediments. Developments in Sedimentology, **8**: 19–89.

Greenwood, R. (1973) Cristobalite: Its relationship to chert formation in selected samples from the Deep Sea Drilling Project. *Jour. Sed. Petrology* **43**: 700–708.

Heezen, B. C., et al. (1973) Diachronous deposits: A kinematic interpretation of the post Jurassic sedimentary sequence on the Pacific plate. *Nature* **241**: 25–32.

Henderson, J. H.; Jackson, M. L.; Syers, J. K.; Clayton, R. N.; and Rex, R. W. (1971) Cristobalite authigenic origin in relation to montmorillonite and quartz origin in bentonites. *Clays and Clay Minerals* **19**: 229–238.

Hummel, K. (1922) Die Entstehung eisenreicher Gesteine durch Halmyrolyse (="submarine Gesteinsersetzung"). *Geol. Rundschau* **13**: 40–81.

Jones, J. B., and Segnit, E. R. (1971) The nature of opal. *I*-nomenclature and constituent phases. *Geol. Soc. of Australia Jour.* **18**: 57–68.

———. (1975) Nomenclature and the structure of natural disordered (opaline) silica. A comment on the paper "A new interpretation of the structure of disordered α-cristobalite" by M. J. Wilson, J. D. Russell, and J. M. Tait, 1974. *Contr. Mineralogy and Petrology* **51**: 231–234.

Kastner, M., and Keene, J. B. (1975) Diagenesis of pelagic siliceous ooze. 9th Interntl. Sedimentol. Cong., Nice. Theme VII: 8–9.

Keene, J. B. (1975) Cherts and porcellanites from the North Pacific, DSDP leg 32. In Larson, R. L., Moberly, R., et al., *Initial Reports of the Deep Sea Drilling Project,* Vol. 32, 429–507. U.S. Government Printing Office, Washington, D.C.

Klasik, J. A. (1975) High cristobalite and high tridymite in Middle Eocene deep-sea chert. *Science* **189**: 631–632.

Lisitzin, A. P. (1971) Sedimentation in the world ocean. *Soc. Econ. Paleontologists and Mineralogists Spec. Pub. No. 17.*

Mapstone, N. B. (1975) Diagenetic history of a North Sea chalk. *Sedimentology* **22**: 601–614.

Matter, A. (1974) Burial diagenesis of pelitic and carbonate deep-sea sediments from the Arabian Sea. In Whitmarsh, R. B.; Weser, O. E.; Ross, D. A.; et al., *Initial Reports of the Deep Sea Drilling Project,* Vol. 23, 421–469. (U.S. Government Printing Office), Washington, D.C.

Moore, T. C., Jr. (1972) DSDP: Successes, failures, proposals. *Geotimes* (July 1972): 27–31.

Neugebauer, J. (1973) The diagenetic problem of chalk—The role of pressure solution and pore fluid. *Neues Jahrb. Geologie u Paläontologie Abh.* **143**: 223–245.

———. (1975) Foraminiferen-Diagenese in der Schreibkreide. *Neues Jahrb. Geologie u Paläontologie Abh* **150**: 182–206.

Oehler, J. H. (1973) Tridymite-like crystals in cristobalitic "cherts." *Nature Phys. Sci.* **241**: 64–65.

———. (1975) Origin and distribution of silica lepispheres in porcelanite from the Monterey Formation of California. *Jour. Sed. Petrology* **45**: 252–257.

———. (1976) Hydrothermal crystallization of silica gel. *Geol. Soc. America Bull.* **87**: 1143–1152.

Packham, G. H., and van der Lingen, G. J. (1973) Progressive carbonate diagenesis at deep sea drilling sites 206, 207, 208, and 210 in the Southwest Pacific, and its relationship to sediment physical properties and seismic reflectors. In Burns, R. E.; Andrews, J. E.; et al., *Initial Reports of the Deep Sea Drilling Project,* Vol. 21, 495–521. U.S. Government Printing Office, Washington, D.C.

Pettijohn, F. J. (1957) *Sedimentary rocks,* 2d ed. Harper and Row, New York, 719 pp.

Schlanger, S. O.; Douglas, R. G.; Lancelot, Y.; Moore, T. C. Jr.; and Roth, P. H. (1973) Fossil preservation and diagenesis of pelagic carbonates from the Magellan Rise, central North Pacific Ocean. In Winterer, E. L.; Ewing, J. I.; et al., *Initial Reports of the Deep Sea Drilling Project,* Vol. 17, 407–427. U.S. Government Printing Office, Washington, D.C.

Shepherd, W. (1972) *Flint–Its origins, properties, and uses.* Faber and Faber, London, 255 pp.

von Gümbel, C. W. (1868) Geognostische Beschreibung des Ostbayerischen Grenzgebirges oder des bayerischen und oberpfälzer Waldgebirges. Perthes, Gotha, 968 pp.

von Rad, U., and Rösch, H. (1975) Progressive chertification of siliceous sediments in the Cretaceous and Tertiary North Atlantic. 9th Interntl. Sedimentol. Cong., Nice, Thème I: 18.

Weaver, F. M., and Wise, S. W., Jr. (1972) Ultramorphology of deep sea cristobalitic chert. *Nature Phys. Sci.* **237:** 56–67.

Wilson, M. J.; Russell, J. D.; and Tait, J. M. (1974) A new interpretation of the structure of disordered α-cristobalite. *Contr. Mineralogy and Petrology* **47:** 1–6.

Wise, S. W., Jr., and Kelts, K. R. (1972) Inferred diagenetic history of a weakly silicified deep sea chalk. *Gulf Coast Assoc. Geol. Socs. Trans.,* 22nd Annual Convention: 177–203.

Wise, S. W., Jr., and Weaver, F. M. (1974) Chertification of oceanic sediments. In Hsü, K. J., and Jenkyns, H. C. (eds.), *Pelagic sediments: On land and under the sea. Internat. Assoc. Sedimentologists' Spec. Pub.* **1:** 301–326.

Part I

CARBONATE DIAGENESIS

Editors' Comments
on Papers 1, 2, and 3

1   BATHURST
Problems of Lithification in Carbonate Muds

2   SCHLANGER and DOUGLAS
The Pelagic Ooze-Chalk-Limestone Transition and Its Implications for Marine Stratigraphy

3   ADELSECK, GEEHAN, and ROTH
Experimental Evidence for the Selective Dissolution and Overgrowth of Calcareous Nannofossils During Diagenesis

The first three papers deal exclusively with carbonate diagenesis. Bathurst's paper (Paper 1) forms a link between the pre-DSDP and DSDP eras. He discusses problems in understanding the lithification of carbonate muds into biomicrite. Though he considers shallow-marine carbonate muds only—study of the lithification of deep-sea muds still being outside the reach of scientists—the basic problems are similar to most pelagic carbonate sediments. He notices that unconsolidated carbonate muds have porosities between 50 and 70 percent, while the lithified end product can have porosities as low as 2 to 3 percent. The baffling aspect is that the lithified muds do not show any signs of compaction. Delicate tests of foraminifera survive uncrushed. From this he concludes that "cement was precipitated in the pores in sufficient quantity to form a resistant framework before the overburden was great enough to cause detectable compaction" (p. 20). However, Bathurst accepts that some initial compaction by dewatering takes places, until the carbonate grains are in contact.

Several later studies support Bathurst's suggestion that a supporting framework is established early in diagenesis. According to Mapstone (1975) such a framework is established by spot-welding of adjacent grains. Such spot-welding has been explained theoretically by Neugebauer (1973; see also Paper 14).

The source of the cement filling the pore spaces poses a problem. Bathurst suggests that it probably comes in part from outside the system. On page 22 he states: "Observations that yield useful evidence of the processes of cementation and neomorphism (= "lithification") of car-

16

bonate oozes are extremely scarce." Since 1969 when Bathurst's paper was submitted, such observations have changed from "extremely scarce" to "abundant," thanks to the DSDP. The next paper is a good example of this progress.

Paper 2 by Schlanger and Douglas is based on results from DSDP drill site 167 in the central North Pacific and enlarges on an earlier paper by Schlanger et al. (1973). Compared with Bathurst's paper, it clearly shows the advantages of both being able to study complete sedimentary sequences showing the transitions from ooze to chalk to limestone and using the SEM. The main aspects treated in this paper are the preservation of fossils, calculations of changes in porosity and volume, the introduction of the diagenetic potential concept, and the relationship between diagenetic potential and seismic reflectors (acousti-stratigraphy).

At site 167 there is a gradual break-up of foraminifera with depth. The authors emphasize that this is not due to crushing under overburden but is caused by dissolution. Thin-walled planktonic foraminifera are the first to be affected. Thicker-walled benthonic foraminifera are preserved longer but are doomed to eventually disappear as well. Nannofossils are subject to selective dissolution and overgrowth during diagenesis.

The transition from ooze to limestone is accompanied by a decrease in porosity from 80 to 40 percent and by a decrease in volume to about one-third. These decreases are due to dewatering (early stage), a change in packing, loss of intrabiotic porosity (due to the dissolution of foraminifera), the selective dissolution of nannofossils, and the reprecipitation of calcite. The authors presume that no calcite has been introduced from outside the system.

It it clear that the observations and conclusions by Schlanger and Douglas differ fundamentally from those by Bathurst, especially with respect to compaction and the source of the secondary calcite (see also Paper 13). There is agreement between the two papers, however, in the observation that foraminifera are not being crushed under overburden. Schlanger and Douglas do not elaborate on this point.

The porosities in the (Upper Jurassic) limestones of site 167 are about 40 percent. Similar values have been reported from other DSDP sites (e.g., Paper 5). Lower porosities are generally due to silicification. To explain the low porosities for limestones (2 to 3 percent) mentioned by Bathurst still poses problems.

Schlanger and Douglas introduced the concept of diagenetic potential. It tries to explain why on a smaller scale there are many fluctuations (reversals) in lithification, even though there is an overall change from ooze to chalk to limestone with increasing depth and time. According to this concept, different "starting" assemblages may have different "lengths of diagenetic pathways" to reach equal stages in lithification. These differences in "diagenetic potential" may reflect changes in paleo-oceanography, changes in paleoclimatology, or changes in spreading rates of oceanic plates. As differences in diagenetic potential eventually result in differences in lithification, they may show up as seismic reflectors. Because of the fact that the diagenetic potential is influenced by regional or even global changes, it should be possible to correlate seismic reflectors over large distances, making it possible to establish an acoustistratigraphy.

The selective dissolution and overgrowth of nannofossils, as observed in DSDP cores (Papers 2, 4, 5, 13, and 14), have been reproduced experimentally by Adelseck et al. (Paper 3). They suggest that size (free energy), morphology, and crystal orientation are important parameters. From their experiments they establish an "order of diagenetic evolution."

The influence of morphology, crystal size, and shape on dissolution and reprecipitation has also been discussed by Matter et al. (Paper 13) and Neugebauer (Paper 14). From these papers it is clear that the original species composition of a biogenic ooze has an important influence on its diagenetic potential.

REFERENCES

Mapstone, N. B. (1975) Diagenetic history of a North Sea chalk. *Sedimentology* **22:** 601–614.

Neugebauer, J. (1973) The diagenetic problem of chalk—The role of pressure solution and pore fluid. *Neues Jahrb. Geologie u. Paläontologie Abh.,* **143:** 223–245.

Schlanger, S. O.; Douglas, R. G.; Lancelot, Y.; Moore, T. C., Jr.; and Roth, P. H. (1973) Fossil preservation and diagenesis of pelagic carbonates from the Magellan Rise, central North Pacific Ocean. In Winterer, E. L.; Ewing, J. I.; et al., *Initial Reports of the Deep Sea Drilling Project,* Vol. 17, 407–427. U.S. Government Printing Office, Washington, D.C.

1

Reprinted from *Geologists' Assoc. Proc.* **81**, Pt. 3:429–440 (1970)

Problems of Lithification in Carbonate Muds

by R. G. C. BATHURST

Received 18 November 1969

CONTENTS

ABSTRACT: The lithification of carbonate muds is examined on the basis of their high primary porosities of 50 to 70 per cent and the characteristic absence of de-watering-compaction structures in the hardened micrite. Some cementation is presumed, therefore, to have been early (pre-compaction), forming a rigid, load-resistant framework. This first generation of cement may have been locally derived, either from marine pore-water or by fresh-water dissolution-precipitation of the more soluble sedimentary particles. The main bulk of the cement is a late second generation and was either allochthonous or produced locally by the development of stylolites. Lithification involved such various processes as loss of water while pores were occluded by cement, the wet transformation of aragonite to calcite, recrystallisation of calcite, dissolution of small supersoluble particles, transfer of Mg^{2+}, the production of secondary voids, influx of allochthonous $CaCO_3$ and pressure-solution. Some micrites have been lithified on the sea-floor as hardgrounds during prolonged exposure to sea-water, super-saturated for $CaCO_3$, for hundreds of thousands of years. Other micrites have been hardened more rapidly as a result of exposure to meteoric groundwater. The evolution of clotted limestones is considered. Finally, it is emphasised that neomorphic fabrics in micrites were produced not by alteration of the dense micrite as we see it now but by neomorphism of the highly reactive primary mud—porous, wet and multimineralogical.

1. THE PROBLEM

IT IS nowadays common knowledge that the lithification of a carbonate mud, like the lithification of carbonate sediments in general, involves a change from a squelchy mixture of solid carbonate phases, bathed liberally in an aqueous pore solution, to a rock composed of low-magnesian calcite with a porosity of, perhaps, 2 or 3 per cent. The problems raised by this transformation are much the same whether we are dealing with a micrite or a calcarenite. The central difficulty is tantalisingly familiar—how to cement a carbonate mud while it is still largely uncompacted.

Holocene carbonate muds have porosities mainly of 50 to 70 per cent (Ginsburg, 1964; Pray & Choquette, 1966). Most ancient micrites[1] and biomicrites (Dunham's, 1962, mudstones and wackestones) show no sign of having been compacted (Pray, 1960). Delicate tests are uncrushed, thin skeletal structures have not been broken as a result of grain-to-grain movement, the crystal size distribution of micrite is the same inside shells where the overburden load was effectively zero as in the intergranular micrite, cavities in the mud remained open until filled with cement. The conclusion is inevitable—cement was precipitated in the pores in sufficient quantity to form a resistant framework before the overburden was great enough to cause detectable compaction. Presumably some compaction by dewatering took place until the particles were in contact, as in Florida Bay, where the rate of decrease of porosity with depth falls off rapidly at about a porosity of 70 per cent (Ginsburg, 1957). Dewatering compaction of this kind would not, however, have proceeded on the same scale in shell chambers where the overburden load was practically zero, yet their filling of micrite is petrographically indistinguishable from the intergranular micrite. That the load in the shell chamber was zero is particularly obvious where the micrite is geopetal and only partly fills the chamber. Again we must infer that compaction of the sediment was very slight. This early cementation is in dramatic contrast with the situation in terrigenous clays (illite, montmorillonite, etc.). In the Carboniferous Limestone of North Wales, layers of biomicrite containing uncompacted crinoid columnals (circular cross-section) are interbedded with terrigenous shales in which the columnals have been squashed flat. The *Inoceramus* and White Chalk lithofacies of Northern Ireland, which show 15 to 31 per cent compaction (Wolfe, 1968), are very rare exceptions.

As the porosities of most carbonate muds have *not* been reduced by *compaction* from initial values of 50 to 70 per cent to their present values of 2 to 3 per cent, then the pores were *filled by cement*. This conclusion in turn presents difficulties. Where was the source of such an enormous quantity of $CaCO_3$—more than half the volume of the limestone? How was it transported and precipitated? These problems are still unsolved.

2. PRIMARY COMPOSITION

The initial composition of past muds is not generally known, although studies with the electron microscope are revealing increasing detail in the less altered muds, particularly coccolith oozes (Honjo & Fischer, 1964; Fischer, Honjo & Garrison, 1967). It is likely that they ranged from

[1] A short glossary of terms is given on p. 438.

aragonite needle-muds like those on the Bahamas–Florida platform (Purdy, 1963; Cloud, 1962; Ginsburg, 1957) and in the Persian Gulf (Evans, 1966; Kinsman, 1969) through mixed detrital skeletal carbonate muds such as are known off British Honduras (Matthews, 1966), to nearly pure oozes of low-magnesian calcite made of coccoliths, as in the Chalk (Black, 1953; Wolfe, 1968). All muds would have contained a number of solid phases, aragonite, various low-magnesian calcites ($MgCO_3$ 2 to 3 mole per cent) and high-magnesian calcites ($MgCO_3$ 12 to 17 mole per cent). Particle size and shape would have varied with the origin (Folk, 1965, 29). The well-known tendency for organic matter to be concentrated in the finer sediments means that the silt and clay grade carbonates will have been rendered even more complex by the addition not merely of organic matter but of the algae, bacteria, fungi and yeasts that accompany it. The loss of this non-carbonate material from muds by oxidation is not as efficient a process as in sand-grade, relatively well-circulated calcarenitic sediments, and the long-term influence of residual organic products may, therefore, be more important in the lithification of carbonate oozes than of calcarenites.

3. PROCESSES

It is plain that, for many calcilutites (lithified muds) the lithification was contrapuntal, a weaving of two melodies, on the one hand the influx and precipitation of externally derived cement, on the other the neomorphism of the original crystals. Not only did the muds undergo early cementation but, associated with this process went extensive development of neomorphic sparry calcite. This is apparent, for example, in the Pleistocene limestones of Guam (Schlanger, 1964) and of Funafuti (Cullis, 1904).

The many processes that, it must be assumed, proceeded simultaneously during lithification were certainly wet ones. As the reduction of porosity during cementation continued, the water content fell and it is sensible to expect that the consequent reduction in the availability of solvent was responsible for a logarithmic slowing of the various processes. The relative importance of such processes is something we cannot yet judge. They include the wet transformation of aragonite to calcite, the dissolution of tiny supersoluble particles and prominences on grains, the transfer of Mg^{2+} from magnesian calcites, dissolution yielding voids, the influx of allochthonous $CaCO_3$, pressure-solution, and the precipitation of cement —in fact, all the paraphernalia of aggrading neomorphism (Folk, 1965).

The results of these processes are nevertheless clear. The calcilutite consists now of low-magnesian calcite. Many of the smallest primary particles (crystals), with shortest diameters as small as 0.1μ, have been lost, and the final calcilutite has a new lower limit of crystal diameter probably at about $0.5\ \mu$ and an upper limit of about 3 to 4 μ (Bathurst, 1959, 367; Folk, 1965, 29). There is commonly a mode around 1 to 2 μ (Schwarzacher,

21

1961; Flügel, 1967). The restricted crystal size-range of the calcilutites is combined with a tendency to equigranular texture, and, surprisingly often, plane intercrystalline boundaries, features that have been amply demonstrated with the aid of the electron microscope.

A note of caution is necessary here regarding the description of the crystal mosaics of micrites—and, indeed, of crystal mosaics, in general, which are (1) monomineralic, (2) have no visible porosity and (3) are equigranular. It is obvious, in these circumstances, that any *plane* interface between two adjoining crystals must be either (a) a face of one of them, or (b) a compromise boundary (Buckley, 1951; Schmidegg, 1928; Bathurst, 1958). Published descriptions of micrites (and commonly of dolomites) as idiomorphic or hypidiomorphic are based, it would seem, on the frequency of plane intercrystalline boundaries and not, as the terms imply, on the frequency of identified crystal faces. A simple arithmetic calculation will show that it is impossible theoretically for more than half the crystals in an equigranular non-porous mosaic to be euhedral. In natural conditions, where the crystals vary somewhat in size, in shape and orientation and are bounded by numerous plane surfaces, the likelihood is that the proportion of euhedral crystals will be much lower, probably less than 10 per cent. Subhedral crystals would not be so rare. The reason for dwelling on this matter of the polygonality of equigranular, monomineralic, non-porous mosaics, is that the *exact* nature of the intercrystalline boundary is a question of profound importance in diagenetic studies. It matters in fabric analysis whether an intercrystalline boundary is a crystal face or simply an unidentified plane interface.

(a) Inferences from the Forms of Intercrystalline Boundaries

Observations that yield useful evidence of the processes of cementation and neomorphism (= lithification) of carbonate oozes are extremely scarce. Intercrystalline boundaries may be plane surfaces or curved. It is necessary to note that plane crystal interfaces can form either by passive growth of cement at crystal-solution interfaces (i.e. syntaxial cement overgrowths) or by syntaxial growth *in situ* during neomorphism at the interface between a crystal face and a solution film, as one crystal enlarges at the expense of adjacent crystals.

There is here an awkward complication. It is well known that intercrystalline boundaries in metamorphosed rocks have simple forms which tend toward plane surfaces as equilibrium is approached. Illustrations in Fischer and others (1967) are instructive. It is necessary to learn to distinguish, therefore, between the fabrics of a micrite that has only experienced low temperature and pressure and of one that has undergone metamorphism. This should not be too difficult, but the appropriate research is awaited.

Other intercrystalline boundaries are amoeboid and thus it is probable that some crystal contacts are pressure-welded. Schwarzacher (1961, 1500) has investigated the possibility that pressure-solution could have been responsible for the fabric of some silty micrites (low-magnesian calcite) in the Carboniferous Limestone of north-western Ireland. He made acetate peels of etched, ground surfaces and in this way produced photomicrographs with exceptionally fine detail for light microscope preparations. He measured the longest axis within each crystal (in two dimensions) and found a strong maximum perpendicular to the bedding with a small submaximum in the bedding. This fabric is not what would be expected in a micrite lithified by pressure-solution under a vertically applied load, but it could, Schwarzacher suggested, have evolved in the presence of a pore solution moving vertically, along a hydrostatic pressure gradient. Pressure-solution might, he wrote, account for the submaximum. A difficulty in the way of interpreting any micrite fabric in terms of pressure-solution is the general paucity of compaction structures in these rocks. The early cementation which this implies seems to rule out the possibility of particle-to-particle pressure-solution, at least as a major lithifying process. On the other hand pressure-welded contacts between micrite crystals are apparent in illustrations by Fischer and others (1967), so the process is not unimportant.

(b) Source of the Cement

It is obvious that the change from aragonite to calcite, by whatever train of processes, with the accompanying 8 per cent increase of volume, yields a definite but very small amount of surplus $CaCO_3$ for precipitation as cement (Harris & Matthews, 1968; Pingitore (1970))—but nothing like enough to fill the porosity of 50 per cent or more. Indeed, even in a purely aragonitic sediment, primary porosities of 60 per cent and 40 per cent would be reduced by polymorphic exchange only to 56.8 and 36.4 per cent. Yet at the moment of writing no one has been able to discover whence the extra quantity of $CaCO_3$ was derived.

Some light on the source of $CaCO_3$ for cementation in general has been shed by the studies of hardgrounds, where submarine cementation of these lithified sea-floors can be demonstrated (for example, Lindström, 1963; Bromley, 1967, 1968; Taylor & Illing, 1969; Shinn, 1969). It is becoming increasingly clear that, in the Swedish Arenig sea, the Chalk sea of northern Europe, the Holocene Persian Gulf, very slow rates of sedimentation have allowed grains to lie undisturbed near the sediment surface for vast periods of time while sea-water, supersaturated for $CaCO_3$, has been pumped through the sediment pores by the action of tides and waves. Given this unchanging diagenetic environment for hundreds of thousands of years, while sedimentation is practically zero, the cementation of lime-

23

stone crusts can be achieved. Other relatively early cementation results from the early exposure of carbonate sediments to the subaerial meteoric environment: this type of early lithification has been demonstrated on Eniwetok and Guam (Schlanger, 1963, 1964), on Bermuda (Friedman, 1964; Land, 1967) and on Barbados (Matthews, 1968; Pingitore (1970).

Recent work on the evolution of stylolites gives the impression that these structures may have played a much greater part in releasing $CaCO_3$ for cementation than has generally been realised. Porosity tends to increase upward and downward, away from the stylolite, as if $CaCO_3$ had moved outward from the stylolitic seam (Harms & Choquette, 1965; Dunnington, 1967). Many stylolites have certainly been syndiagenetic in origin, developing gradually during the process of lithification (Park & Schot, 1968). The release of $CaCO_3$ for cementation during pressure-solution has also been estimated by Barrett (1964) and Oldershaw & Scoffin (1967), in a loosely quantitative manner.

(c) Experiments of Hathaway & Robertson

The experiments of Hathaway & Robertson (1961, 301) are of uncertain relevance to the problem of the lithification of micrite—despite their inevitable fascination—because they imply a degree of compaction not generally found in nature. These authors subjected wet aragonite mud from the Bahamas to various temperatures and pressures, in a cylinder from which surplus pore water could escape as the mud compacted. Temperatures ranged up to 400° C. (equivalent to a depth of about 20 km.) and pressure to 3450 bars (equivalent to an overburden of about 10 km. of average crust). Times ranged up to sixty-three days. Their series of electron photomicrographs (the mineralogy checked by X-ray diffraction) shows a change from aragonite to calcite, accompanied by rounding of needles and the appearance of increasingly larger globular-shaped masses of calcite. With their maximum pressure and time an equigranular mosaic of crystals was formed, with many plane intercrystalline boundaries. Photographs of the end-product (fracture surface) are embarrassingly like those of natural micrites, such as those from the famous Upper Jurassic Solnhofen Limestone of Bavaria, the Triassic Halstätter Limestone of Austria and the Carboniferous Limestone of the British Isles. Nevertheless, it is inconceivable that micrites in general were constructed with the heat and violence lavished upon the Bahamian muds by Hathaway & Robertson, but it is significant, as they point out, that an artificial calcite micrite can be produced in this way. Transformation of the aragonite was completed early in the process, so the remainder of the evolution consisted of wet recrystallisation of calcite, dissolution and cementation (the pore filling) of an accumulation of calcite crystals, giving a final specific gravity for the artificial rocks of 1.9 (Solnhofen Limestone has sp. gr. of 2.6). The end-product retains, therefore, a porosity of about 30 per cent, so that, to complete the lithification, a further 30 per cent of $CaCO_3$ would have to be exotic in origin.

(d) Significance of the Upper Crystal Size of Calcite Micrites

The widespread upper crystal diameter for the groundmass of lithified micrites at 3 to 4 μ, ignoring the included coarser skeletal debris, is intriguing. As I wrote earlier (1959, 366), this 'points to the existence of a universal threshold state at which fabric evolution stops and beyond which it can, but need not, continue'. A possible reason for this, which would bear further investigation, is that a stage is reached in the combined neomorphism and cementation, when the porosity and permeability are so reduced that the transport of Ca^{2+} and CO_3^{2-} from one crystal face to another becomes slow even on a geological time-scale. This stage would represent virtual stability. Some new driving force would be needed to induce further progress in neomorphism, such as elastic strain induced during deformation. Probably a more plausible explanation has been made by Folk (1965, 36). He suggested that the crystal size of the micrite represents the *long* axes of the original crystals which 'have mainly expanded in volume by fattening out rather than lengthening'. This, of course, implies the introduction of allochthonous $CaCO_3$.

(e) Structure Grumeleuse

In connexion with the question of micrite lithification it is useful to glance for a moment at this clotted limestone, named and lucidly described by Cayeux (1935, 271) thus: ' ... elle montre de tout petits éléments calcaires, à pâte extrêmement fine, se détachant en gris sombre, de forme générale globuleuse ou irrégulière, dont les contours ne sont jamais franchement arrêtés, et sans différenciation d'aucune sorte. Ces matériaux, dont la microstructure est invariablement cryptocristalline, sont plongés dans une gangue de calcite incolore et grenue.' The fabric studied by Cayeux in the Carboniferous Limestone of France and, above all, of Belgium is known in limestones of many other places and ages. The two-component fabric, consisting of patches of micrite embedded in a matrix of microspar (Folk, 1965), occupies a central position in a spectrum of limestone fabrics. In one direction, from the *classificational* point of view this clotted limestone passes into micrite, as the proportion of microspar falls and the boundaries between micrite patches and matrix become less obvious. In the other, it passes into *structure pseudoolithique* (Cayeux, 1935), the familiar pelsparite of Folk (1959). Cayeux noted particularly that the *grumeaux* are of silt grade. (*Grumeau* = clot or lump = peloid = grain of micritic composition with no special origin implied.) The passage from *structure grumeleuse* to micrite, again from a purely classificational aspect, is also marked by a merging of the micrite patches. Cayeux's words cannot be improved: 'Dans un stade de différenciation moins prononcé, les grumeaux, toujours, séparés par de la calcite pure, contractent entre eux des adhérences multiples' (1935, 271). These various types of clotted calcilutites have been described and illustrated by Beales (1958) from the Palaeozoic limestones of Alberta.

The existence of a range of petrographic types, from micrite through *structure grumeleuse*, or clotted limestone, to pelsparite, does not by itself mean that this variation represents the different stages in a continuous diagenetic evolution. Indeed, the two leading writers on the question have expressed opposing views on the course of the diagenetic evolution. Cayeux believed that *structure grumeleuse* evolved by the growth of calcite crystals throughout the mass of an originally homogeneous micrite, and the gradual differentiation, thereby, of a more coarsely crystalline, continuous matrix separating *residual* clots of microcrystalline (micritic) calcite. Beales, on the other hand, whose extensive researches into pellet limestones are an indispensable introduction to this field (1956, 1958, 1965), thought that the processes operated the other way about, in that 'Many closely packed grains [peloids] appear to have merged on recrystallization into a homogeneous microcrystalline rock differentiated with difficulty from calcilutite'. The outstanding characteristic, and the most puzzling aspect, of this fabric, is the merged patches of micrite, with 'des adhérences multiples'. This obviously cannot be a primary fabric of mechanically deposited peloids: it must be a secondary feature. Once this is granted, the field is open to speculation on the diagenetic evolution.

A factor that must have an important bearing on the development of any working hypothesis is the universality of clotted texture, the widespread occurrence of some degree of clotting in micrites of many ages and localities, despite differences of both depositional and diagenetic environments. Two additional factors also need to be kept in view. One is the ubiquitous formation of faecal pellets in modern carbonate sediments allied to the common occurrence of heavily micritised skeletal debris (Bathurst, 1966). The other is the well-known nature of crystal growth fabrics. Where crystal growth starts at a number of points in a homogeneous crystal mosaic and spreads outward from these points (Folk's 'porphyroid aggrading neomorphism', 1965, 22), the resultant fabric is likely to contain radial elements (radial fibres or centrifugal increase in crystal size) and to yield, in ideal cases, a spherical growth front. An obvious example is the growth of spherulitic, radial-fibrous ooids, or needle bundles. Having regard to these three factors, my own, purely intuitive, conclusion as to the origin of *structure grumeleuse* runs thus: the growth of sparry calcite, beginning at a number of points in a homogeneous carbonate mud (or mudstone) would produce a collection of relatively coarsely crystalline patches in a matrix of unaltered finely crystalline ooze. This is the reverse of what we find in *structure grumeleuse*. Furthermore, because of the ubiquitous occurrence of primary peloids, diagenetic processes are more likely to take the form of a reduction of the individuality of peloids rather than a fabrication of new ones. Such reduction may follow the merging of soft faecal pellets or the action of pressure-solution. *Structure grumeleuse* is common in algal stromatolites and it may well be that, in a peloidal calcarenite or calcisiltite trapped by algal filaments, photosynthesis can lead to the precipitation of additional aragonitic micrite which, attaching itself to the existing peloids, will obscure their boundaries and cause merging. Grapestone (Illing, 1954; Purdy, 1963), in which calcarenite peloids, intensely bored by algae, are cemented by micrite (Bathurst, 1966, 21), may have formed in this way. These are, nevertheless, only tentative conclusions, and the elucidation of individual cases remains a matter of extraordinary difficulty, requiring patience and, maybe, the virtue for the time being of suspended judgment.

(f) Fossilised Stages

In micrites which have been partly replaced by neomorphic sparry calcite, it is important that we should not mistake the micrite as it is now for the oozy stuff that was originally altered to spar.

In diagenetically immature limestones (e.g. the Cainozoic limestones of Guam (Schlanger, 1964) or Funafuti (Cullis, 1904)), the present fabric and mineralogy of the unaltered, barely consolidated, micrite may truly reflect the condition normally obtaining while neomorphism is active. However, in diagenetically mature limestones (e.g. Carboniferous of Europe, Pennsylvanian of the United States), which are low-magnesian calcite, the present micrite must be different from its mineralogically heterogeneous precursor which existed at the time when neomorphism was operating. In other words, when we see, in a thin section, patches of secondary, sparry low-magnesian calcite associated with residual low-magnesian micritic calcite, we cannot logically infer that the spar is simply a product of recrystallisation of the micrite. I think we are bound to assume, in view of the foregoing discussion, that the spar is the neomorphic product of a material that no longer exists, and that the original micrite (more finely crystalline, more porous, mineralogically heterogeneous) has since changed to low-magnesian calcite with porosity 2 to 3 per cent. This is, of course, another way of saying that neomorphism goes on in some rocks during lithification.

4. PROGRESS

In the last fifteen years much has been discovered about the diagenesis of carbonate sediments. We have, first of all, learned how to look at them, to take advantage of the fabric evidence they display. In 1947 Professor Read

27

438 R. G. C. BATHURST

spoke to me before I went into the field to do my independent mapping. He urged me to bring back a map—even if it were only the size of a postage stamp! Since then much of my work and that of others in the field of carbonate sedimentology has, indeed, been concerned with the 'mapping' of areas of philatelic dimensions—as observed with the light and electron microscopes. More recently the introduction of quantitative chemical and isotopic data has deepened our conception of diagenesis. The effects of biological agents are only now beginning to be widely appreciated. Yet the situation is profoundly different from that of, say, ten years ago. We are at last in a position to pose many of the critical questions which future research workers should be able to answer.

5. GLOSSARY

Biomicrite. A limestone composed of skeletal grains in a matrix of micrite (Folk, 1959).

Cement. A non-skeletal void-filling, precipitated on an intragranular or intrasedimentary free surface. This convenient working definition, necessarily a compromise, was informally adopted by the Bermuda Seminar on Carbonate Cementation, September 1969.

Compromise boundary. A plane interface between two crystals which evolved by mutual interference of their respective growing faces. This interface is a face of neither crystal (Buckley, 1951).

Micrite. An abbreviation of 'microcrystalline ooze', it refers to finely crystalline carbonate sediments with upper crystal diameter of 4 μ (Folk, 1959).

Microspar. Carbonate crystal mosaic with crystal diameters from 4 to 10 μ, or even higher, to about 50 μ (Folk, 1959).

Mudstone. A micrite containing less than 10 per cent grains (Dunham, 1962).

Neomorphism. More strictly 'aggrading neomorphism' as used here. A complex of processes whereby a mosaic of finely crystalline carbonate is replaced by a coarser (sparry) mosaic without the development of visible porosity (Folk, 1965). Dominant reactions are the wet transformation of aragonite to calcite and recrystallisation. The process is 'porphyroid' where some of the neomorphic crystals are conspicuously larger than those which surround them.

Peloid. A sedimentary grain formed of micritic carbonate irrespective of origin (McKee & Gutschick, 1969).

Pelsparite. A limestone composed of pellets (peloids) in a matrix of cement (Folk, 1959).

Wackestone. A micrite containing more than 10 per cent of grains: these grains are not in three-dimensional contact with each other. They are 'mud-supported' (Dunham, 1962).

REFERENCES

BARRETT, P. J. 1964. Residual Seams and Cementation in Oligocene Shell Calcarenites, Te Kuiti Group. *J. sedim. Petrol.* **34**, 524–31.

BATHURST, R. G. C. 1958. Diagenetic Fabrics in Some British Dinantian Limestones. *Lpool Manchr geol. J.*, **2**, 11–36.

———. 1959. Diagenesis in Mississippian Calcilutites and Pseudobreccias. *J. sedim. Petrol.*, **29**, 365–76.

———. 1966. Boring Algae, Micrite Envelopes and Lithification of Molluscan Biosparites. *Geol. J.*, **5**, 15–32.

BEALES, F. W. 1956. Conditions of Deposition of Palliser (Devonian) Limestone of Southwestern Alberta. *Bull. Am. Ass. Petrol. Geol.*, **40**, 848–70.

———. 1958. Ancient Sediments of Bahaman Type. *Bull. Am. Ass. Petrol. Geol.*, **42**, 1845–80.

———. 1965. Diagenesis in Pelletted Limestones. *In* L. C. PRAY & R. C. MURRAY (Editors): *Dolomitization and Limestone Diagenesis: a Symposium.* Soc. Econ. Paleontologists Mineralogists, Spec. Publ., 13, 49–70.

BLACK, M. 1953. The Constitution of the Chalk. *Proc. geol. Soc., Lond.* 1499, lxxxi–lxxxii.

BROMLEY, R. G. 1967. Some Observations on Burrows of Thalassinidean Crustacea in Chalk Hardgrounds. *Q. Jl geol. Soc. Lond.*, **123**, 157–77.

———. 1968. Burrows and Borings in Hardgrounds. *Meddr dansk geol. Foren.*, **18**, 247–50.

BUCKLEY, H. E. 1951. *Crystal Growth.* Wiley, New York, 571 pp.

CAYEUX, L. 1935. *Les Roches Sédimentaires: Roches Carbonatées.* Masson, Paris, 463 pp.

CLOUD, P. E. 1962. Environment of Calcium Carbonate Deposition West of Andros Island, Bahamas. *U.S. Geol. Surv. Profess. Paper*, 350, 1–138.

CULLIS, C. G. 1904. The Mineralogical Changes Observed in the Cores of the Funafuti Boring. *In* T. G. BONNEY (Editor): *The Atoll of Funafuti.* Royal Society, London, 392–420.

DUNHAM, R. J. 1962. Classification of Carbonate Rocks According to Depositional Texture. *In* W. E. HAM (Editor): *Classification of Carbonate Rocks.* Am. Assoc. Petrol. Geologists, Tulsa, Okla., 108–121.

DUNNINGTON, H. V. 1967. Aspects of Diagenesis and Shape Change in Stylolitic Limestone Reservoirs. *World Petrol. Cong. Proc. 7th Mexico*, **2**, 339–52.

EVANS, G. 1966. Persian Gulf. *In* R. W. FAIRBRIDGE (Editor); The Encyclopedia of Oceanography. 1. Reinhold, New York, 689–95.

FISCHER, A. G., S. HONJO & R. E. GARRISON. 1967. *Electron Micrographs of Limestones.* Princeton Univ. Press, 141 pp.

FLÜGEL, E. 1967. Elektronenmikroskopische Untersuchungen an mikritischen Kalken. *Geol. Rdsch.*, **56**, 341–58.

FOLK, R. L. 1959. Practical Petrographic Classification of Limestones. *Bull. Am. Ass. Petrol. Geol.*, **43**, 1–38.

———. 1965. Some Aspects of Recrystallization in Ancient Limestones. *In* L. C. PRAY & P. C. MURRAY (Editors); *Dolomitization and Limestone Diagenesis: a Symposium.* Soc. Econ. Paleontologists Mineralogists, Spec. Publ., 13, 14–48.

FRIEDMAN, G. M. 1964. Early Diagenesis and Lithification in Carbonate Sediments. *J. sedim. Petrol.*, **34**, 777–813.

GINSBURG, R. N. 1957. Early Diagenesis and Lithification of Shallow-water Carbonate Sediments in South Florida. *In* R. J. LEBLANC & J. G. BREEDING (Editors); *Regional Aspects of Carbonate Deposition.* Soc. Econ. Paleontologists Mineralogists, Spec. Publ. 5, 80–99.

———. 1964. South Florida Carbonate Sediments. *Guidebook for Field Trip No. 1.* Geol. Soc. Am. Convention, 1–72.

HARMS, J. C. & P W. CHOQUETTE. 1965. Geologic Evaluation of a Gamma-Ray Porosity Device. *Trans. Soc. Profess. Well Log Analysts, 6th Annual Logging Symposium, Dallas, Texas*, 1–37.

29

HARRIS, W. H. & R. K. MATTHEWS. 1968. Subaerial Diagenesis of Carbonate Sediments: Efficiency of the Solution-Precipitation Process. *Science*, **160**, 77–9.

HATHAWAY, J. C. & E. C. ROBERTSON. 1961. Microtexture of Artificially Consolidated Aragonite Mud. *U.S. Geol. Surv. Profess. Paper*, 424–C, 301–4.

HONJO, S. & A. G. FISCHER. 1964. Fossil Coccoliths in Limestone Examined by Electron Microscopy. *Science*, **144**, 837–9.

ILLING, L. V. 1954. Bahaman Calcareous Sands. *Bull. Am. Ass. Petrol. Geol.*, **38**, 1–95.

KINSMAN, D. J. J. 1969. Interpretation of Sr^{2+} Concentrations in Carbonate Minerals and Rocks. *J. sedim. Petrol.*, **39**, 486–508.

LAND, L. S. 1967. Diagenesis of Skeletal Carbonates. *J. sedim. Petrol.*, **37**, 914–30.

LINDSTRÖM, M. 1963. Sedimentary Folds and the Development of Limestone in an Early Ordovician Sea. *Sedimentology*, **2**, 243–75.

McKEE, E, D. & R. C. GUTSCHICK, 1969. History of the Redwall Limestone of Northern Arizona. *Mem. geol. Soc. Am.*, **114**, 726 pp.

MATTHEWS, R. K. 1966. Genesis of Recent Lime Mud in Southern British Honduras. *J. sedim. Petrol.*, **36**, 428–54.

———. 1968. Carbonate Diagenesis: Equilibration of Sedimentary Mineralogy to the Subaerial Environment; Coral Cap of Barbados, West Indies. *J. sedim. Petrol.*, **38**, 1110–19.

OLDERSHAW, A. E. & T. P. SCOFFIN. 1967. The Source of Ferroan and Non-Ferroan Calcite Cements in the Halkin and Wenlock Limestones. *Geol. J.*, **5**, 309–20.

PARK, W. C. & E. H. SCHOT. 1968. Stylolites: Their Nature and Origin. *J. sedim. Petrol.*, **38**, 175–91.

PINGITORE, N. E. 1970. Diagenesis and Porosity Modification in *Acropora palmata*: Pleistocene of Barbados, West Indies. *J. sedim. Petrol.*, **40**, 712–21.

PRAY, L. C. 1960. Compaction in Calcilutites. *Bull. geol. Soc. Am.*, **71**, 1946 (abstract).

——— & P. W. CHOQUETTE. 1966. Genesis of Carbonate Reservoir Facies. *Bull. Am. Ass. Petrol. Geol.*, **50**, 632 (abstract).

PURDY, E. G. 1963. Recent Calcium Carbonate Facies of the Great Bahama Bank 1 and 2. *J. Geol.*, **71**, 334–55, 472–97.

SCHLANGER, S. O. 1963. Subsurface Geology of Eniwetok Atoll. *U.S. Geol. Surv. Profess. Paper*, 260 BB, 991–1066.

———. 1964. Petrology of the Limestones of Guam. *U.S. Geol. Surv. Profess. Paper*, 403–D, 1–52.

SCHMIDEGG, O. 1928. Über geregelte Wachstumsgefüge. *Jb. geol. Bundesanst., Wien.*, **78**, 1–52.

SCHWARZACHER, W. 1961. Petrology and Structure of Some Lower Carboniferous Reefs in Northwestern Ireland. *Bull. Am. Ass. Petrol. Geol.*, **45**, 1481–503.

SHINN, E. A. 1969. Submarine Lithification of Holocene Carbonate Sediments in the Persian Gulf. *Sedimentology*, **12**, 109–44.

TAYLOR, J. M. C. & L. V. ILLING. 1969. Holocene Intertidal Calcium Carbonate Cementation, Qatar, Persian Gulf. *Sedimentology*, **12**, 69–107.

WOLFE, M. J. 1968. Lithification of a Carbonate Mud: Senonian Chalk in Northern Ireland. *Sediment. Geol.*, **2**, 263–90.

2

Copyright © 1974 by the International Association of Sedimentologists

Reprinted from pp. 117–148 of *Pelagic Sediments: On Land and Under the Sea,*
Internat. Assoc. Sedimentologists Spec. Publ. No. 1, K. J. Hsü and
H. C. Jenkyns, eds., 1974, 447 pp.

The pelagic ooze-chalk-limestone transition and its implications for marine stratigraphy*

SEYMOUR O. SCHLANGER *and* ROBERT G. DOUGLAS

*Department of Earth Sciences, University of California, Riverside,
California, and Department of Geology, Case Western Reserve University,
Cleveland, Ohio, U.S.A.*

ABSTRACT

Recovery of long sequences of cores, at Deep Sea Drilling Project sites, from Recent to Upper Jurassic pelagic ooze-chalk-limestone sections has shown that in general lithification increases with age and depth of burial. However, the relationship between degree of lithification and depth of burial in any core is not a direct one. A diagenetic model is presented that accounts for the major reduction in porosity and foraminiferal content with depth and age and the development of cement and overgrowth on those microfossils which are not dissolved. The primary diagenetic mechanism functions through the solution of less stable, very small, calcite crystals such as make up small coccolith elements and the walls of Foraminifera, and reprecipitation of calcite upon large crystals such as make up discoasters and large coccoliths. The concomitant decrease in surface energy probably provides the driving force for this process. The variation in the degree of cementation of ooze-chalk-limestone sequences when plotted as a function of depth is ascribed to initial variations in the diagenetic potential of the sediments as they are buried. The diagenetic potential of a sediment is defined as the length of the diagenetic pathway the sediment has left to traverse before it becomes a crystalline aggregate. In marine acoustistratigraphy, the concept of the diagenetic potential relates seismic reflectors to original intrastratal differences in microfossil content. Seismic reflectors in this context therefore record palaeo-oceanographic events such as changes in the calcite compensation depth, surface water temperature, plankton productivity and glacio-eustatic sea level.

INTRODUCTION

The success of the Deep Sea Drilling Project (DSDP) in recovering cores from thick sequences of strata of Recent to Jurassic age comprising oozes, chalks and limestones made up almost entirely of the remains of calcareous plankton has accelerated the study of the diagenesis of oceanic pelagic carbonates. Many chapters in the Initial Reports of the Deep Sea Drilling Project and dozens of articles in journals

*Contribution No. IGPP–UCR–73–43, Institute of Geophysics and Planetary Physics, University of California, Riverside.

31

have appeared covering all aspects of the diagenesis of pelagic carbonates. A brief review of the results of Legs I through IX was given by Davies & Supko (1973). Much data and many observations on the variation with depth of burial and age, of (1) physical properties—density, porosity, and sonic velocity—; (2) fossil preservation and (3) sediment textures have been published (e.g. Gealy, 1971; Moberly & Heath, 1971; Roth & Thierstein, 1972; Wise & Kelts, 1972; Douglas, 1973a, b; Bukry, 1973; Wise, 1973). One of the results of the Deep Sea Drilling Project has been the finding that in relatively pure pelagic carbonate sections there is a general progression from ooze to chalk to limestone with increasing depth of burial and age of the sediment. The exact relationship between lithification, age and depth of burial is, however, far from clear.

In this paper we develop a diagenetic model for the ooze-chalk-limestone transition using data on the physical properties, fossil preservation and textures and the geological setting of DSDP Site 167 (Schlanger *et al.*, 1973). At this Site, on the crest of the Magellan Rise in the central North Pacific, the drilling bit penetrated to what was then a record sub-bottom depth of 1185 m, 1172 of which were in a purely pelagic sequence of Quaternary to Berriasian-Tithonian (Cretaceous) strata dominated by carbonates (Figs 1, 5). From the diagenetic model we have developed the concept of a 'Diagenetic Potential' (Table 2); we have then applied this concept to the problem of the development of acoustic reflectors in the thick sections of pelagic carbonates deposited in the equatorial Pacific Basin since the period of widespread chert development in Middle to Late Eocene time.

TERMINOLOGY

As used in this paper

Lithification refers to the total effect of all diagenetic processes that serve to convert a mass of loose grains into a coherent rock; these include but are not limited to gravitational compaction, grain interpenetration and cementation.

Cementation refers to the single process in which calcite, precipitated from an intergranular solution, connects and binds grains into a rigid framework.

As stated in the lithological nomenclature conventions of the Deep Sea Drilling Project (see Winterer, Ewing *et al.*, 1973, pp. 9–10) 'Oozes have little strength and are readily deformed under the finger or the broad blade of a spatula. Chalks are partly indurated oozes; they are friable limestones that are readily deformed under the fingernail or the edge of a spatula blade. Chalks more indurated than that are simply termed limestones.' This nomenclature for oozes, chalks and limestones is followed in the present paper.

A DIAGENETIC MODEL FOR THE OOZE-CHALK-LIMESTONE TRANSITION

Any diagenetic model must be based on observations and set into a reasonable geological framework. In developing the model (pictured in Fig. 6 and discussed below) we considered the following attributes of pelagic oozes—using the carbonate sequence on Magellan Rise as an example.

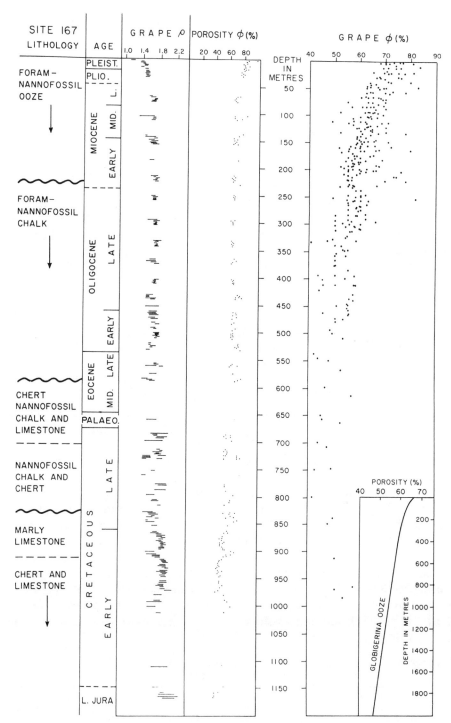

Fig. 1. Porosity-depth relations in pelagic carbonates. Plots to the left of the depth scale are GRAPE density and porosity data for DSDP Site 167 (Winterer, Ewing *et al.*, 1973). The points plotted to the right of the scale are based on averaged porosity data from DSDP Sites, 62, 63, 64, 71, 77, 78, 158, 161, 214, 216, 217. These sections are all characterized by high CaCO₃ (80% or higher) content. Inset is from Hamilton (1959) based on experimental data on the compaction of a *Globigerina* ooze containing 54% CaCO₃. Note the rapid decrease of porosity with depth in the upper 100–200 m.

33

(1) Physical properties.
(2) State of fossil preservation.
(3) Textural changes with depth.
(4) The geological setting and sedimentological history.

Physical properties

Perhaps the most important property relevant to diagenetic studies is porosity. In pure carbonate sequences where grain density may be considered a constant, the porosity is inversely proportional to the density and the seismic velocity. Early on in the progress of the Deep Sea Drilling Project, as pure carbonate pelagic sequences were drilled (Hays *et al.*, 1972; Tracey *et al.*, 1971; Winterer *et al.*, 1971), it became apparent that there was a rough correlation between depth of burial (and therefore age) and degree of lithification of such sections. This correlation was predicted by Hamilton (1959) and by the experimental work of Laughton (1954, 1957). Depth-porosity data from a number of DSDP Sites characterized by high calcium carbonate contents (above 80%) are summarized in Fig. 1. The depth-porosity data for Site 167 are also shown as is the curve generated by Laughton's (1954) experimental compact-tion of a *Globigerina* ooze (inset on Fig. 1). Space limitations do not permit a comprehensive review of porosity data from other DSDP Sites. However, certain Pacific sites at which thick sections of relatively pure carbonates were drilled are of interest. Cook & Cook (1972) noted that porosity values ranged from 94% to 41% and that they generally decreased with depth in the holes. They attributed this decrease to compaction, to incipient cementation, or to both. Tracey *et al.* (1971, p. 22) reported a relationship between the depth of burial and the degree of induration of pelagic oozes sampled on Leg 8. They noted (p. 23) a decrease in porosity and an increase in sediment stiffness which might have occurred from gravitational compaction and partial lithification. Sequences of nannofossil ooze-chalk-limestone as deep as 985 m and as old as Middle Eocene were penetrated at Sites 62, 63 and 64. At all three sites the porosity decreases irregularly and the rate of decrease appears to be related to age as well as depth of burial (Gealy, 1971, pp. 1103–1104). The saturated bulk density and porosity of Upper Oligocene and Middle Miocene nannofossil ooze, respectively, are nearly identical, despite different depths of burial (Table 1). Gealy noted (p. 1104) that porosity reversals, where porosity decreases with depth and then increases, occur on both the centimetre and decimetre scale. The change in porosity with depth and the density distributions at Site 167 (Fig. 1) are similar to those of Leg 8 and Sites 62, 63 and 64. The change from nannofossil ooze to chalk at Site 167 does not seem to be

Table 1. Comparison of bulk density (ρB) and porosity (φ) with age and depth of burial at three Deep Sea Drilling sites in the Western Pacific (Gealy, 1971)

	Site 62			Site 63			Site 64		
	Depth (m)	ρB	φ	Depth (m)	ρB	φ	Depth (m)	ρB	ρ
Recent	0	1·50	72	0	1·50	72	0	1·50	72
Middle Miocene	340	1·75	57	140	1·75	57	300	1·71	59
Upper Oligocene	520	1·90	49	350	1·91	49	560	1·85	51

caused by a change in porosity and occurs within a thick interval of almost constant densities of 1·60–1·70 (Winterer, Ewing *et al.,* 1973).

From the above the following can be seen.

(a) There is indeed a trend, over long stratigraphic intervals, towards decreasing porosity and increasing lithification in carbonate sections; the ooze to chalk to limestone transition appears, over long stratigraphic intervals, to be an irreversible one.

(b) Over shorter intervals, however, the correlation between depth and porosity is inexact.

(c) Furthermore, ooze-chalk-ooze-chalk sequences can occur in sections through which there is no discernible change in porosity. This last conclusion has important implications for acoustistratigraphy as discussed below. In terms of a diagenetic model there appear to be two stages in the reduction of porosity within a long section.

(i) An early dewatering stage in which, as may be seen in Fig. 1, half of the total porosity reduction takes place in the upper 200 m. This part of the section is termed in this paper the Shallow-burial Realm (Realm V of Table 2) where porosity is reduced from approximately 80% to 60% in most sections. Considering a sedimentation rate of 20 m/million years to encompass most sections drilled, this stage is completed in at most 10 million years. The dominant mechanism in this realm is gravitational compaction; cementation is a subordinate process.

(ii) A slower dewatering stage (lasting on the order of tens of millions of years). This is the Deep-burial Realm (Realm VI of Table 2) where porosity is reduced from approximately 65% to approximately 40% at a depth of 1000 m. The dominant process in this realm is cementation; gravitational compaction is the subordinate process.

Another important related physical property to be considered in any diagenetic model is the compressional velocity defined as:

$$V_c = \left(\frac{\lambda+2\mu}{\rho}\right)^{\frac{1}{2}}$$

where

V_c=compressional velocity; ρ=density; λ and μ are the Lamé stress constants:

$$\lambda = \frac{\sigma}{1-2\sigma} \cdot \frac{E}{1+\sigma}; \; \mu = \frac{1}{2} \cdot \frac{E}{1+\sigma}$$

σ and E being Poisson's ratio and Young's modulus respectively.

The values of the elastic constants, E and σ, are influenced in part by the degree of cementation of the rock whereas ρ may simply be a function of the degree of compaction of the sedimentary column. Thus, if a foraminiferal-nannofossil ooze were merely compacted without cementation, the velocity-depth function would be smooth (Fig. 7). However, there are marked deviations of V_c from such a curve because the processes of cementation occurring over an extended time interval should have the effect of increasing velocity (Nafe & Drake, 1963, pp. 811–812). The triangles on Fig. 7 are the interval velocities for the major lithological units at Site 167 (Winterer, Ewing *et al.,* 1973): from 0 to 220 m, 1·82 km/s; 220–600 m, 2·11 km/s: 600–820 m, 2·38 km/s; and 820–1185 m, 3·26 km/s. The divergence of these measured values from the values predicted by the smooth compaction curve indicates the degree of cementation the various intervals of the sedimentary column have undergone.

Fossil preservation

Foraminifera

The two types of Foraminifera, benthonic and planktonic, present in these pelagic carbonates contribute in somewhat different ways to the diagenetic process. On the average, benthonic Foraminifera are more preservable than are planktonic Foraminifera. Benthonic species usually have thicker walls and fewer pores than planktonic Foraminifera which tend to make the former more resistant to dissolution and, in some cases, to mechanical breakage. The preferential loss of planktonic species relative to benthonic species is well illustrated in the down-hole changes in Foraminifera at Site 167 (Fig. 2). The changes which occur in benthonic Foraminifera with increased depth of burial, dissolution and lithification (ooze to chalk to limestone) are many and include the following.

A deterioration in surface lustre and transparency. This seems to be the initial sign of dissolution. Hyaline calcareous benthonic Foraminifera from the Quaternary have smooth surfaces and a shiny lustre and are translucent. However, by Late Miocene these lustres have been lost or are considerably duller in appearance. Miliolids, which had a porcelaneous lustre, have a chalky appearance and most shells are broken or contain holes. Many hyaline shells have acquired a cloudy appearance.

A preferential removal of thin-walled, more porous rotaline species and especially miliolids. No miliolids were found below Core 8 which corresponds to the beginning of the transition from ooze to chalk.

A rapid down-hole increase in the percentage of broken tests. The increased fragmentation of planktonic species is most noticeable in the upper four cores, thereafter the percentage is fairly constant or decreases as the fragments become too small to identify or are dissolved. The percentage of broken benthonic species gradually increases from about 20% in the ooze to over 50% in the chalk. Below about Core 27 (Upper Eocene) the percentage of broken tests decreases as the total number of benthonic Foraminifera decreases. In the firm to hard chalks and limestones of Early Tertiary and Cretaceous age foraminiferal loss increases noticeably.

Calcite overgrowth and chamber infilling. Beginning in the Oligocene chalks (around Core 13) calcite overgrowth was noted on the inner chamber wall and pores of some species. Chamber infilling of calcite occurs in the lower part of the chalk sequence but was more common in the limestone and chalks of the Cretaceous. The percentage of shells with calcite overgrowth or infillings increased rapidly in the Cretaceous cores below Core 50 (Upper Campanian) and all shells were affected below Core 55.

The main effect of diagenesis on planktonic Foraminifera is a continued down-hole decrease in abundance due to dissolution. The minimum loss in foraminiferal sediment assemblages due to dissolution can be estimated from the equation $L=100(1-R_0/R)$, where L is the loss necessary to increase the insoluble residue R_0 to R %, and by assuming benthonic Foraminifera are an insoluble component of the sediment assemblage and their fraction is altered only by dissolution, not by changing productivity (Berger, 1971). The estimate supposes there is no dissolution of benthonic Foraminifera, an assumption that can be shown to be incorrect although most benthonic species appear to be more resistant to dissolution than planktonic species. Arrhenius (1952) and Berger (1973b) assumed that benthonic species dissolve three times more slowly than planktonic species. Intuitively this assumption is more attractive than assuming R_0 is insoluble but in fact it has little justification as dissolution

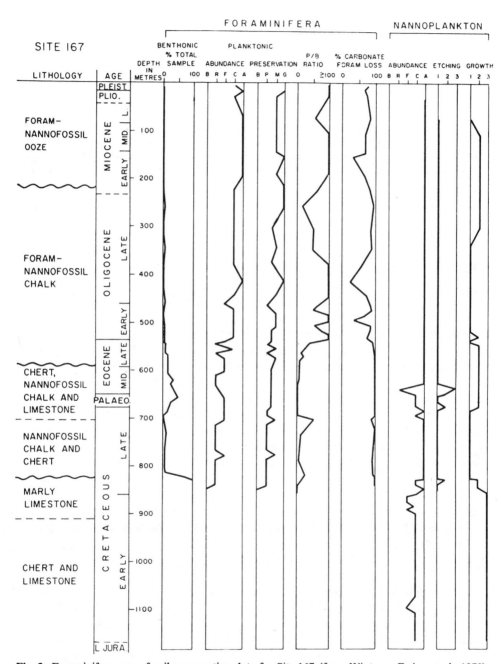

Fig. 2. Foraminifera-nannofossil preservation data for Site 167 (from Winterer, Ewing *et al.*, 1973). The shape of the planktonic/benthonic ratio (*P/B* ratio) curve is of interest in that the two sharp decreases in this ratio at horizons of Late Miocene and Late Oligocene age correspond to the position of reflectors b and d. These decreases show the extent of planktonic test dissolution at these horizons.

rates for benthonic Foraminifera have not been determined. The estimate L provides a useful measure for understanding preservation in pelagic foraminiferal sediments even though the calculation yields values which are probably low. In the calculated foraminiferal carbonate loss (L) values shown in Fig. 2, R_0 is 0.2%, based on the fact that in modern, well-preserved pelagic sediments planktonic Foraminifera compose 99.5–99.9% of the total assemblage (Schott, 1935; Parker, 1954; Thiede, 1972). A small increase in the percentage of benthonic Foraminifera may reflect a large loss of planktonic Foraminifera.

At Site 167 (Fig. 2) the value of L increases with increasing depth below the sea floor and with increasing lithification. Apparent loss values range from 30% to 85% for oozes, 60–96% for Tertiary chalks, and over 90% for Cretaceous chalks and marly limestones. According to this estimate, even well-preserved sediments in the upper seven or eight cores have had on the average more than 50% of the planktonic Foraminifera dissolved to account for the increased percentage of benthonic species. The L values for ooze are in good agreement with L values in Recent sediments from comparable water depth and latitude (Parker & Berger, 1971) and similar sediment cores (Berger, 1971; Berger & von Rad, 1972). Planktonic foraminiferal loss for hard chalks and marly limestones is on the average over 75% and may be as high as 98% of the original sediment assemblage. In this calculation, the actual destruction of planktonic species may be underestimated since benthonic species have been destroyed by dissolution and disintegration.

Site 167 is located near the crest of the Magellan Rise; there is no source for redeposited benthonic species. The large percentage of less-than-44 µm fraction suggests there has been no winnowing. Thus the estimated foraminiferal loss based on the value of L is a reasonable measure of the diagenetic destruction that has occurred in the planktonic foraminiferal population.

Thus for the purpose of developing a diagenetic model the benthonic Foraminifera are important inasmuch as they indicate the amount of dissolution the planktonic populations have undergone. The very large losses of the planktonics show that these forms are altered during the diagenetic process due to direct dissolution of their framework-supporting structure. Dissolution and collapse of weakened tests can account for most of the porosity loss within the sedimentary column (Fig. 1). The amount of calcite dissolved from the tests almost equals the amount of new calcite formed as interstitial cement, calcite overgrowths and infillings of remaining tests and nannofossil overgrowths.

Nannofossils

Since 1969 the availability of both Scanning Electron Microscopes (SEM) and cores recovered by the Deep Sea Drilling Project have literally fostered an explosion in the study of the progressive diagenesis of nannofossils. In going from Bramlette's pioneering work (1958) to the experimental evidence presented by Adelseck, Geehan & Roth (1973) it has become abundantly clear that nannofossils are an important factor in the diagenesis of pelagic carbonates. According to Roth & Thierstein (1972): 'a slight degree of secondary calcite overgrowth is found in most carbonate oozes that have been buried under about 100 metres of sediment'. Thus the initial stages of subsurface nannofossil diagenesis are within the Shallow-burial realm. As shown by Fischer, Honjo & Garrison (1967) the final stages of pelagic-carbonate diagenesis involve the remaining nannofossils. The trends in nannofossil diagenesis that have become clear through the studies of, among others, Fischer *et al.* (1967) and Adelseck *et al.* (1973) are as follows.

(i) Small coccoliths tend to dissaggregate along sutures due to dissolution, thus producing in the sediment large numbers of micron-sized crystals that are very susceptible to solution.

(ii) Discoasters tend to grow by the precipitation on them of highly euhedral calcite overgrowths; the overgrown discoaster may contain a volume of calcite several times greater than the original.

(iii) Larger coccoliths show overgrowths and persist in highly recrystallized limestones.

(iv) The very abundant micron-sized elements of coccoliths can supply some of the calcite for the overgrowths on discoasters and larger coccoliths, the interstitial cement and the calcite infillings of Foraminifera.

Textural considerations

The fact that these pelagic sediments are virtually 100% biogenic calcite means that the textural changes observed in their diagenetic transformation from ooze to chalk to limestone are the direct result of the response of the foraminiferal tests and nannofossils to burial. Numerous scanning electron micrographs in many papers have illustrated these changes (e.g. Wise, 1973; Bukry, 1973; Adelseck *et al.*, 1973; Wise & Kelts, 1972). For the purposes of this paper four scanning electron micrographs from the Magellan Rise Deep Sea Drilling site summarize the pertinent textural changes for which a diagenetic model must account (Fig. 3a–d).

(a) A Middle Miocene specimen (DSDP core 167–6, depth 103–112 m) showing intact but slightly etched tests of planktonic Foraminifera in a matrix of generally well-preserved nannofossils, mainly coccoliths. The still-whole chambers enclose large volumes of intrabiotic pore space. The abundant tests form a supporting framework. The scattered anhedral to subhedral grains of calcite are the result of disintegration of some small coccoliths and possibly Foraminifera tests. The porosity in this interval ranged from 62·8% to 67% (Winterer, Ewing *et al.*, 1973).

(b) An Upper Oligocene chalk (DSDP core 167–11–4, depth 297–306 m) shows the whole spectrum of fossil-preservation phenomena; corroded foraminiferal tests (lower right), discoasters with massive euhedral overgrowths (upper right) and severely etched coccoliths. The porosity in this interval ranged from 61% to 64%.

(c) A Valanginian-Hauterivian limestone (DSDP core 167–81CC, depth 1040 m). This shows a mass of calcite crystals, that appears to be filling a void or replacing a foraminiferal test, in a matrix of coccolith fragments and fine-grained cement. Porosity was not determined on this sample.

(d) A Berriasian-Tithonian limestone (DSDP core 167–94–2, depth 1165–1168 m) shows very severely etched coccoliths in a matrix of euhedral to subhedral cement; an etched nannoconid is seen in the lower left corner. Porosity in this interval ranged from 33% to 48%.

Geological considerations

The Magellan Rise is a relatively smooth-topped structure that reaches to within 3200 m of the sea surface and stands above a plain at a depth of 5000–6000 m (Fig. 5). The seismic profile shows that the internal stratigraphy is quite uniform. Basement

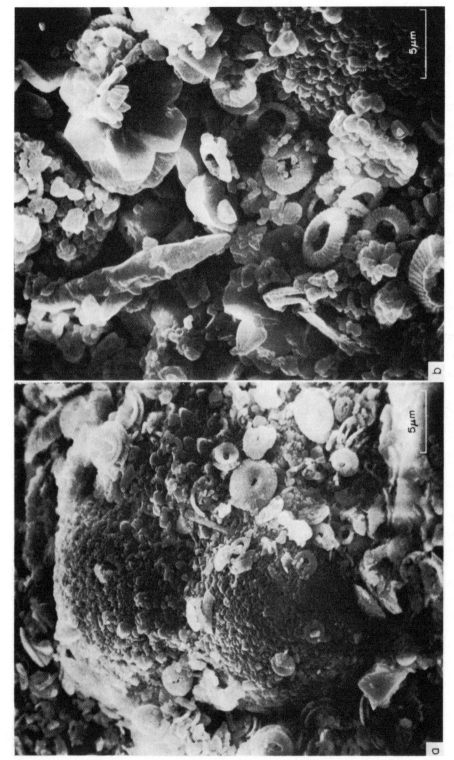

Fig. 3. Scanning electron micrographs of representative sediments from Site 167. See text for descriptions.

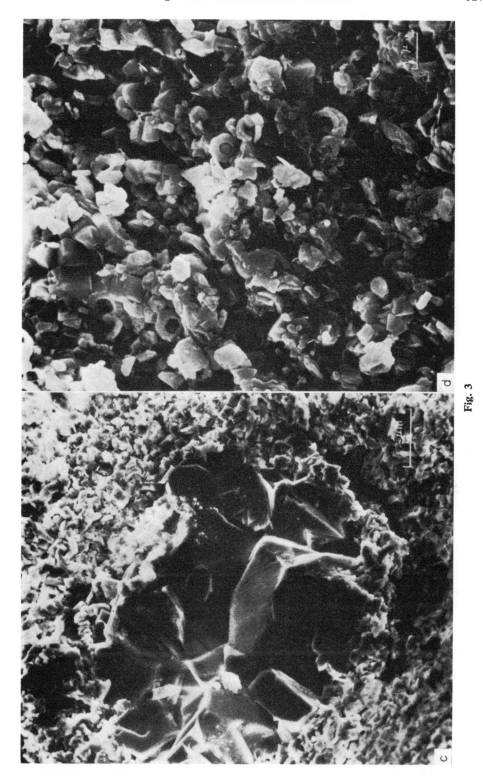

Fig. 3

ridges, that trend parallel to the long axis of the Rise, are local features where the sediments show draping effects perhaps accentuated by compaction. The Rise has, quite simply, been the recipient of pelagic sediments over the past 135 million years (Winterer, Ewing *et al.*, 1973). There is no evidence in the cores that the surface of the Rise was ever below the calcite compensation depth. The geometry of the sedimentary cap suggests that water flow within it has been primarily upward and outward as the accumulating sediments slowly compacted. It is difficult to see how much water could have flowed into the cap. Thus a diagenetic model must provide for a cement source within the sediments, i.e. a conservative system exists in which the original biogenic calcite is the sole diagenetic source material.

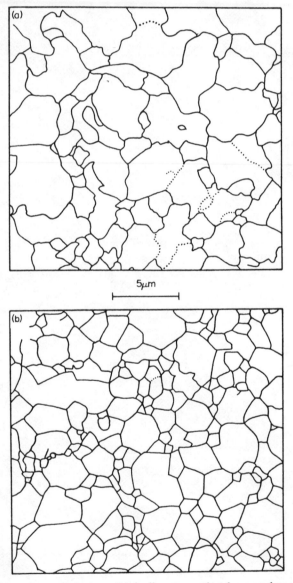

5μm

Fig. 4. Ameboid (a) and mosaic textures (b) in limestones that have undergone some degree of metamorphism (from Fischer *et al.*, 1967).

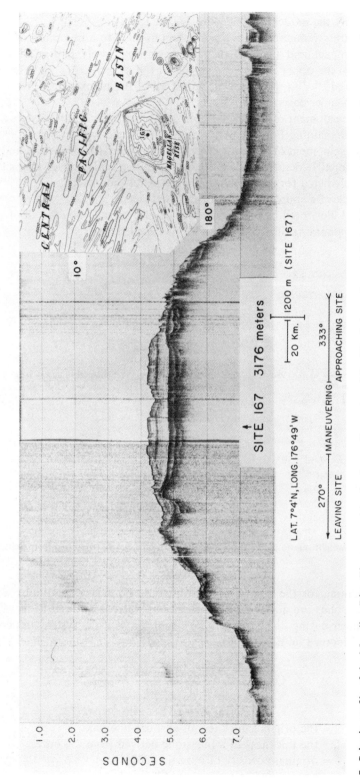

Fig. 5. Seismic profile of the Magellan Rise. The apparent 'graben' near the centre of the Rise is in part an artefact of the course changes during manoeuvring.

In summary the model must account for the following.

(a) A porosity reduction from a maximum of 80% to a minimum of 40% as the sediments compact and undergo diagenesis, approximately half of this reduction taking place in the upper 200 m.

(b) The gradual disappearance of almost all of the planktonic Foraminifera and many of the smaller coccoliths.

(c) The development of interstitial cement, overgrowths on discoasters and calcite fillings in the remaining Foraminifera (largely benthonic types).

(d) The development of cement to a degree that the elastic properties and therefore the compressional velocities of the more deeply buried sediments diverge significantly from values predicted for sediments that were merely compacted.

(e) The lack of a source of extra-formational calcite.

The model illustrated in Fig. 6 was developed by accounting for all of the calcite in the system assuming little or no loss to the expelled waters.

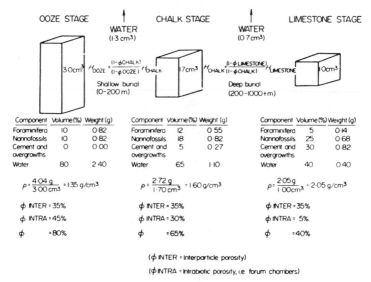

Fig. 6. Volume-weight relations in the ooze-chalk-limestone system (modified from Schlanger *et al.*, 1973).

In accounting for the components, volumetric boundary conditions were established. If no solids are added to the system and water is allowed to leave, then the volume changes during diagenesis can be calculated as if the system was compacting as follows (Moore, 1969):

$$\text{Let } H_{\text{ooze}} = \frac{(1 - \phi_{\text{limestone}})}{(1 - \phi_{\text{ooze}})} H_{\text{limestone}};$$

where

H_{ooze}	= the original thickness of an ooze interval,
$H_{\text{limestone}}$	= the thickness of a limestone derived from the ooze,
ϕ_{ooze}	= original porosity of ooze (i.e. 80%) and
$\phi_{\text{limestone}}$	= porosity of limestone (i.e. 40%).

Substituting the appropriate values and setting $H_{\text{limestone}}$ equal to 1 cm:

$$H_{\text{ooze}} = \frac{(1 - 0{\cdot}40)}{(1 - 0{\cdot}80)} \cdot 1 \text{ cm}$$

$H_{\text{ooze}} = 3$ cm.

Thus, approximately 3 cm³ of foraminiferal-nannofossil ooze with a porosity of 80% will reduce to 1 cm³ of limestone with a porosity of 40% in going from ooze to limestone.

By using a petrographic and textural approach based on scanning electron micrographs of pelagic sediments and tests of individual Foraminifera, a model of ooze diagenesis was developed.

Scanning electron micrographs of foraminiferal tests show that the three types of primary intrabiotic porosity (chambers, foramen, and interwall) make up 80% of the volume occupied by an individual foraminiferal test. Further, a loosely packed aggregate of tests would have primary interparticle porosity. Considering a pure foraminiferal ooze as an aggregate of spheres and considering that the interparticle porosity of spherical aggregates of uniform size ranges from 26% (rhombohedral packing) to 48% (cubic packing) an original interparticle porosity of 45% is reasonable. Thus, 80% (intrabiotic porosity) of 55% (sphere-enclosed volume) equals 44% and this, plus the interparticle porosity, equals approximately 90%—that of a pure foraminiferal ooze. If an equal amount, by volume, of nannofossil calcite is added to the interparticle void space, the resulting ooze is volumetrically: 10% foraminiferal calcite, 10% nannofossil calcite, and 80% intrabiotic and interparticle pore space; the sediment would have a honeycomb structure. The ooze stage in Fig. 6 was constructed on the above considerations.

The major changes in going from ooze to limestone involve the dissolution, and eventual almost complete destruction of the framework-supporting tests of the Foraminifera and the etching and dissolution of smaller coccoliths. This decrease in the volume of biogenic calcite by dissolution is balanced by a build-up of calcite overgrowths on remaining fossils, particularly discoasters, and formation of calcite cement. The foraminiferal loss amounts to almost 50% going from ooze to chalk and nearly 100% from ooze to limestone (Fig. 6). As shown above, approximately 44% of the original ooze was intrabiotic porosity. Wholesale dissolution of the tests with reprecipitation of the test calcite as cement and nannofossil overgrowths in going from ooze to limestone would result in a porosity reduction approaching the 40% demanded by this model (Fig. 6).

THE DIAGENETIC POTENTIAL

The shallow- and deep-burial realms, discussed above, can be placed into a scheme covering all stages of diagenesis of pelagic carbonate sediments (Table 2). Such a scheme is convenient to the discussion of the concept of a diagenetic potential.

The life history of a pelagic limestone can be thought of as beginning in the upper 200 m of the ocean (Realm I) where planktonic organisms produce calcite as a highly dispersed suspension which has a very high potential for future diagenetic change. As this dispersed calcite aggregates by sedimentation to the sea floor and lithifies by compaction and cementation its diagenetic potential decreases; the sediment matures.

Table 2. Diagenetic realms

Depth	Realm	Residence Time	Petrography	Porosity (ϕ) % / Velocity (V_c) km/s	Diagenetic potential 0 → ∞
0–200 m (surface water)	I Initial production	Weeks	Highly dispersed calcite-sea water system; $10–10^2$ forams/m^3, $10^4–10^6$ nannoplankton/m^3 (Lisitzin, 1971; Berger, 1971)		
200 m to sea floor	II Settling	Days to weeks for forams; months to years for coccoliths depending on pelletization (Smayda, 1971)	Pelletized coccoliths, ratio of broken to whole nannoplankton increases downward, ratio of living to empty foram tests decreasing during settling (Lisitzin, 1971; Berger, 1971)		
3000–5000 m (see Diagenetic Potential)	III Deposition	Inversely proportional to sedimentation rate and dissolution rate	'Honeycombed' structure (Tschebotarioff, 1952). Large foram tests supported by chains of coccolith discs. This surface is actually part of Realm IV	$\phi \simeq 80\%+$ $V_c \simeq 1·45–1·50$ (Nafe & Drake, 1963)	← slope at 3000 m ← Slope at 5000 m.
0–1 m sub-bottom)	IV Bioturbation	50,000 years (at 20 m/10^6 years sedimentation rate)	Remoulded 'honeycomb', slight compaction, burrowing, destruction by ingestion and solution	$\phi \simeq 75–80\%$ $V_c \simeq 1·45–1·6$	
1 200 m sub-bottom)	V Shallow-burial	10×10^6 years (at 20 m/10^6 years sedimentation rate)	Ooze affected by gravitational composition, establishment of firm grain contacts; dissolution of fossils and initiation of overgrowths	$\phi \simeq 75–60\%$ $V_c \simeq 1·6–1·8$	
200–1000 m + (sub-bottom)	VI Deep-burial	Up to $\simeq 120 \times 10^6$ years (by then either subducted or uplifted)	Chalk with strong development of interstitial cement and overgrowths; transition down to limestone with dissolution of forams, pervasion by cement and overgrowths—grain interpenetration, welding and 'ameboid mosaics' (Fischer et al., 1967)	$\phi \simeq 60\%$ down to 35–40% V_c 1·8 increasing to 3·3 km/s	
1–10 km sub-surface	VII Metamorphic	$10^6–10^7$ years	Recrystallization trending to 'pavement mosaic' (Fischer et al., 1967) of completely interlocking crystals	$\phi \simeq 40\%$ down to < 5% V_c 3+ up to 6 km/s	

The life history of a pelagic limestone ends as the rock approaches a final state in which further diagenetic change is not probable, e.g. an aggregate of pure, unstrained crystals at a minimum free energy level. As Byrne points out (1965, p. 105):

'In three dimensions, all space can be filled so as to satisfy surface tension requirements if all grains are cubo-octahedrons of minimum surface-to-volume ratio. A cubo-octahedron is made up of eight faces which are regular hexagons and six faces which are squares . . . . If all grains had this shape, all grain boundaries would be in equilibrium and no further growth would occur.'

In discussing lithified micrites, Bathurst (1971, p. 511) states that the widespread upper crystal diameter for the groundmass of lithified micrites at 3–4 µm, 'points to the existence of a universal threshold state at which fabric evolution stops and beyond which it can, but need not, continue. A possible reason for this, which would bear further investigation, is that a stage is reached in the combined neomorphism and cementation, when the porosity and permeability are so reduced that the transport of Ca^{2-} and CO_3^{2+} from one crystal face to another becomes slow even on the geological time scale. This stage would represent virtual stability'.

He also states (Bathurst, 1971, p. 502) that:

' . . . lithification of micrite and the growth of neomorphic spar yield between them, a range of calcite fabrics which, once evolved, appear to resist strongly any further diagenetic change. These are arbitrarily classified as micrite (0·5–4 µ), microspar (5–50 µ) and pseudospar (50–100 µ) . . . . '

Fischer *et al.* (1967, Fig. 3a, b, Fig. 4a, b, p. 20) show the apparently penultimate and final life stages of nannofossil limestones from the Franciscan Formation of California as 'ameboid mosaics' and 'pavement mosaics' respectively, see Fig. 4, and Table 2. Byrne's space-filling cubo-octahedrons, Bathurst's micrite and microspar, which are the right size range from limestone made up of overgrown large coccoliths and discoasters, and the ameboid and pavement mosaics of Fischer *et al.* (1967) all represent states of near-zero diagenetic potential. Bramlette (1958, p. 126) recognized this trend to the holocrystalline aggregate stage in his studies of outcrop material from France and North Africa. He stated that the absence of coccoliths in older fine-grained limestone would be inevitable with the recrystallization that has produced the aplite texture.

Diagenetic potential may thus be defined as the length of the diagenetic pathway left for the original dispersed foraminiferal-nannoplankton assemblage to traverse before it reaches the very low free-energy level of a crystalline mosaic.

The concept of the diagenetic potential can be used to explain deviations from a strict depth-of-burial/lithification dependence demanded by a simple gravitational compaction model of diagenesis. The diagenetic potential remaining to a sediment after it passes through the critical boundary between Realms IV and V (Table 2) will determine how far cementation will proceed per unit time. Thus, if layers of very different diagenetic potential are buried sequentially, the amount of cementation in a layer per unit time will be proportional to the diagenetic potential of that layer so that chalks can form above oozes and limestones above chalks in the sedimentary column.

The diagenetic potential (DP) can be expressed as follows:

$$DP = f \text{(water depth, sed. rate, Temp (surface), Productivity (surface), foraminifer}$$
$$+ \text{coccolith: discoaster ratio; } Size_{(max)} : Size_{(min)} \text{ ratio, predation rate . . . .)}$$

The depth of water and the sedimentation rate are very important factors as these affect the degree to which the calcite is dissolved while at the sediment/water interface; calcite dissolved at this interface is the more soluble portion of the sediment, and is not available for further diagenesis in buried strata.

The diagenetic potential of a sediment is enhanced in zones of high productivity, high lysocline and low calcite compensation depths which favour the mobilization and redeposition of skeletal calcite (Wise, 1972). The surface water temperature, and salinity and biological factors not understood, affect the $MgCO_3$ content of the pelagic sediment in a complex manner because the temperature-salinity structure of the upper water layers apparently influences the contributions made to the general population by various species and genera of Foraminifera, each of which have characteristic $MgCO_3$ contents. As Parker & Berger (1971) and Savin & Douglas (1973) point out, the $MgCO_3$ content of Foraminifera affects the solubility of these tests as they sink and as they reside at the sea floor. Since discoasters appear to be the most favoured receptors of calcite overgrowths, and foraminiferal tests and coccoliths the most called-upon donors for calcite in the cement, it seems possible that the original Foraminifera +coccolith/discoaster ratio is an important factor in the diagenetic potential because a high content of $MgCO_3$-rich tests in a buried sediment would give it a high diagenetic potential. A sorting measure should also be included since an abundance of very small grains, mixed with larger ones such as make up discoasters, appear to promote calcite transfer as discussed by Adelseck et al. (1973). These investigators subjected an Upper Pliocene nannofossil ooze from Site 167 to temperatures and pressures of up to 300°C and 3 kb for 1 month in order to simulate the diagenetic changes such a sediment might undergo with long and deep burial. They found that the smaller and more delicate coccoliths were more easily destroyed than the larger forms with strongly overlapping elements (crystallites). Most larger forms showed formation of secondary overgrowth though some underwent slight etching. Discoasters displayed well-developed overgrowths with crystal faces.

They further pointed out that the relatively rapid disaggregation of smaller coccoliths produced large quantities of less than 1 μm size crystals that evidently supplied the calcium carbonate for the overgrowths on the discoasters and the remaining larger coccoliths.

This growth of the larger crystals at the expense of the smaller ones, in the buried foraminiferal-nannofossil ooze, is evidently taking place through a dissolution-diffusion-precipitation mechanism. The abundant, less than 1 μm crystals, present as disaggregated coccoliths and in the tests of Foraminifera (see Fig. 3b, lower right) represent the phase that provides the calcite and the up-to-10 μm-size crystals present as discoasters and large coccoliths represent the phase that receives the calcite. The driving force for the dissolution-precipitation process is dependent on the large difference in surface area between a population of very small grains and a population of larger grains. In a dispersed system a term containing the surface energy must be taken into account in expressing the Gibbs free energy G of the system:

$$dG = -Sdt + PdV + \alpha dw + \xi \mu dn$$

where α represents the surface energy per cm² and dw is an element of surface area. As the small particles of calcite dissolve and the same volume of calcite is reprecipitated on larger crystals the free energy change will be proportional to the decrease in surface area of the small particles minus the increase in the surface area of the large crystals.

Further, edges and sharp corners contribute to the surface free-energy component of the total free energy of the calcite; the dissolution of very small grains lessens this contribution. In essence the appearance of the adw term causes the dispersed system to be more soluble than the bulk system. Evidently, based on petrographic and experimental data, the apparent irreversible trend in foraminiferal-nannofossil sediments towards the ameboid and pavement mosaics of Fischer *et al.* (1967) follows a trend of decreasing free energy in the calcite system.

The predation rate of plankton-feeders may be important as Smayda (1971) showed that the accumulation of nannoplankton in faecal pellets at near-surface levels hastened the sinking of this material, lessening the chances of its being dissolved as it descended into deep waters. Faecal pellets charged with nannofossil debris sink at the rate of approximately 100 m/day whereas naked fragments of coccoliths would take 1 or more years to sink 5000 m. At this slow slow sinking rate micron-sized crystals would totally dissolve (Peterson, 1966) and these ready donors of overgrowth calcite would not be available in the buried sediment.

We postulate that variations of the diagenetic potential with depth of burial in the sedimentary column are related to original variations in basin depth, the calcite compensation depth (CCD), and the calcareous-plankton productivity of the upper water layers. This relationship is important to the discussion of diagenetic potential and acoustistratigraphy developed below.

THE DIAGENETIC POTENTIAL AND ACOUSTISTRATIGRAPHY

If the velocity-depth function in foraminiferal-nannofossil sub-sea sections was smooth and continuous, such sections would be acoustically transparent. Such would be the case if the lithification of carbonate-rich sediments were solely the result of gravitational compaction. This transparency would be due to the lack of sharp acoustic impedance gradients and discontinuities within the sedimentary column. However, refraction and Deep Sea Drilling data (see Fig. 7) show that deviations in velocity values from those predicted by compaction models are the rule rather than the exception. These deviations have generally been ascribed to cementation effects which drastically influence the values of the elastic constants. Cementation, which markedly increases the rigidity of the sediment, increases the compressional velocity. Since acoustic impedance = density × compressional velocity (Hamilton, 1959, 1971), a marked change between two layers, in compressional velocity, without marked changes in density, will cause a reflection of acoustic energy.

Pure pelagic carbonate sections are characterized by an abundance of closely spaced reflectors (acoustic artefacts aside) caused by discontinuities, marked gradients and reversals in acoustic impedance values. Data presented by Gealy (1971) for DSDP Site 64 show that both positive and negative impedance differences over distances of only a few metres, within apparently homogeneous sediments, are very common and these appear to result from very subtle changes in degree of lithification. Thus, instead of seeing a non-reversible gradual transition from ooze to chalk to limestone with increasing depth of burial, there are short range fluctuations, of ooze - chalk - ooze - chalk - ooze - chalk - limestone - chalk - limestone sequences, within the long-range trend. These alternations show that the ooze-chalk-limestone transition is not strictly time- and depth-of-burial-dependent. The fact that

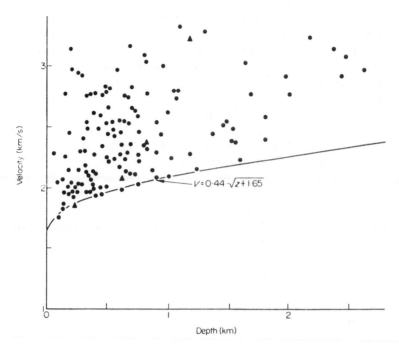

Fig. 7. Compressional velocity versus depth of burial plot. (Modified from Nafe & Drake, 1963, by addition of DSDP Site 167 interval velocity data (▲), see text for details.) The solid line is based on data from Laughton (1954) derived through compaction of *Globigerina* ooze. Other data from seismic refraction work.

uncemented oozes are found below hard chalks indicates that time and depth of burial taken alone cannot account for degree of cementation. According to arguments presented in a preceding section the degree of cementation exhibited by a carbonate pelagic sediment is in large part correlative with the diagenetic potential the sediment possessed when it finally passed below the bioturbation realm of diagenesis.

The diagenetic potential of the sediment, as discussed previously, is in large part pre-determined by oceanographic conditions prior to burial and even prior to the arrival of the sediment at the sea floor. We believe the abundant reflectors characteristic of carbonate sequences are related to the degree of cementation and that the degree of cementation is controlled by the diagenetic potential. Thus, since diagenetic potential is determined by palaeo-oceanographic conditions, it follows that an acoustistratigraphic event should correlate with a palaeo-oceanographic event.

This relationship of acoustistratigraphy to time-stratigraphy was noted by geologists during drilling operations at DSDP Site 64 on the Ontong Java Plateau (Winterer *et al.*, 1971) where many of the reflectors can be traced on profiles for tens or even hundreds of kilometres; further, the evidence of close parallelism of reflectors on the profiles suggested a time-stratigraphic control on induration. If the reflectors in these ooze-chalk-limestone sequences, so abundant in the post-Eocene-chert stratigraphy of the Pacific Basin, preserve palaeo-oceanographic events they should be correlatable and the strength of the correlation should be proportional to the magnitude and length of the event. To test this argument we compared reflection profiles and data on travel times to prominent reflectors from the Ontong Java Plateau (near DSDP Site

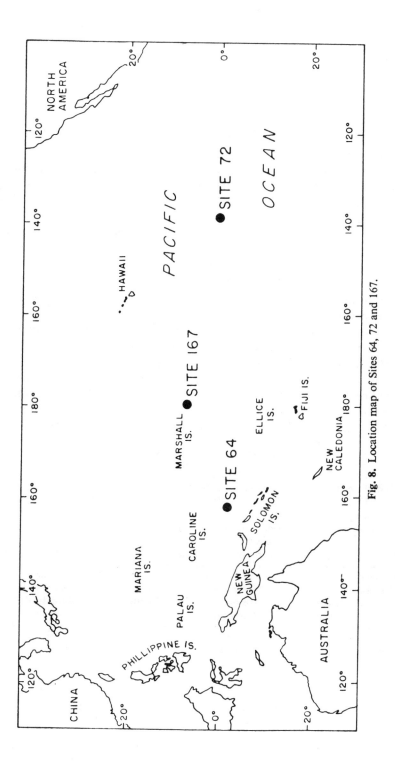

Fig. 8. Location map of Sites 64, 72 and 167.

64), the Equatorial Pacific (DSDP Site 72), and the Magellan Rise (DSDP Site 167), (Fig. 8). These three Deep Sea Drilling sites were picked because they have the following features in common.

(a) Pure carbonate sections above Eocene chert.

(b) Good seismic profiles near the drill sites.

(c) They are in areas free from turbidite contamination and represent as pure pelagic sedimentation as one can expect.

(d) The sections are well documented biostratigraphically.

(e) They are all in the central latitudes.

The results of this comparison between DSDP Sites 64, 72 and 167 are summarized on Tables 3 and 4 and in Figs 9–11 (the reflector terminology of 'a' to 'e' is for this paper only). Examination of (1) the reflection travel-time data published for Site 64 (Winterer *et al.*, 1971) and Site 72 (Tracey *et al.*, 1971) and (2) the average accumulation rates for the sites revealed that reflector travel time is proportional to the average post-middle Eocene accumulation rate. The proportionality between reflector travel time and accumulation rate provides a tool for looking back and forth between seismic profiles and arriving at correlations with respect to reflections. Thus the presence of a reflector at one site suggested where to look for a correlative reflector at another site. For example, the 0·035 reflector noted at Site 72 had no counterpart picked at Site 64. However, changes in the drilling rate were noted at that site as having a high degree of correlation with reflectors. Inspection of the drilling rate graph (Winterer *et al.*, 1971, Site 64, Fig. 5) shows a sharp drilling rate decrease at 85 m sub-bottom depth. This depth corresponds to the depth to a reflector at approximately 0·1 s (using the Site 64 $\bar{V}$ of 1·7 km/s for the upper part of the section). According to the accumulation rate proportionality argument this reflector should correspond to the 0·035 reflector at Site 72 (accumulation rate ratio of 2·5 : 1 yields $0·035 \times 2·5 = 0·09$ s).

Table 3. DSDP data for Sites 64, 72 and 167

Site 64 (Winterer *et al.*) Ontong Java Plateau 1°45'N, 158°37'E Water depth: 2052 m Depth to Middle Eocene chert: 985 m (total depth, 985 m)		Site 72 (Tracey *et al.*) Equatorial Pacific 00°26'N, 138°52'W Water depth: 4326 m Depth to Middle-Upper Eocene chert: 460 m (in adjacent basin)		Site 167 (Winterer *et al.*) Magellan Rise 7°04'N, 49'W Water depth: 3176 m Depth to Middle Eocene chert: 600 m (total depth, 1185 m)	
Prominent reflectors (seconds of travel time)	Depth (m)	Prominent reflectors (seconds of travel time)	Depth (m)	Prominent reflectors (seconds of travel time)	Depth (m)
		0·035	30		
0·43	366	0·16	135		
0·71	660	0·30 (0·28)	265	0·24	220
0·97	983	0·46	460	0·60	600
Average post-chert accumulation rate= 25 m/million years		Average post-chert accumulation rate= 10 m/million years		Average post-chert accumulation rate= 15 m/million years Average accumulation rate from 0–220 m=9 m/million years	

Table 4. Acoustistratigraphic correlations—Sites 64, 72 and 167

	Site 64					Site 72					Site 167					Reflecting horizons¶
	Reflector strength	Travel time	Depth	Zone	Age (million years)	Reflector strength	Travel time	Depth	Zone	Age (million years)	Reflector strength	Travel time	Depth	Zone	Age (million years)	Designation
a	Weak	0·10	85 (d.b.)*	N.20 interpol. (N.19–22)	3	Moderate	0·035	30	N.20–N.21	3			Not seen			a
b	Moderate	0·20	170 (d.b.)*	N.18 interpol. (N.17–19)	5–6			Not seen			Weak	0·07 0·08	60–70 §	N.17–N.18	5–6	b
c	Strong	0·43	366	N.12 interpol. (N.10–14)	14	Strong	0·16	135	N.14 interpol. (N.12–16)	13			Not seen			c
d	Strong	0·71	660	NP 25 L. bipes	24–26	Strong	0·30 (0·28)†	265	P.22	23–26	Strong	0·24	220	P.22–N.4	21–26	d
e	Strong	0·97	983	P.14 or older	43 or less	Strong	0·46	460 ‡	P.13 or younger	44 or less	Strong	0·60	600	P.14	43–44	e

* d.b. indicates drilling time break used to locate reflector.
† 0·28 corrected figure used in this paper. (Fig. 12, Site 72 report, Tracey *et al.*, 1971.)
‡ Projected basement depth in basin adjacent to Site 72.
§ Probably at hiatus between cores 3 and 4 in N. 17–18 Zone.
¶ Letter assigned for this paper.

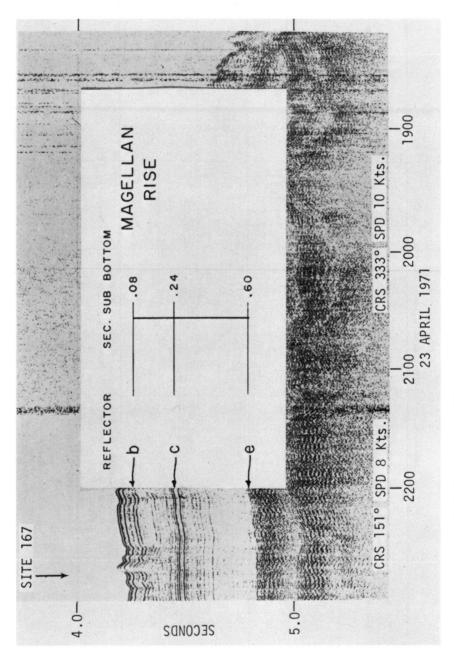

Fig. 9. Seismic profile of Site 167 and vicinity showing reflectors discussed in text.

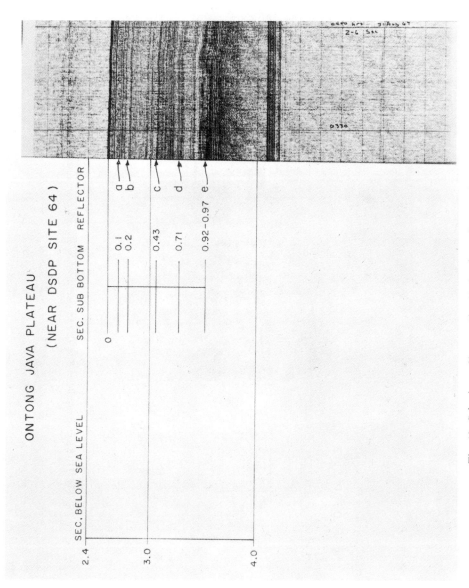

Fig. 10. Seismic profile near Site 64 showing reflectors discussed in text.

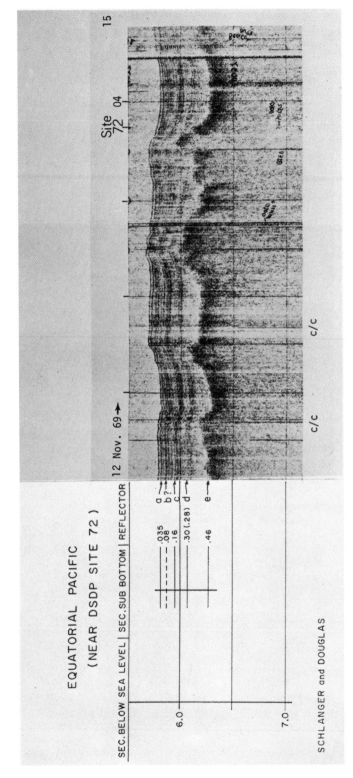

Fig. 11. Seismic profile of Site 72 and vicinity showing reflectors discussed in text.

In this manner the correlations shown on Table 3 were generated. The biostratigraphic zone assignments and ages are taken from the data published in the Deep Sea Drilling Project reports for these sites.

Certain events appear significant with respect to the 'a', 'b', 'c' and 'd' reflectors which date at approximately 3, 5–6, 12–14 and 21–26 million years B.P. respectively. Berger (1973a, Fig. 12); see Fig. 12 of this paper) points to two major fluctuations in the calcite compensation depth (CCD), that stand out above the 'noise' of shorter range events, at approximately 13 and 27 million year B.P. These two dates correlate with major temperature changes in the Pacific (Devereux, 1967; Douglas & Savin, 1971, 1973; Savin, Douglas & Stehli, in press) A rapid temperature decline in the Eocene reached a minimum in the Late Oligocene, about 27 million years B.P. and corresponds to the last and probably major phase of Palaeogene glaciation in the southern hemisphere (Savin *et al.*, in press; Hayes *et al.*, 1973) At approximately 12–14 million years B.P., the first significant temperature decline of the Neogene occurred. This event appears related to initiation of major glaciation in the northern hemisphere, glaciation having started earlier in the southern hemisphere (Hayes *et al.*, 1973). These dates also correlate very well with the periods of emergence of atolls in the Pacific (Fig. 12) described by Schlanger (1963), Emery, Tracey & Ladd (1954) and Ladd, Tracey & Gross (1970). The coincidence of these periods of atoll emergence with the 'deepenings' of the CCD, suggests that these events both correlate perhaps with glacio-eustatic changes in the depth of the Pacific Basin. Thus the 'c' and 'd' reflectors correlate well with glacio-eustatic regressions, the 'a' reflector at 3 million years B.P. correlates with a widespread cool period discussed by Ciaranfi & Cita (1973, pp.

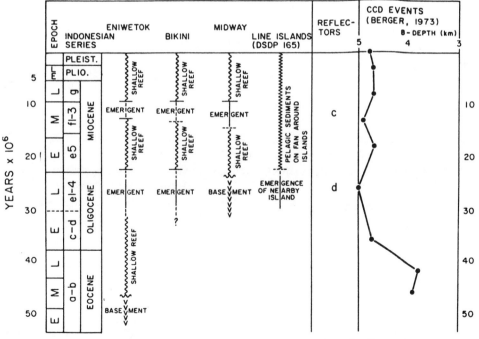

Fig. 12. Geological events, variations in calcite compensation depth and reflector horizons in the Pacific.

1396–98) and Cita & Ryan (1973, pp. 1408–09). The 'b' reflector correlates with the Messinian regression (5·5–7·0 million years B.P.) (see discussion in Ryan, Hsü *et al.,* 1973) and a glacial climax in Antarctica (Hayes *et al.,* 1973). Evidently the sediments deposited during the regression had a high diagenetic potential. Arrhenius (1952) postulated that during glacial intervals there would be a higher carbonate productivity due to increased upwelling. Thus a greater rate of carbonate deposition might account for the higher diagenetic potential and, therefore, the presence of the reflectors.

There are large numbers of minor local reflectors, e.g. the multiplicity of these on the Ontong Java Plateau. These are spaced only 10 m or so apart; at a sedimentation rate of approximately 25 m/million years (at Site 64): these would represent events that took place about 400,000 years apart. These may then represent fluctuations in diagenetic potential due to short-range changes in surface-water conditions. Short-term temperature fluctuations in surface temperatures, with a periodicity of approximately 80,000 years, have been identified in carbonate sequences in the Pacific and Atlantic (Savin *et al.,* in press). Changes of this order of time magnitude suggest correlations with events such as shifts in upper mixed-layer productivity discussed by Arrhenius (1963, Fig. 39).

Another kind of event that appears to correlate with the reflectors is tectonic in nature—although these may be directly related to changes in basin depths as shown by Rona (1973). According to his analysis of sedimentation rates on continental shelves, marine transgression-regression cycles and sea-floor spreading rates, fast spreading correlates with transgression and slow spreading rates correlate with regression. Two major changes in either Pacific plate direction or velocity appear to have taken place, at approximately 25 million years B.P. (Dott, 1969) and at approximately 10 million years B.P. (data summarized by Hays *et al.,* 1972).

There are large numbers of minor local reflectors, e.g. the multiplicity of these on the Ontong Java Plateau. These are spaced only 10 m or so apart; at a sedimentation rate of approximately 25 m/million years (at Site 64): these would represent events that took place about 400,000 years apart. These may then represent fluctuations in diagenetic potential due to short-range changes in surface-water conditions. Short-term temperature fluctuations in surface temperatures, with a periodicity of approximately 80,000 years, have been identified in carbonate sequences in the Pacific and Atlantic (Savin *et al.,* in press). Changes of this order of time magnitude suggest correlations with events such as shifts in upper mixed-layer productivity discussed by Arrhenius (1963, Fig. 39).

NOTE ADDED IN PRESS

On Leg 33 of the Deep Sea Drilling Project which took place during November and December 1973, after the presentation of this paper at the September 1973 European Geophysical Society meeting in Zürich, particular attention was paid to the identification and correlation of acoustic reflectors. At Sites 315 and 316 in the central and southern Line Islands, the 'a', 'b', 'c' and 'd' reflectors were identified and partly cored and at Site 317 on the Manihiki Plateau the 'b', 'c', 'd' and 'e' reflectors were cored and identified—and at all sites the reflectors occurred within strata that occupy the age intervals indicated on Table 3. These reflector sequences have now been identified and dated for the equatorial Pacific (Site 72) at 138°W through the Line Islands, the Magellan Rise, and the Manihiki Plateau to the Ontong Java Plateau (Site 64) at 158°E, a distance of approximately 3840 km (Schlanger, Jackson *et al.,* in preparation).

CONCLUSIONS

The data on the textural, palaeontological and physical properties of relatively pure pelagic ooze-chalk-limestone sequences can be interpreted as indicating the following.

(1) That the general trend towards increasing lithification with length and depth of burial is interrupted by local reversals in the degree of lithification. Thus soft, plastic, oozes occur below stiff but friable chalks and the latter occur below harder, more dense limestones.

(2) That these reversals are due to variations in the amount of calcite cement in the rock and that this amount is somewhat independent of depth of burial.

(3) That pelagic carbonate oozes, because they contain an abundance of both: (a) calcite crystals, less than 1 μm in size, produced by the dissaggregation of small coccoliths and foraminiferal tests and (b) larger crystals such as discoaster segments and large coccolith elements, up to 10 μm in size, become cemented through a dissolution-diffusion-reprecipitation mechanism. The direction of diagenesis is irreversible because the dissolution of the very fine-grained calcite and the reprecipitation of the same volume of calcite on the larger crystals lowers the free energy of the system.

(4) That the degree to which a pelagic carbonate sediment becomes cemented depends on the diagenetic potential the sediment possessed on burial. The diagenetic potential is the measure of how much more diagenesis the sediment can be expected to undergo in the normal course of geological history.

(5) The diagenetic potential is a function of the following.

(a) The depth of water in which the sediment was deposited and the relation of this depth to the CCD level.

(b) The fertility of the upper mixed layers in terms of calcareous plankton production since this affects the sedimentation rate. The sedimentation rate is in part a function of the temperature and salinity structure of the upper mixed layers.

(c) The ratio of Foraminifera + small coccoliths: discoasters + large coccoliths in the sediment; this determines the size distribution characteristics of the sediment.

(d) Predation by plankton feeders as this affects the state of aggregation of the nannoplankton and therefore the sinking rate of this calcite.

(6) Acoustic reflectors are pre-determined in their characteristics and stratigraphic distribution by the diagenetic potential the reflecting horizon had upon burial. Reflectors then are related to palaeo-oceanographic events such as glacio-eustatic sea-level changes, alterations in calcite compensation depth and tectonic events that affected plate motion.

ACKNOWLEDGMENTS

This study was supported in part by the Institute of Geophysics and Planetary Physics, University of California, Riverside and the Marathon Oil Company. The authors wish to thank Professors L. H. Cohen and J. B. Combs and Samuel Savin for their suggestions and criticism. Peter Kolesar and Michael Arthur helped compile and analyse data.

REFERENCES

ADELSECK, C.G., GEEHAN, G.W. & ROTH, P.R. (1973) Experimental evidence for the selective dissolution and overgrowth of calcareous nannofossils during diagenesis. *Bull geol. Soc. Am.* **84,** 2755-2762.

ARRHENIUS, G. (1952) Sediment cores from the East Pacific. *Rep. Swed. deep Sea Exped.* (1947–1948), Parts 1–4, **5**, 1–288.

ARRHENIUS, G. (1963) Pelagic sediments. In: *The Sea*, Vol. 3 (Ed. by M. N. Hill), pp. 655–727. Interscience Publishers, New York.

BATHURST, R.G.C. (1971) *Carbonate Sediments and Their Diagenesis,* pp. 620. Elsevier Publishing Co., Amsterdam, London, New York.

BERGER, W.H. (1971) Sedimentation of planktonic Foraminifera. *Mar. Geol.* **11**, 325–358.

BERGER, W.H. (1973a) Cenozoic sedimentation in the Eastern Tropical Pacific. *Bull. geol. Soc. Am.* **84**, 1941–1954.

BERGER, W.H. (1973b) Deep-sea carbonates. Pleistocene dissolution cycles. *J. Foram. Res.* **3**, 187–195.

BERGER, W.H. & VON RAD, U. (1972) Cretaceous and Cenozoic sediments from the Atlantic Ocean. In: *Initial Reports of The Deep Sea Drilling Project,* Vol. XIV (D. E. Hayes, A. C. Pimm *et al.*), pp. 788–854. U.S. Government Printing Office, Washington.

BRAMLETTE, M.N. (1958) Significance of coccolithophorids in calcium-carbonate deposition. *Bull. geol. Soc. Am.* **69**, 121–126.

BUKRY, D. (1973) Coccolith stratigraphy, eastern Equatorial Pacific, Leg 16, D.S.D.P. In: *Initial Reports of the Deep Sea Drilling Project,* Vol. XVI (Tj. H. van Andel, G. R. Heath *et al.*), pp. 653–712. U.S. Government Printing Office, Washington.

BYRNE, J.G. (1965) *Recovery, Recrystallization and Grain Growth,* pp. 175. Macmillan, New York.

CIARANFI, N. & CITA, M.B. (1973) Paleontological evidence of changes in the Pliocene climates. In: *Initial Reports of the Deep Sea Drilling Project,* Vol. XIII (W. B. F. Ryan, K. J. Hsü *et al.*), pp. 1367–1399. U.S. Government Printing Office, Washington.

CITA, M.B. & RYAN, W.B.F. (1973) Time scale and general synthesis. In: *Initial Reports of the Deep Sea Drilling Project,* Vol. XIII (W. B. F. Ryan, K. J. Hsü *et al.*), pp. 1405–1415. U.S. Government Printing Office, Washington.

COOK, F.M. & COOK, H.E. (1972) Physical properties synthesis. In: *Initial Reports of the Deep Sea Drilling Project,* Vol. IX (J. D. Hays *et al.*), pp. 645–646. U.S. Government Printing Office, Washington.

DAVIES, T.A. & SUPKO, P.R. (1973) Oceanic sediments and their diagenesis: some examples from deep-sea drilling. *J. sedim. Petrol.* **43**, 381–390.

DEVEREUX, I. (1967) Oxygen isotope palaeotemperatures on New Zealand Tertiary fossils. *N. Z. Jl Sci.* **74**, 49–57.

DOTT, R. H. JR (1969) Circum-Pacific Late Cenozoic structural rejuvenation: implications for sea floor spreading. *Science,* **166**, 874–876.

DOUGLAS, R.G. (1973a) Benthonic foraminiferal biostratigraphy in the Central North Pacific. *Initial Reports of the Deep Sea Drilling Project,* Vol. XVII (E. L. Winterer, J. I. Ewing *et al.*), pp. 607–671. U.S. Government Printing Office, Washington.

DOUGLAS, R.G. (1973b) Planktonic foraminiferal biostratigraphy. In: *Initial Reports of the Deep Sea Drilling Project,* Vol. XVII (E. L. Winterer, J. I. Ewing *et al.*), pp. 673–694. U.S. Government Printing Office, Washington.

DOUGLAS, R.G. & SAVIN, S.M. (1971) Isotopic analysis of planktonic Foraminifera from the Cenozoic of the northwest Pacific. In: *Initial Reports of the Deep Sea Drilling Project,* Vol. VI (A. G. Fischer *et al.*), pp. 1123–1127. U.S. Government Printing Office, Washington.

DOUGLAS, R.G. & SAVIN, S.M. (1973) Oxygen and carbon isotope analyses of Cretaceous and Tertiary Foraminifera from the central north Pacific. In: *Initial Reports of the Deep Sea Drilling Project,* Vol. XVII (E. L. Winterer, J. I. Ewing *et al.*), pp. 591–605. U.S. Government Printing Office, Washington.

EMERY, K.O., TRACEY, J.I., JR & LADD, H.S. (1954) Geology of Bikini and nearby Atolls. *Prof. Pap. U.S. geol. Surv.* **260-A**, 1–255.
ment Printing Office, Washington.

FISCHER, A.G., HONJO, S. & GARRISON, R.W. (1967) *Electron Micrographs of Limestones and their Nannofossils,* pp. 137. Princeton University Press.

GEALY, E.L. (1971) Saturated bulk density, grain density and porosity of sediment cores from the Western Equatorial Pacific. In: *Initial Reports of the Deep Sea Drilling Project,* Vol. VII (E. L. Winterer *et al.*), pp. 1084–1104. U.S. Government Printing Office, Washington.

HAMILTON, E.L. (1959) Thickness and consolidation of deep-sea sediments. *Bull. geol. Soc. Am.* **70**, 1399–1424.

HAMILTON, E.L. (1971) Elastic properties of marine sediments. *J. geophys. Res.* **76**, 579–604.

HAYES, D.E., FRAKES, L.A., BARRETT, P., BURNS, D.A., CHEN, P.-H., FORD, R.B., KANEPS, A.G., KEMP, E.M., McCOLLUM, D.W., PIPER, D.J.W., WALL, R.E. & WEBB, P.N. (1973) Leg 28 deep-sea drilling in the Southern Ocean. *Geotimes,* **18** (6), 19–24.

HAYS, J.D. *et al.* (1972) *Initial Reports of the Deep Sea Drilling Project,* Vol. IX, pp. 1205. U.S. Government Printing Office, Washington.

LADD, H.S., TRACEY, J.I., JR & GROSS, G. (1970) Deep drilling on Midway Atoll. *Prof. Pap. U.S. geol. Surv.* **680-A,** A1–A21.

LAUGHTON, A.S. (1954) Laboratory measurements of seismic velocities in ocean sediments. *Proc. R. Soc.* **222,** 336–341.

LAUGHTON, A.S. (1957) Sound propagation in compacted oceanic sediments. *Geophysics,* **22,** 233–260.

LISITZIN, A.P. (1971) Distribution of carbonate microfossils in suspension and in bottom sediments. In: *The Micropalaeontology of the Oceans* (Ed. by B. M. Funnell and W. R. Riedel), pp. 197–218. Cambridge University Press, London.

MOBERLY, R., JR & HEATH, R. (1971) Carbonate sedimentary rocks from the western Pacific: Leg 7, Deep Sea Drilling Project. In: *Initial Reports of the Deep Sea Drilling Project,* Vol. VII (E. L. Winterer *et al.*), pp. 977–986. U.S. Government Printing Office, Washington.

MOORE, D.G. (1969) Reflection profiling studies of the California continental borderland. *Spec. Pap. geol. Soc. Am.* **107,** 142 pp.

NAFE, J.E. & DRAKE, C.L. (1963) Physical properties of marine sediments. In: *The Sea* (Ed. by M. N. Hill), pp. 794–813. Interscience Publishers, New York.

PARKER, F.L. (1954) Distribution of the Foraminifera in the northeastern Gulf of Mexico. *Bull. Mus. comp. Zool.* **III,** 454.

PARKER, F.L. & BERGER, W.H. (1971) Faunal and solution patterns of planktonic Foraminifera in surface sediments of the South Pacific. *Deep Sea Res.* **18,** 73–107.

PETERSON, M.N.A. (1966) Calcite: rates of dissolution in a vertical profile in the Central Pacific. *Science,* **154,** 1542–1544.

RONA, P.A. (1973) Relations between rates of sediment accumulation on continental shelves, sea-floor spreading and eustasy inferred from the Central North Atlantic. *Bull. geol. Soc. Am.* **84,** 2851–2872.

ROTH, P.H. & THIERSTEIN, H. (1972) Calcareous nannoplankton. In: *Initial Reports of the Deep Sea Drilling Project,* Vol. XIV (D. E. Hayes, A. C. Pimm *et al.*), pp. 421–486. U.S. Government Printing Office, Washington. '

RYAN, W.B.F. & HSÜ, K.J. *et al.* (1973) *Initial Reports of the Deep Sea Drilling Project,* Vol. XIII, pp. 1447. U.S. Government Printing Office, Washington.

SAVIN, S.A. & DOUGLAS, R.G. (1973) Stable isotope and magnesium geochemistry of Recent planktonic Foraminifera from the South Pacific. *Bull. geol. Soc. Am.* **84,** 2327–2342.

SAVIN, S.M., Douglas, R.G. & STEHLI, F.G. (in press) Tertiary marine paleotemperatures. *Bull. geol. Soc. Am.*

SCHLANGER, S.O. (1963) Subsurface geology of Eniwetok Atoll. *Prof. Pap. U.S. geol. Surv.* **260-BB,** 901–1066.

SCHLANGER, S.O., DOUGLAS, R.G., LANCELOT, Y., MOORE, T.C. & ROTH, P. (1973) Fossil preservation and diagenesis of pelagic carbonates from the Magellan Rise, Central North Pacific Ocean. In: *Initial Reports of the Deep Sea Drilling Project,* Vol. XVII (E. L. Winterer, J. I. Ewing *et al.*), pp. 467–527. U.S. Government Printing Office, Washington.

SCHLANGER, S.O., JACKSON, E.D. *et al.* (in preparation) *Initial Reports of the Deep Sea Drilling Project,* Vol. XXXIII. U.S. Government Printing Office, Washington.

SCHOTT, W. (1935) Die Foraminiferen in dem äquatorialen Teil des Atlantischen Ozeans. *Dt. Atl. Exped. Meteor* 1925–1927, **3,** 43–143.

SMAYDA, T.J. (1971) Normal and accelerated sinking of phytoplankton in the sea. *Mar. Geol.* **11,** 105–122.

THIEDE, J. (1972) Planktonische Foraminiferen in Sedimenten vom ibero-marokkanischen Kontinentalrand. *'Meteor' Forsch. Ergebn.* **7,** 15–102.

TRACEY, J.I., JR *et al.* (1971) *Initial Reports of the Deep Sea Drilling Project,* Vol. VIII, pp. 1037. U.S. Government Printing Office, Washington.

TSCHEBOTARIOFF, G.P. (1952) *Soil Mechanics, Foundations and Earth Structures*, pp. 645. McGraw-Hill, New York.

WINTERER, E.L. *et al.* (1971) *Initial Reports of the Deep Sea Drilling Project*, Vol. VII, pp. 1756. U.S. Government Printing Office, Washington.

WINTERER, E.L., EWING, J.I. *et al.* (1973) *Initial Reports of the Deep Sea Drilling Project*, Vol. XVII, pp. 930. U.S. Government Printing Office, Washington.

WISE, S.W., JR (1972) Calcite overgrowths on calcareous nannofossils: a taxonomic irritant and a key to the formation of chalk. *Abst. geol. Soc. Am.* **4,** 115–116.

WISE, S.W., JR (1973) Calcareous nannofossils from cores recovered during Leg 18, Deep Sea Drilling Project: biostratigraphy and observations of diagenesis. In: *Initial Reports of the Deep Sea Drilling Project*, Vol. XVIII (L. D. Kulm, R. von Heune *et al.*), pp. 565–615. U.S. Government Printing Office, Washington.

WISE, S.W., JR & KELTS, K.R. (1972) Inferred diagenetic history of a weakly silicified deep sea chalk. *Trans. Glf-Cst Ass. geol. Socs*, **22,** 177–203.

Copyright © 1973 by the Geological Society of America

Reprinted from *Geol. Soc. America Bull.* **84**:2755–2762 (Aug. 1973)

Experimental Evidence for the Selective Dissolution and Overgrowth of Calcareous Nannofossils During Diagenesis

CHARLES G. ADELSECK, JR.
GREGORY W. GEEHAN } *Scripps Institution of Oceanography, La Jolla, California 92037*
PETER H. ROTH

ABSTRACT

Well-preserved samples of upper Pliocene calcareous nannofossil ooze were subjected to elevated temperatures and pressures in order to simulate diagenetic effects. The samples were then studied and compared to an untreated control sample using both a light microscope and a scanning electron microscope. The order of alterations, progressing from the original sample to the sample exposed to highest temperatures and pressures, was observed to be (1) minor etching and overgrowth formation; (2) continued etching and secondary overgrowth formation, and the disaggregation of the elements of smaller coccoliths; and (3) dominant overgrowth on whole coccoliths with particularly massive overgrowth formation on the discoasters. Diversity of the nannofossil assemblage decreased under increasingly severe conditions.

INTRODUCTION

Calcareous nannofossils are important constituents of pelagic carbonate sediments. The effects of diagenetic solution and overgrowth have been recognized by Bramlette (1958), Bramlette and Sullivan (1961), Bukry and others (1971), and Wise and Hsü (1971). Electron microscopy of limestone and chalk (Fischer and others, 1967; Noël, 1970) indicates that much of the fine carbonate fraction in these rocks consists of complete or fragmented nannofossils which were subsequently altered during diagenesis. An excellent carbonate section more than 1,000 m thick was drilled on the Magellan Rise in the central Pacific Ocean (Deep Sea Drilling Project, Leg 17, Site 167). In this section, a gradual transition occurs from soft ooze in the top 200 m to chalk between 200 m and 600 m and to limestone in the basal 400 m. Unlike those in well-preserved sediments, nannofossil assemblages in some of the chalk beds and especially in the limestone showed reduced diversity and a relative increase in species more resistant to secondary calcite deposition. The percentage of small calcite fragments also increased considerably in the more altered carbonate sediments. Scanning electron microscopy of some of the Upper Cretaceous and Tertiary nannofossil assemblages (Roth, 1973) has shown that solution and secondary overgrowths on coccoliths can occur in the same sample.

This study simulated, in a short-term experiment, the long-term diagenetic changes observed in the continuous section of calcareous ooze drilled at Site 167. The original sample was selected from the upper Pliocene ooze cored at this site (Deep Sea Drilling Project, Leg 17, Site 167, Core 3, Section 6 at 51 cm, 26 m below the sea floor). It was chosen because of its good state of preservation and the presence of both coccoliths and discoasters.

Normal in situ temperatures and pressures in oceanic sediments probably do not exceed 30°C and 0.6 kb. The temperatures and pressures used in this experiment (up to 300°C and 3 kb) were higher in order to produce in one month effects similar to those produced during millions of years of natural diagenetic conditions. After equilibrium is reached, the high temperatures and pressures accelerate the rates of calcite overgrowth and solution reactions.

Figure 1. Comparisons of selected species. All bar scales equal 1 μ. A. *Cyclococcolithina macintyrei*, a distal view of the distal shield from the 300°C–3 kb bomb. B. Disaggregation of the elements of an undetermined coccolith, from the 300°C–1 kb bomb. C. *Discoaster brouweri*, distal view from the original sample. D. *Discoaster brouweri*, distal view from the 200°C–1 kb bomb. E. *Discoaster brouweri*, distal view from the 300°C–1 kb bomb. D. *Discoaster brouweri*, distal view from the 300°C–1 kb bomb.

Figure 2. Comparisons of selected species. All bar scales equal 1 μ. A. *Coccolithus pelagicus*, distal view of the distal shield from the original sample. B. *Coccolithus pelagicus*, distal view of a distal shield from the 300°C–3 kb bomb. C. *Cyclococcolithina leptopora*, distal view of a distal shield from the original sample. D. *Cyclo-coccolithina leptopora*, distal view of a distal shield from the 300°C–1 kb bomb. E. *Cyclococcolithina leptopora*, proximal view from the original sample. F. *Cyclococco-lithina leptopora*, proximal view of both shields from the 300°C–1 kb bomb.

METHODS

Three subsamples were sealed into silver tubing along with artificial sea water with a pH of 8.0. Each subsample was then placed in a hydrothermal "bomb" for 32 days. One sample was held at a temperature of 200°C and a pressure of 1 kb, another at 300°C and 1 kb, and the third at 300°C and 3 kb. Smear slides of the original sample and the three experimental samples were made for examination with both light microscope (LM) and scanning electron microscope (SEM).

The light microscope was used to obtain counts of the various nannofossil species present in each sample along with estimates of the degree of preservation of each species. The samples were viewed under cross-polarized and phase-contrast illumination, and counts of at least 300 specimens of each sample were made before relative abundances were calculated.

Whereas the LM was used to evaluate changes in relative proportions of the species present, the SEM was employed to observe in detail the results of solution and(or) overgrowth on individual members of the various nannofossil species. Several individuals of a species were studied before taking photographs of selected specimens that illustrated the characteristic appearance of that species resulting from the experimental conditions imposed on it.

EXPERIMENTAL RESULTS

The resistance to destruction of the calcareous nannofossil species during the experiment varied considerably. Smaller and more delicate coccoliths were more easily destroyed than the larger forms with strongly overlapping elements (crystallites). Some larger forms underwent slight etching but generally showed formation of secondary overgrowth. Discoasters displayed well-developed overgrowths having the appearance of crystal faces with constant interfacial angles. The changes observed in comparing the three samples from the hydrothermal bombs to the original sample are summarized in Table 1. For individual samples, the variability of induced changes among members of the same species was low.

In the original sample, *Discoaster brouweri* showed thin fragile rays with little evidence of overgrowth. The central area had a uniform arrangement of smooth planar surfaces (Fig. 1C). Specimens from the 200°C-1 kb bomb

showed a slight thickening of the rays with distinct etching and pitting of the central area. Such solution effects along the sutures during early diagenesis may have contributed to fragmentation of the discoaster. At later stages of diagenesis, calcium carbonate precipitation, resulting in overgrowth formation, dominated. No etching was visible, and the discoasters became massive, with thick rays (Fig. 1E, F). There were numerous partially overgrown *D. brouweri* with no calcium carbonate precipitation on the outward tips of the rays (Fig. 1E). This indicates that the predominance of early overgrowth formation occurs in and around the central area of the discoaster and then proceeds outward along the rays. Ultimately the complete ray becomes overgrown (Fig. 1F).

The coccoliths of *Cyclococcolithina leptopora* in the original sample were well preserved (Fig. 2C, E); however, the attachment between the proximal and distal shields was weak and most of the specimens were isolated shields. The abundances of each shield in the samples are shown in Table 1. No changes were observed in the 200°C-1 kb bomb sample. In the two high-temperature bombs, the distal side of each shield showed selective overgrowth on a limited number of elements, accompanied by dissolution of the central area. Subsequently, there was uneven overgrowth of the outer cycle of elements, which tended to grow into the central area (Fig. 2F). The proximal sides of both shields developed greater overgrowth formation, which was more evenly distributed upon all the elements (Fig. 2F). Figure 2F also shows the central area extensively dissolved. It appeared that the interlocking of the overgrowths was holding the coccolith together.

Cyclococcolithina macintyrei (Fig. 1A) reacted analogously to *C. leptopora* in the hydrothermal experiments. Figure 1A shows a good example of the selective nature of calcite overgrowths on the elements of the distal sides of the distal shields. Notice the overgrown element at the 3 o'clock position in comparison to the numerous unaltered elements.

Coccolithus pelagicus showed the least alteration of all species in the sample. The specimens from the two low-pressure bombs showed no observable change, and those from the 300°C-3 kb bomb developed minor overgrowth on specific elements of the distal shield (Fig. 2A, B).

In the original sample and the 200°C-1 kb

TABLE 1. SPECIES ABUNDANCE AND PRESERVATION

ABUNDANCE (%)   "P" = Present

Sample	*Discoaster brouweri* Tan	*Cyclococcolithina leptopora* (Murray & Blackman) whole coccoliths	*Cyclococcolithina leptopora* (Murray & Blackman) – proximal shields	*Cyclococcolithina leptopora* (Murray & Blackman) – distal shields	*Coccolithus pelagicus* (Wallich)	*Cyclococcolithina macintyrei* (Bukry & Bramlette)	*Umbilicosphaera mirabilis* Lohmann	*Helicopontosphaera kamptneri* Hay & Mohler	*Pseudoemiliania lacunosa* (Kamptner)	*Syracosphaera pulchra* Lohmann	*Ceratolithus rugosus* Bukry & Bramlette	*Discoaster pentaradiatus* Tan	*Discoaster perplexus* Bramlette & Riedel	*Pontosphaera discopora* Schiller	*Coccolithus doronicoides* Black & Barnes & *Cyclolithella* sp.	TOTAL COUNTED
original sample	18	9	12	5	3	1	1	18	2	P	1	P	P	P	30	417
200°C-1 kb	19	9	9	1	2	1	1	18	2	P	1	1	1	1	36	391
300°C-1 kb	43	20	18	4	4	1				1	1	1	1	P	8	358
300°C-3 kb	40	19	20	4	4	1				1	1	P	P	P	11	317

PRESERVATION

Sample	*Discoaster brouweri*	*C. leptopora* whole	*C. leptopora* proximal	*C. leptopora* distal	*Coccolithus pelagicus*	*C. macintyrei*	*U. mirabilis*	*H. kamptneri*	*P. lacunosa*	*S. pulchra*	*C. rugosus*	*D. pentaradiatus*	*D. perplexus*	*P. discopora*	*C. doronicoides* & *Cyclolithella*
original sample	G	G	G	G	G	G	M / E-1	G	M / E-1	M / O-1	P	G	G	P	G
200°C-1 kb	E-1 / O-1	G	G	G	G	G	M / E-1	G	M / E-1	O-2 / B	P	G	G	M	M
300°C-1 kb	O-3	E-1 / O-2	E-1	G	E-1 / O-1	M				O-3 / B	O-2 / B	G / B	G / B	M	P
300°C-3 kb	O-3	E-1 / O-2	E-1 / O-2	O-1	E-1 / O-1	M				O-3 / B	O-2 / B	G / B	G / B	M	P

KEY TO PRESERVATION:
"G" = Good
"M" = Moderate
"P" = Poor
"B" = Broken
"O" = Overgrown
"E" = Etched
"-1", "-2", or "-3" = Degree of Etching or Overgrowth,
1 = least to 3 = greatest

67

bomb sample, *Helicopontosphera kamptneri* showed advanced solution etching in the central area; similarly, *Syracosphaera pulchra* had a well-preserved outer cycle of elements but a greatly dissolved center.

The specimens of *Ceratolithus rugosus* in the original sample exhibited definite overgrowths. In progressively higher temperature and pressure bomb samples, an increase occurred in both the amount of overgrowth and the fragmentation of the specimens. Considering the developments in the other species, the fragmentation was unexpected. The presence of overgrowths on all the other nannofossils seems to favor their preservation and resistance to breakage. In the 300°C–3 kb bomb, *C. rugosus* was in such poor condition that identification was difficult.

Coccolithus doronicoides and a few *Cyclolithella* species were dominant among the smaller coccoliths in the sample. Their abundance counts (Table 1) are not based upon the strictest identification of species, but instead represent all interference patterns that appear to fit this grouping reasonably. These species were destroyed in the two high-temperature bomb samples (Table 1) by solution etching along sutural contacts, resulting in the eventual disaggregation of the coccolith's elements (Fig. 1B).

DISCUSSION

In approaching this project, little was known of pressures, temperatures, and time durations necessary to simulate diagenetic effects. A wide range of temperatures and pressures was eventually chosen in order to maximize the chances of success in at least one bomb. It is evident that only minor changes were produced at 200°C–1 kb, and that the pressure increase from 1 kb to 3 kb in the 300°C bombs had little effect. The major changes were observed to have occurred in the range from 200°C to 300°C at 1 kb pressure.

The disaggregation of the coccolith elements showed that solution etching attacks the sutures between the elements. Coccoliths with greatest resistance to destruction had strongly overlapping elements in their distal shields (*Cyclococcolithina*, *Coccolithus*, and so forth), and so their sutural contacts were relatively larger than those of the smaller coccoliths; therefore, they were the last. to disaggregate. Thus, initial disaggregation occured mainly among the smaller coccoliths.

Berner (1971, p. 28) stated that larger crystals have a lower free energy; therefore, "larger crystals can grow at the expense of smaller crystals during the aging of a precipitate." For crystals smaller in size than 1 μ, the excess free energy of the surface ions and atoms becomes important, and the total free energy of the crystal is increased. This indicates that in a sample containing a range of crystal sizes, from 1 μ to several, the smallest particles will be selectively destroyed while the larger ones will form secondary overgrowths. It is evident from Figures 1, 2, and 3 that most of the disaggregated elements in this experiment were smaller than 1 μ. Therefore, they underwent dissolution and thus provided the calcium carbonate for precipitation onto larger elements as secondary overgrowths. These overgrowths on the larger coccoliths apparently then bound the elements and shields together, preventing their disaggregation (Fig. 2F).

The overgrowths formed on the resistant species displayed a controlled pattern of carbonate precipitation. Preferential calcium carbonate precipitation occurred upon the proximal sides of the shields (Fig. 2D, F). Individual elements on the distal side of the distal shields showed well-developed overgrowths, but the other elements displayed neither etching nor overgrowth. Organic compounds can inhibit normal calcium carbonate reaction rates (Suess, 1970), and organic coverings are formed around coccoliths (Franke and Brown, 1971; McIntyre and McIntyre, 1971, p. 259); such coverings will exert controls on overgrowth formation. These effects certainly warrant a more detailed study to investigate the controls of calcium carbonate overgrowth on nannofossils.

Bain (1940, p. 7) showed that the different crystal faces of calcite have varying solubilities. He found that the faces oriented perpendicular to the c-axis are most resistant to solution while those that are parallel to the c-axis are least resistant. Bukry (1971, p. 304) related the orientation of the optic axis of the elements to their susceptibility to dissolution and ultimately to the solution resistance of the species. The orientation of the axis may also be related to the greater susceptibility to etching in the central area of many coccoliths. The distal shields of *C. leptopora*, *C. macintyrei*, and other coccoliths that remain dark in cross-polarized illumination have a c-axis oriented perpendicular to the stage of the microscope. The faces of the elements on the distal shields are therefore

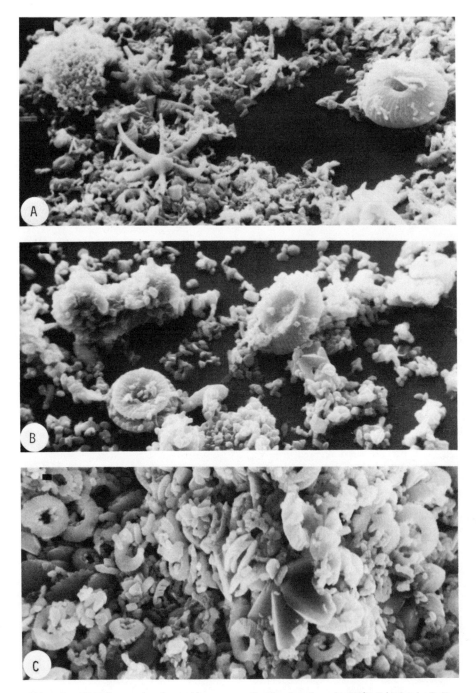

Figure 3. SEM photographs of assemblage overviews. All bar scales equal 1 μ. A. Original sample, a well-preserved upper Pliocene calcareous ooze. B. Sample taken from the 300°C–3 kb bomb. C. Upper Oligocene chalk from Deep Sea Drilling Project Site 167.

close to being oriented perpendicular to the c-axis, and the faces forming the central area are more closely oriented to the plane of the c-axis. Thus, solution etching would occur primarily in the central area of these coccoliths and along the outer edges.

The significance of this experiment depends on how much the results mirror the observable changes from natural diagenetic conditions. Figure 3 compares photographic overviews of the standard sample, its condition after removal from the 300°C–3 kb bomb, and an upper Oligocene chalk at 441-m penetration from DSDP Site 167. The experiment showed the following order of diagenetic evolution: initial minor etching and minor formation of overgrowths; the disaggregation of the smaller coccolith elements with a corresponding decrease in diversity and continued overgrowth and etching on larger specimens; and a dominance of overgrowth formation on all remaining whole coccoliths, with discoasters becoming particularly massive.

Comparing these effects with the observed nature of the Oligocene chalk (Fig. 3C), it is evident that similar processes occurred. The chalk lacked a reasonable distribution of smaller coccoliths, and there appeared to be an abundance of disaggregated elements, similar to that observed in a Cretaceous chalk and attributed to diagenetic processes by Bramlette (1958). The larger coccoliths were best preserved, and the discoaster showed massive overgrowth, similar to that seen in the experiment. Although the chalk exhibited a lesser degree of diagenetic change, the similarities between it and the experimental results are strongly evident.

ACKNOWLEDGMENTS

We thank John T. Whetten for his helpful suggestions and James W. Hawkins for providing the hydrothermal equipment. F. B Phleger and W. H. Berger furnished encouragement and support during the period of the experiment. W. H. Berger, D. Bukry, F. B Phleger, and E. L. Winterer reviewed the manuscript and provided many helpful comments and ideas. We also thank Ellen Flentye, Phyllis Helms, David Wirth, Larry Lauve, and Thomas Walsh for their technical assistance in preparing this paper.

The National Science Foundation supplied the sample, and the research was supported in part by the Oceanography Section, National Science Foundation, NSF Grant GA–35451 and by the American Petroleum Institute, API Research Project 143.

REFERENCES CITED

Bain, G. W., 1940, Geological, chemical and physical problems in the marble industry: Am. Inst. Mining and Metall. Engineers, Tech. Pub. no. 1261, 16 p.
Berner, R. A., 1971, Principles of chemical sedimentology: New York, McGraw-Hill, 240 p.
Bramlette, M. N., 1958, Significance of coccolithophorids in calcium-carbonate deposition: Geol. Soc. America Bull., v. 69, p. 121–126.
Bramlette, M. N., and Sullivan, F. R., 1961, Coccolithophorids and related nannoplankton of the early Tertiary in California: Micropaleontology, v. 7, p. 129–188.
Bukry, D., 1971, Cenozoic calcareous nannofossils from the Pacific Ocean: San Diego Soc. Nat. History Trans., v. 16, p. 303–327.
Bukry, D., Douglas, R. G., Kling, S. A., and Krasheninnikov, V., 1971, Planktonic microfossil biostratigraphy of the northwestern Pacific Ocean: Deep Sea Drilling Project Initial Repts., v. 6, p. 1253–1300.
Fischer, A. G., Honjo, S., and Garrison, R. E., 1967, Electron micrographs of limestones and their nannofossils: Princeton Mon. Geology and Paleontology, no. 1, 141 p.
Franke, W. W., and Brown, R. M., Jr., 1971, Scale formation in chrysophycean algae: Archiv für Mikrobiologie, v. 11, p. 12–19.
McIntyre, A., and McIntyre, R., 1971, Coccolith concentrations and differential solution in oceanic sediments, in Funnell, B. M., and Riedel, W. R., eds., The micropalaeontology of oceans: London, Cambridge Univ. Press, 828 p.
Noël, D., 1970, Coccolithes crétacés de la craie campanienne du Bassin de Paris: Paris, Editions Centre National Récherche Scientifique, 129 p.
Roth, P. H., 1973, Calcareous nannoplankton: Leg 17 of the Deep Sea Drilling Project: Deep Sea Drilling Project Initial Repts., v. 17 (in press).
Suess, E., 1970, Interaction of organic compounds with calcium carbonate—I. Association phenomena and geochemical implications: Geochim. et Cosmochim. Acta, v. 34, p. 157–168.
Wise, S. W., Jr., and Hsü, K. J., 1971, Genesis and lithification of a deep-sea chalk: Eclogae Geol. Helvetiae, v. 64, p. 273–278.

ERRATUM

Page 2758, line 33 in the second column should read: "the central area (Fig. 2D)..."

Part II

CARBONATE AND SILICA DIAGENESIS

Editor's Comments
on Papers 4 and 5

4   WISE and HSÜ
Genesis and Lithification of a Deep Sea Chalk

5   VAN DER LINGEN and PACKHAM
Relationships Between Diagenesis and Physical Properties of Biogenic Sediments of the Ontong-Java Plateau (Sites 288 and 289, Deep Sea Drilling Project)

Papers 4 and 5 deal with both carbonate and silica diagenesis. Wise and Hsü (Paper 4) describe aspects of lithification of an Oligocene chalk, cored by the DSDP at many sites in the South Atlantic. This chalk has an exceptional composition, consisting almost entirely of the nannofossil *Braarudospheara rosa* (see also Wise and Kelts 1972). SEM photographs show that the sediment has been cemented by calcite and by spherulitic aggregates. Basing their study on X-ray analysis of the insoluble residue, the authors first thought that these aggregates were clinoptilolite. However, they later realized (note added in proof) that a platy mineral (also present in the insoluble residue) is clinoptilolite and that the spherical aggregates are probably amorphous silica.

The historic interest of their paper is that, for the first time, the suggestion has been made that such spherical aggregates are an important mineral phase in the silica diagenesis of deep-sea sediments. Later studies (e.g., Wise et al. 1972; Wise and Kelts 1972) showed that these aggregates are very common in siliceous deep-sea sediments. The spheres are made up of blades, and Wise and Kelts therefore called them "lepispheres" ("spheres of blades").

Wise and Hsü were not the first to publish SEM photographs of lepispheres. Earlier in 1971 (February) SEM photographs of lepispheres were published in volume 6 of the *Initial Reports of the DSDP* (Pimm et al. 1971, Plates 27 and 28). However, these authors thought that the lepispheres were "hemispherical clusters of aragonite crystals." In the same month a paper by Buurman and van der Plas (1971) appeared, showing pictures of lepispheres from chert-nodules in chalks from the Netherlands, which they called "botryoidal structures." They consider these structures to be "amorphous silica that has crystallized towards the above mentioned 'badly crystalline tridymite'" (p. 22). Their mineralogical analysis of the lepispheres was quite advanced as compared with later studies by others (see "Introduction"; Papers 6 through 9; Wilson et al. 1974; Jones and Segnit 1975).

Van der Lingen and Packham (Paper 5) describe progressive diagenesis at two DSDP drill sites on the Ontong-Java Plateau, north of the Solomon Islands. They discuss both carbonate and silica diagenesis. The sediments at both sites consist of biogenic sediments. Site 288 was cored intermittantly. The oldest sediment sampled was of Aptian age. No igneous basement was reached. Site 289 was cored continuously, and igneous basement was reached after penetrating 1271m of sediment. The oldest sediment was also of Aptian age. Diagenetic changes were studied in great detail with the aid of an SEM. These data were then compared with various physical parameters such as induration, porosity, bulk density, sonic velocity, and seismic reflectors, as well as with mineralogical data. The authors review and discuss important publications by other workers.

At the end of the paper the authors try to delineate several problem areas, one of which is the interaction between silica diagenesis and carbonate diagenesis, a problem generally ignored in the literature and which will require more attention in the future. This problem is highlighted by differences between sites 288 and 289. At site 288, chert is present in sediments as young as Late Oligocene (at about 270 m subbottom depth), while at site 289 the youngest chert is of Eocene age (at about 1030 m subbottom depth). There seems to be a correlation between the presence of chert and the lithification of the carbonate sediment (see also the volume editor's comments on Paper 14). Another problem van der Lingen and Packham mention is the influence of heatflow, as it relates to sea-floor spreading, on diagenesis. Both silica diagenesis and heatflow could cause a false impression of progressive carbonate diagenesis with time and depth of burial.

The Ontong-Java Plateau has a geologic setting comparable to other oceanic plateaus in the Pacific. Some of these have also been drilled by the DSDP, e.g., the Magellan Rise (Paper 2) and the Shatsky Rise (Paper 13). With all three papers reproduced together in this volume, it is possible to compare the data from these plateaus in detail.

REFERENCES

Buurman, D. A., and van der Plas, L. (1971) The genesis of Belgian and Dutch flints and cherts. *Geol. Mijnbouw* **50:** 9–27.

Jones, J. B., and Segnit, E. R. (1975) Nomenclature and the structure of natural disordered (opaline) silica. A comment on the paper "A new interpretation of the structure of disordered α-cristobalite" by M. J. Wilson, J. D. Russell, and J. M. Tait, 1974: *Contr. Mineralogy and Petrology* **47:** 1–6. *Contr. Mineralogy and Petrology* **51:** 231–234.

Pimm, A. C.; Garrison, R. E.; and Boyce, R. E. (1971) Sedimentology synthesis: Lithology, chemistry, and physical properties of sediments in the northwestern Pacific Ocean. In Fischer, A. C.; Heezen, B. C.; et al., *Initial Reports of the Deep Sea Drilling Project,* Vol. 6, 1131–1252. U.S., Government Printing Office, Washington, D.C.

Wilson, M. J.; Russell, J. D.; and Tait, J. M. (1974) A new interpretation of the structure of disordered α-cristobalite. *Contr. Mineralogy and Petrology* **47:** 1–6.

Wise, S. W., Jr.; Buie, B. F.; and Weaver, F. M. (1972) Chemically precipitated cristobalite and the origin of chert. *Eclogae Geol. Helvetiae* **65:** 157–163.

Wise, S. W., Jr.; and Kelts, K. R. (1972) Inferred diagenetic history of a weakly silicified deep sea chalk. *Gulf Coast Assoc. Geol. Socs. Trans.* 22nd Annual Convention: 177–203.

4

Reprinted from *Eclogae Geol. Helvetiae* **64**(2):273–278 (1971)

Genesis and Lithification of a Deep Sea Chalk

By Sherwood W. Wise, Jr., and K. Jinghwa Hsü

Department of Geology, Swiss Federal Institute of Technology, Zürich, Switzerland

ABSTRACT

Lithification of a deep sea Oligocene chalk in the South Atlantic has been accomplished by chemical precipitation of two types of cement: 1. calcite, which occurs as secondary overgrowths and euhedral crystals up to 10 μm in diameter, and 2. spherical aggregates (about 3 μm diameter), insoluble in HCI, which were deposited as a late stage pore filling. Results suggest that Tertiary strata of present day ocean basins are ideal for studying early phases of diagenesis and lithification in carbonate rock.

Introduction

Although lithified chalks are rare in the younger Tertiary sediments of present day ocean basins, a few isolated chalk horizons have been sampled within unconsolidated pelagic ooze sequences during the recent exploratory operations of the Deep Sea Drilling Project. Studies of these chalks may provide keys for interpreting the early lithification and diagenetic histories of more widespread but more highly lithified chalks of Mesozoic age. We report here our preliminary analyses of an Oligocene braarudosphaerid chalk and the discovery of an interstitial cement or pore filling of unusual character which we hold to be responsible in part for the consolidation of this rock.

Results

The chalk was cored at Glomar Challenger Stations 14, 17, 19, 20 and 22 in the South Atlantic (Maxwell et al 1970a, 1970b) and was sampled by one of us (KJH) aboard ship. It is unusual in several respects: 1. It is restricted geographically to the South Atlantic and in time to rock units of Oligocene age. 2. It is sufficiently lithified to form an acoustical reflector and, as such, was of special interest as a key bed during on site drilling operations. No explanation could be given, however, for its lithification. 3. Although monospecific oozes are extremely rare in the geologic column, this chalk consists almost entirely of isolated skeletal fragments derived from a single species of golden brown algae (*Braarudosphaera rosa* Levin and Joerger 1967; see Maxwell et al. 1970a, 1970b). The pentalith construction of the coccospheres of this calcareous planktonic species is well known (Levin and Joerger 1967; Roth 1970; see also Gaarder 1954; Fischer, Honjo and Garrison 1967), and each plate of five distinct segments (Fig. 1) is normally found preserved intact in shallow water sediments; however, within our material, most of the pentaliths are themselves disaggregated, so that the sediment consists primarily of wedge-shaped segments. In view of these facts, this peculiar lithified material has been singled out for intensive laboratory study by scanning and transmission electron microscopy, X-ray diffraction, microprobe and oxygen isotope analysis.

Electron microscope analyses of *Braarudosphaera* chalk sample 3/22/4/1 from Station 22 (Rio Grande Rise) reveal that the pentalith segments and the relatively few specimens of other nannofossil species present show some evidence of solution, and this appears to have caused the disaggregation of the pentaliths. In addition, two types of cement are observed: 1. calcite, which is the principal lithifying material in the rock, and 2. an insoluble substance which occurs as a late stage pore filling to form an accessary cement[1]).

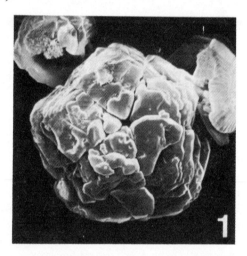

Fig. 1. *Braarudosphaera rosa*, a calcareous nannofossil composed of five wedge-shaped segments which, in a disaggregated state, are the principal constituents of the Oligocene chalk reported here.
2,900 ×

The calcite occurs as secondary overgrowths and as euhedral crytals of a variety of shapes and sizes up to 10 μm in diameter. The latter often envelope biogenic particles, and the large prismatic/rhombohedral crystal in Figure 2 has enveloped a portion of placolith. Similar crystals have been reported in the Jurassic Solenhofen Limestone (LAFFITTE and NoëL 1967, Pl. 8, Fig. 2), and our findings support LAFFITTE and NoëL's contention that the Solenhofen Ls is not inorganic in origin, but represents a lithified nannofossil ooze which has undergone diagenesis.

The second cementing material in our sample consists of fine platelets which grow in spherical aggregates about 3 μm in dia (Fig. 3) within the void areas of the chalk. Because the cementation process has not been carried to completion, there is still considerable void space within the rock, and the aggregate material typically shows spherical surfaces of unhindered growth. On fracture surfaces, however, it often shows

---

[1]) *Braarudosphaera* chalk samples from stations 14 and 20 have also been analyzed under the electron microscope. In there chalk samples only the calcite cement is present; the pore-filling silicate was not observed.

smooth flat impressions which result from a close contact against larger sedimentary particles (arrow, Fig. 2). This material, therefore, has gone through an active growth stage within the interstices of the sediment, and is an important accessary cement. In transmitted light the aggregates exhibit high refringence and complete extinction under crossed nichols. These features, together with their small size and fibrous construction, give them a slight superficial resemblance to palynomorphs. The aggregates, however, are insoluble in cold HCl, and X-ray analyses of the insoluble residue from the chalk (Rex 1970; K. Kelts 1970, personal communication) suggest that they are the mineral clinoptilolite, a silicate belonging to the zeolite group, or an amorphous substance.

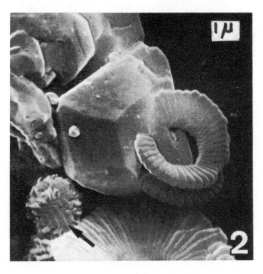

Fig. 2. Euhedral calcite crystal (center) which, through accretion, has overgrown a portion of a placolith.   7,200 ×

Morphologically, the aggregates are unlike any late Tertiary or Quaternary shallow water cements reported in the literature (example, see Hathaway and Degens 1969; Ffiedman, Amiel and Schneidermann 1970) or any of the fine constituents of unconsolidated shallow water carbonate sediments observed during extensive electron microscope studies (Hay, Wise and Stieglitz 1970; Stieglitz 1971). The peculiar form and habit of this precipitant are undoubtedly a product of its deep sea environment of deposition. In this respect, the potential importance of interstitial cements of this type as sedimentary environmental indicators should not be overlooked.

X-ray analyses of bulk samples of the chalk indicate that the rock consists of calcium carbonate in the form of low-magnesium calcite (about 1 mole per cent $MgCO_3$). It is not possible to say how much this composition differs from normal skeletal calcite of calcareous phytoplankton because comparative data on modern nannofloras is lacking. Recent studies by Gomberg and Bonatti (1970), however, suggest that diagenetic alteration of high magnesium calcite to low magnesium calcite can be very rapid in deep sea environment, and this factor should be kept in mind when

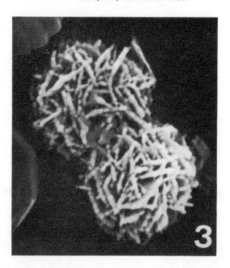

Fig. 3. Spherulitic aggregates composed of small platelets which constitute a second type of cement in the chalk. The flat surface of the aggregate in Figure 2 (arrow) was formed by close contact against a placolith shield.   12,600 ×

interpreting the trace element data. The oxygen isotope analysis for our material (LLOYD and Hsü 1971) indicates a paleotemperature of crystallization of 6° Centrigrade (assuming that isotope equilibrium has been achieved). This figure, however, is below the life tolerance limits of most modern calcareous phytoplankton (MCINTYRE and BE 1967), and no live representatives of modern *Braarudosphaera* have been sampled from sub-polar waters lying beyond the 10 °C isotherm. The low temperature readings for our sample probably reflect isotopic re-equilibration during the diagenetic alteration of the original skeletal material. In addition, solution and reprecipitation of calcite within this rock have been significant, and the calcareous cement was apparently formed in equilibrium with sea bottom temperatures which, during the Oligocene, should not have been much higher than the present day value of 4 °C. Both the paleotemperature reading and the trace element analysis, therefore, are considered to be indicative of conditions on the sea bottom rather than at the surface where the skeletal calcite was originally formed by planktonic organisms.

Discussion

The Tertiary braarudosphaerid chalk reported here has had an unusual depositional and diagenetic history, but one which should provide important new data on the little understood problems of deep sea diagenesis and lithification. Certainly it would be more difficult to follow the early phases of these processes in older, more tightly consolidated chalks and limestones. At most of the drilling sites, several laminae of chalk were encountered within an otherwise normal sequence of unconsolidated Oligocene ooze (MAXWELL et al. 1970b). Although the chalk is nearly monospecific, it does not have an unusually high terreginous content which, in the South Atlantic, is a reliable index of degree of dissolution (MAXWELL et al. 1970a). Selective solution of species other than *B. rosa*, therefore, cannot be invoked to explain the monospecificity of the deposit. Extensive selective solution would have produced high insoluble residue contents and would have required inordinantly long periods of time to concentrate the pentaliths if production rates remained constant. Unusual conditions, therefore, must have been responsible for unusually high production rates of *B. rosa*[2]) and the rarity of other species. We postulate that special environmental conditions which came into play only periodically over the South Atlantic during Oligocene time caused the accumulation of the unusually thick laminae (up to 90 cm) of concentrated braarudosphaerid ooze in its highly disaggregated state. These conditions are thought to be related to the Antarctic current system or temporary climatic cooling. Reduced surface water temperatures would tend to exclude many tropical and semitropical phytoflagellate species from the planktonic flora. Cold bottom temperatures would promote in situ dissolution of the more soluble calcareous nannofossils at the sediment/water interface. The massive, compact construction of the braarudosphaerid pentaliths, however, would make them relatively resistant to solution, although not sufficiently so to prevent their disaggregation into isolated segments.

During the cold intervals (times of *Braarudosphaera* deposition), the organic calcite which formed at surface water temperatures would be in a non-steady state on the sea bottom, and much of the calcium carbonate derived from its dissolution would be held in solution by the cold interstitial fluids. The warming which would accompany a return to normal conditions would cause supersaturation of the interstitial fluids and the inorganic precipitation of calcite cement. It is also possible that circulating supersaturated bottom waters could have supplied additional quantities of calcium carbonate for the precipitation of cement. Until recently, it was generally thought the inorganic precipitation of calcite and the lithification of carbonate sediments could not occur on the sea floor at bathyal depths; however, the compelling evidence presented by FISCHER and GARRISON (1967) indicates that these processes do occur on the sea bottom. This gives us additional reason to believe that our material was indurated shortly after deposition while still in the vicinity of the sediment-water interface.

The origin of the clinoptilolite in the insoluble residue is not certain. REX (1970) reported the occurrence of this species of zeolite in several other samples of the Rio Grande Rise. We suspect that its origin may be related to the particular paleo-oceanographical milieu of the Rise.

Acknowledgments

We are indebted to Dr. Katherine Gaarder for helpful discussion and unpublished ecological data on *Braarudosphaera*. Scanning electron micrographs were made by SWW on a Cambridge model IIA instrument kindly made available by Dr. Hans-Ude Nissen of the Swiss Federal Institute of Technology (ETH); P. Cattori, H.E. Franz and R. Wessicken provided valuable technical assistance. We thank Mr. Kerry R. Kelts (ETH) and Dr. Johan DeVilliers (National Physical Research Laboratory, Pretoria) for the X-ray analyses and Mr. Juerg Sommerauer (ETH) for contributing the microprobe data. All samples were supplied by the National Science Foundation. SWW was supported at the ETH by an NSF Postdoctoral Fellowship; Prof. Dr. Hans M. Bolli (host professor) kindly provided laboratory facilities.

---

[2]) The possibility of high production rates of *B. rosa* in the open marine environment was suggested by MAXWELL et al. (1970b) and is favored by us because abundant populations of modern *Braarudosphaera* in open ocean water have been reported by GAARDER (1954).

Note added in proof: Our electron micrographs show that the insoluble residue of the chalk sample consits largely of the spherical aggregates and a platy mineral; the X-ray pattern indicates the presence of both clinoptilolite and an amorphous substance (amorphous silica?), as well as traces of various detrital minerals. Comparison with the newly published electron micrograph by Gibson and Towe (Science, v. 172, p. 153) identifies the platy mineral as clinoptilolite; the aggregate would then be the amorphous material. This interpretation is in accord with our observation that the spherical aggregates show signs of having been compacted, or squashed (Figure 2, arrow), as one would expect from spherules of silica gels prior to their hardening. Research is currently underway to determine the exact mineralogy and chemical composition of the insoluble residues and to interpret their genetic significance.

REFERENCES

Fischer, A.G., and Garrison, R.E. (1967): *Carbonate Lithification on the Sea Floor*. J. Geol. *75*, 488–496.

Fischer, A.G., Honjo, S., and Garrison, R.E. (1967): *Electron Micrographs of Limestones and Their Nannofossils* (Princeton University Press, Princeton, New Jersey). Monographs in Geology and Paleontology, No. 1, 141.

Friedman, G.M., Amiel, A.J., and Schneidermann, N. (1970): *Submarine Cements in Modern Red Sea Reef Rock*. Geol. Soc. Am., Abstracts with Programs *2/7*, 554–555.

Gaarder, K.R. (1954): *Coccolithineae, Silicoflagellatae, Pterosperma taceae and Other Forms.* In: Michael Sars: *North Atlantic Deep-Sea Expedition, 1910* (John Grieg, Bergen), Vol. II, No. 4, 20 p.

Gomberg, D.N., and Bonatti, E. (1970): *High-Magnesian Calcite: Leaching of Magnesium in the Deep Sea*. Science *168*, 1451–1453.

Hathaway, J.C., and Degens, E.T. (1969): *Methane-Derived Marine Carbonates of Pleistocene Age*. Science *165*, 690–692.

Hay, W.W., Wise, S.W., and Stieglitz, R.D. (1970): *Scanning Electron Microscope Study of Fine Grain Size Biogenic Particles*. Trans. Gulf-Coast Ass. geol. Soc. *20*, 287–302.

Lafitte, R., and Noël, D. (1967: *Sur la Formation des Calcaires Lithographiques*. C.r. Acad. Sci. Paris *265*, 1379–1382.

Levin, H.L., and Joerger, A.P. (1967): *Calcareous Nannoplankton from the Tertiary of Alabama*. Micropaleontology *13*, 163–182.

Lloyd, R.M., and Hsü, K.J. (1971): *Isotope Investigations of Lithified Sediments from the Joides III Cruise to South Atlantic*. Sedimentology (in press).

Maxwell, A.E. et al. (1970a): *Initial Reports of the Deep Sea Drilling Project*. US Government Printing Office, Washington, D.C., Vol. 3, 806 p.

–   (1970b): *Deep Sea Drilling in the South Atlantic*. Science *168*, 1047–1059.

Rex, R.W. (1970): *X-ray Mineralogy Studies – Leg 13*. In: Maxwell, A.E. et al.: *Initial Reports of the Deep Sea Drilling Project* (US Government Printing Office, Washington, D.C.), Vol. 3, p. 509–581.

Roth, H.P. (1970): *Oligocene Calcareous Nannoplankton Biostratigraphy*. Eclogae geol. Helv. *63*, 799–881.

5

Reprinted from pp. 443–481 of *Initial Reports of the Deep Sea Drilling Project,*
Volume 30, J. E. Andrews *et al.,* Washington, D.C.: U.S. Government Printing Office,
1975, 753 pp.

RELATIONSHIPS BETWEEN DIAGENESIS AND PHYSICAL PROPERTIES
OF BIOGENIC SEDIMENTS OF THE ONTONG-JAVA PLATEAU
(SITES 288 AND 289, DEEP SEA DRILLING PROJECT)

Gerrit J. van der Lingen, Sedimentation Laboratory, New Zealand Geological Survey, Christchurch, New Zealand
and
Gordon H. Packham, Department of Geology and Geophysics, University of Sydney, NSW, Australia

ABSTRACT

The Ontong-Java Plateau is a large oceanic plateau north of the
Solomons Island, covered with a thick sequence of subhorizontal
strata of predominantly biogenic sediments. The sediments sampled
at the two sites drilled on this plateau during Leg 30 of the Deep Sea
Drilling Project are well suited for progressive-diagenesis studies.

At Site 288, 988.5 meters of sediment were penetrated. Oldest
sediments are Aptian limestones. No igneous basement was reached.
Stratigraphic hiatuses occur at several levels. Chert horizons are
present from exceptionally young sediments (early Miocene, 267 m)
downwards. At Site 289 basaltic basement was reached after drilling
through 1271 meters of sediment. Oldest sediments are Aptian
limestones. Chert is present below 1000 meters, in Eocene and older
sediments.

For both sites the stratigraphic columns are compared with
downhole plots of lithologic and physical data, viz, drilling disturbance, induration, percentage of siliceous fossils, calcium carbonate
content, seismic reflectors, porosity, bulk density, and sonic velocity.
Drilling disturbance correlates reasonably well with the overall
changes in induration from ooze to chalk to limestone. At Site 288
the boundary between ooze and chalk is well defined, while at Site
289 the induration of ooze to chalk is intermittent right up to the
change to limestone at about 1000 meters. The later chalk ooze
relationship is probably due to the absence of chert at this level at
Site 289. At Site 288 there is a transition zone between chalk and
limestone, while at Site 289 the transition from ooze/chalk to
limestone is sharp. At both sites there is a reversal in induration to
ooze in the Maestrichtian, though a difference of about 600 meters
subbottom depth for this level exists between the two sites.

At Site 288 the porosity decreases, and the bulk density and sonic
velocity increase with depth. There are several sudden changes in all
three parameters at levels related to induration and composition
(especially silica content) of the sediments. In contrast, at Site 289
these three parameters change gradually, without any jumps, to
about 1000 meters depth. The sudden change in values below 1000
meters is due mainly to the presence of chert, and, the associated
abrupt change in induration.

The presence of silica at Site 288 significantly complicates the picture of progressive carbonate diagenesis, as shown in scanning electron photomicrographs. The relatively silica-free carbonate
sediments in the top thousand meters at Site 289 provide a better picture. The aspects of progressive carbonate diagenesis are broadly
similar to those described in earlier papers by the present and other
authors. Dissolution and precipitation of calcite seems to take place
simultaneously. Reprecipitation of secondary calcite begins on discoasters. Subsequently, new calcite is formed on the edges and in the
central parts of coccoliths, as well as on micarb particles (fossil
fragments). In the next stage, secondary calcite bridges the space
between proximal and distal shields of coccoliths, and euhedral
calcite crystals grow inside foraminiferal chambers. The secondary
calcite overgrowth is the major agent in welding particles together.
Further calcite precipitation is indicated by more and more euhedral
crystal faces on overgrown fossil particles. Porosity decreases
gradually. In the most advanced diagenetic stage all fossil particles

are covered with secondary calcite, foraminiferal chambers are completely filled with new calcite, and most of the original pore space is taken up by secondary calcite.

Silica diagenesis takes place in several ways: the formation of quartzose chert nodules; the formation of silica spherules; the replacement and infill of microfossil tests (silicosphere); and the coating and bridging of microfossil particles by smooth "icing sugar" silica.

The correlation of seismic reflectors with diagenetic changes is not always straightforward. A list of possible causative factors is compiled from the present data and from data obtained by the authors during Leg 21.

Several fundamental questions in the understanding of diagenesis of biogenic sediment still remain: (1) is quartzose-chert formation an early diagenetic process?; (2) are (cristobalite-tridymite) silica spherules an intermediate stage in the formation of quartzose chert, or are they formed as an independent phase, late diagenetic and associated with volcanic material?; (3) how much is carbonate diagenesis influenced by silica diagenesis?; (4) can paleo-oceanographic chemistry explain reversals in progressive carbonate diagenesis?; (5) how much is diagenesis influenced by heatflows, decreasing in time, etc?

INTRODUCTION

Deep-sea drilling has made it possible to study diagenetic processes in deep-sea sediments in great detail. Scattered through the Initial Reports series of the Deep Sea Drilling Project are many observations as well as specific papers on diagenesis. In the early stages most of the observations and papers dealt with specific aspects of diagenesis only (Pimm et al., 1971; Moberly and Heath, 1971; Heath and Moberly, 1971; von der Borch et al., 1971; Lancelot and Ewing, 1972; Kastner and Siever, 1973; von Rad and Rösch, 1972; Berger and von Rad, 1972; Heath, 1973; Lancelot, 1973). Since it became possible, because of improved drill bits, to drill through thick chert-bearing sequences, more complete studies of progressive diagenesis with depth (up to 1300 m) and in time (up to Late Jurassic) could be made (Schlanger et al., 1973; Packham and van der Lingen, 1973; Matter, 1974). At the time of writing this paper, Volume 23 of the Initial Reports of the Deep Sea Drilling Project had just been published. Some papers on diagenesis, based on DSDP material, also appeared outside the Initial Reports series.

Most diagenesis studies have concerned themselves with biogenic sediments. Matter (1974), however, also describes progressive diagenesis in a thick (1300 m) terrigenous pelitic sediment sequence, drilled in the Indus Cone in the Arabian Sea (Site 222).

It was a lucky coincidence that the first commercial scanning electron microscope came on the market in 1965, only three years before the beginning of the Deep Sea Drilling Project. Scanning electron microscopy has been essential in the study of diagenesis.

The present authors studied the relationships between the progressive diagenesis of biogenic sediments, their physical properties, and seismic reflectors, from material and data collected during drilling in the southwest Pacific (Leg 21; Packham and van der Lingen, 1973). The same line of investigation is followed in the present paper. For this study, Sites 288 and 289, on the Ontong-Java Plateau, were chosen. The sedimentary sequences at these sites are almost purely biogenic.

EARLIER STUDIES OF PROGRESSIVE DIAGENESIS IN BIOGENIC SEDIMENTS

In dealing with the diagenesis of biogenic sediments, carbonate and silica diagenesis should be discussed separately. The two processes are largely independent, physicochemically, as well as time-wise.

Carbonate Diagenesis

Schlanger et al. (1973) studied progressive diagenesis in biogenic sediments at Site 167. This site is situated on the Magellan Rise, a large oceanic plateau in the central Pacific. The hole bottomed in extrusive oceanic basalts after penetrating 1172 meters of sediment. The age of the sediment immediately overlying basalt is Late Jurassic (Tithonian-Berriasian). The diagenetic changes from ooze to chalk to limestone were studied in detail. The authors proposed a diagenetic model for these changes (carbonate component only): 3.0 cc of foram-nannofossil ooze (density 1.35; porosity 80%) changes to 1.7 cc of foram-nannofossil chalk (density 1.60; porosity 65%), which changes to 1.0 cc of nanno-bearing limestone (no forams; density 2.0; porosity 40%). The changes take place through tighter packing (early stages), mechanical breakdown of some fossils (a very minor contributor to compaction), and by progressive dissolution of an "easily dissolved calcite phase" (mainly foram calcite, but also some nannofossil calcite), and simultaneous reprecipitation on a more stable calcite phase. For their model, they accept that the system is "calcite conservative," that is, no introduction of carbonate from an outside source is required.

Packham and van der Lingen (1973) studied progressive carbonate diagenesis in sediments at four sites in the southwest Pacific: Site 206 in the New Caledonia Basin, Sites 207 and 208 on the Lord Howe Rise, and Site 210 in the Coral Sea Basin. The bulk of the sediments are biogenic, apart from a silty claystone unit at the base of the sediment sequence at Site 207, and a terrigenous turbidite sequence (intercalated with nannofossil layers) in the upper part of Site 210. The oldest sediments studied are Maestrichtian (Sites 207 and 208), and the deepest sediment penetration was 734 meters (Site 206). An effort was made to correlate scanning electron microscope observations with physical sediment properties such as drilling disturbance (or deformation of cores), induration (degree of lithification), bulk density, and sonic velocity. Correlation of all these data with seismic reflectors was only partly successful. Possible correlations are: (1) where the biogenic ooze changes from "creamy" to "stiff" (induration); (2) at levels where the sediments become "crumbly" in the "stiff" zone (induration); (3) at the stiff-semilithified boundary, where welding of particles begins; (4) at the level where the bulk density starts to increase more rapidly with depth, a transition also characterized by a pronounced decrease in porosity and the beginning of granular calcite overgrowth.

The presence of a regional hiatus in all four holes, straddling the Eocene-Oligocene boundary, enabled the relative importance of time and depth of burial in relation to diagenesis to be evaluated. No simple relationship could be detected. Diagenesis seems to be dependent on both the depth and the duration of burial.

Matter (1974) studied the diagenesis in carbonate (Leg 23, Sites 220 and 223) and noncarbonate (Site 222) sediments in the Arabian Sea. Maximum age of the carbonate sediments is early Eocene (Site 220), and maximum depth of carbonate sediments is 657 meters (Site 223). His findings on progressive carbonate diagenesis supports and widens the findings of both Schlanger et al. (1973) and Packham and van der Lingen (1974). Matter also stresses the opinion that there is no need for the introduction of carbonate from an outside source. The calcite for cementation comes from the dissolution of "supersoluble" grains and from pressure solution at grain contacts. He calls this dissolution-reprecipitation process "autolithification."

Silica Diagenesis

In most biogenic sediments there is a siliceous-fossil component of varying magnitude. As pointed out by many authors, biogenic opal seems to be the main source for the ubiquitous deep-sea cherts (e.g., Heath and Moberly, 1971; Wise and Kelts, 1972; Heath, 1973; Lancelot, 1973; van der Lingen et al., 1973). Fundamental, yet unsolved problems exist in the understanding of silica diagenesis. One of these is the question whether cristobalite is an intermediate phase in the formation of chert between opaline silica and quartz. Another problem is whether chert formation is an early diagenetic process or not. These two problems will be discussed shortly

The Cristobalite Problem

Early in the Deep Sea Drilling Project tiny (3 to 12 μm spherical authigenic silica blade aggregates were discovered in Oligocene chalks of the South Atlantic (Wide and Hsü, 1971). X-ray data of insoluble residues suggested that the "silica" aggregates might be clinoptilotite. Similar spherical aggregates ("botryoidal structures") from silicified chalks in the Netherlands and Belgium were described earlier in the same year by Buurman and van der Plas (1971). From analytical data they concluded that the aggregates are "badly crystalline tridymite." Since then, similar spherical aggregates were

found to be a common diagenetic silica mineral in deep-sea sediments. They have been given various morphological names, such as "spherules" (Wise et al., 1972; "microspheres" (Oehler, 1973); and "lepispheres" (= spheres of blades, Wise and Kelts, 1972).

It should be noted, however, that not all silica spherules are spheres of blades. At least two other morphological types have been noted. Matter (1974) depicts spherules consisting of vermiform crystal elements. Another form shows indistinct crystallinity, or at most a "sugary" surface (Lancelot, 1973; this study, Figure 31).

Quite a lot of analytical work has now been done on these spherules. Some of the mineralogical terms given are: "poorly crystallized cristobalite" (Heath and Moberly, 1971); "disordered cristobalite" (Lancelot, 1973); "low cristobalite" (Matter, 1974); "α-cristobalite" (Weaver and Wise, 1972); "lussatite" or "opal-CT" (Berger and von Rad, 1972; Heath, 1973; Greenwood, 1973). Heath (1973) suggested that the α-cristobalite structure can be interposed with tridymite layers. Oehler (1973) synthesized silica microspheres. He concluded that the crystal habit resembles tridymite rather than cristobalite, and suggests that "both the synthetic microspheres and the deep-sea microspheres may be composed of hybrid crystals of interlayered cristobalite and tridymite" (p. 64). Packham and van der Lingen (1973) described silica spherules from a noncalcareous Maestrichtian silty claystone (p. 504, Plate 4, fig. 2 and 3). As the bulk X-ray data for this silty claystone indicate, apart from other components, 69% cristobalite and 12% tridymite, it seems likely that the spherules are a combination of these two minerals. From the available data it can therefore be concluded that the silica spherules are hybrid aggregates of (α- or low-) cristobalite and tridymite. It may well be that these minerals form a series with pure α-cristobalite and tridymite as end members.

Because of the variations in morphology and in mineralogy, the present authors will use the more general term "silica spherules."

Regardless of their morphology or mineralogy, the silica spherules are often associated with euhedral crystals of clinoptilolite. This explains why Wise and Hsü (1971) detected clinoptilolite in the insoluble residue of Oligocene chalk containing silica spherules. Clinoptilolite, a high-silica member of the heulandite zeolite group, is generally found as an alteration product of volcanic ash and tuff (Deer et al., 1963). This raises the question whether both clinoptilolite crystals and silica spherules can be derived from volcanic material. This question is discussed by Wise and Kelts (1972). They argue that some excess silica becomes available during the alteration of volcanic glass to clinoptilolite, and that this excess could precipitate as cristobalite. However, they reason that such a reaction could not produce enough silica to account for the large volumes of chert present in deep-sea sediments. They agree with many other authors that most silica reprecipitated as chert (cristobalitic or quartzitic) is derived from the tests of siliceous organisms.

Cristobalite has generally been considered as an intermediate stage in the formation of chert in deep-sea sediments (Heath and Moberly, 1971; Weaver and

Wise, 1972; Wise et al., 1972; Wise and Kelts, 1972; Heath, 1973; Greenwood, 1973). However, in a recent paper Lancelot (1973) questions this theory. In an excellent study of silica diagenesis in sediments at Site 167 of the Deep Sea Drilling Project (Leg 17), he puts forward an alternative theory. He distinguishes two main types of chert, porcelanitic chert, which is composed predominantly of disordered cristobalite, and quartzose chert, composed of quartz and/or chalcedony. The first type occurs in marly or clayey sediments, often as thin beds or as sediment impregnation. It forms where foreign cations find a place in the disordered structure of the cristobalite. Clay minerals are considered a likely source for such cations. Low permeability also favors the formation of disordered cristobalite. The second type is found in relatively pure (siliceous fossil-bearing) carbonate sediments. The absence of foreign cations and high permeability seem to favor the precipitation of quartz (chalcedony) directly from dissolved biogenic opal and in the form of nodules. Some quartzose chert nodules have a rim of cristobalite. In this transition zone between chert and matrix, impurities are concentrated , creating a microenvironment conducive to the formation of cristobalite. Lancelot thinks that "maturation," the process of converting metastable disordered cristobalite into stable quartz, plays only a minor role in the diagenesis of deep-sea sediments. It could take place under exceptional conditions of high temperature and great overburden.

This interesting theory warrants further verification and study. Because of the common association of silica spherules and clinoptilolite crystals, the importance of volcanic material as a source for foreign cations (and free silica) should be investigated. The question should also be put whether clinoptilolite can be an alteration product of nonvolcanic material.

Early or Late Diagenesis

Because Neogene sediments hardly ever contain chert, it is generally thought that chert formation is dependent on time (the earlier mentioned "maturation process"). It was again Lancelot (1973) who questions this opinion and puts forward strong arguments in favor of chert being an early diagenetic product.

It does not seem that this question can be given a final answer at the present time. Careful sorting and synthesizing of all DSDP data no doubt will bring us close to solving this problem. However, because of the above discussion of the cristobalite problem, the present authors are of the opinion that the question of whether chert formation is early diagenetic or not should be answered separately for quartzose chert and for cristobalitic (tridymitic) chert.

THE ONTONG-JAVA PLATEAU SITES

Lithological, chemical, and physical data of the sediments drilled at Sites 288 and 289 are summarized in Figures 1 and 2. These data will be discussed first, followed by a detailed description of the scanning electron microscopic features of the sediments. Subsequently, a correlation with seismic reflectors will be attempted. And finally tentative generalizations on the diagenesis of deep-sea biogenic sediments will be presented and problem areas outlined, incorporating information from earlier publications.

The regional geologic setting of the Ontong-Java Plateau is discussed in detail elsewhere in this volume, and will not be repeated here.

Lithology

At Site 288, 988.5 meters of sediment were penetrated. No igneous basement was reached. Coring was discontinuous. The oldest sediments cored are Aptian. Stratigraphic hiatuses were encountered at several levels. Eocene sediments are probably missing altogether. The Tertiary sequence is thinner than at Site 289, while a thicker, fairly complete Late Cretaceous sequence is present. From the base of the sequence up to the Coniacian, volcanic ash and tuff beds are intercalated. Glass shard ash is present in early Miocene sediments and younger. Chert is present from the bottom of the hole up to early Miocene.

At Site 289, the drill hole bottomed in extrusive oceanic basalt, after penetrating 1271 meters of sediment. Coring was continuous. The oldest sediments cored have the same age as those at Site 288. The Neogene sequence appears to be complete. Several hiatuses were encountered in the Paleogene and in the Cretaceous. Ash-bearing (or tuffaceous) horizons are present in the Cretaceous and the Paleocene. The volcanic debris forms a minor component only, with the exception of two fairly pure tuff beds near the base of the sedimentary sequence. Chert is present in the lower part of the sequence up to the middle Eocene.

SEM Samples

Twenty-nine samples from Site 288 and 38 from Site 289 were studied with the scanning electron microscope. Their positions are indicated on Figures 1 and 2. The results will be described further on.

Drilling Disturbance

Drilling disturbance is mainly a function of the induration of the sediments. However, drilling procedures, such as different rates of water circulation, also affect the disturbance of the cores. Drilling disturbance is estimated onboard the ship during visual core description. Such estimates are rather subjective, and only the broad pattern should be taken into account. The drilling disturbance is given numerical values from 1 to 4: 1 means no deformation of internal sedimentary structures; 2 means gently deformed structures; 3 means strong deformation, but some structures are still visible; and 4 means almost total destruction of structures through mixing and homogenization. The results can best be discussed in combination with the induration data.

Induration

This column is separated into two parts, the first giving the induration adjectives to the rock names in the second part. As with the drilling disturbance, estimating induration is somewhat subjective. The sediment is considered creamy and stiff when the cores can be cut with a wire cheesecutter. The term ooze is used for unindurated biogenic sediment. When the splitting of cores requires the use of a bandsaw, the sediment is considered to be semilithified and is called a chalk when dealing with predominantly carbonate sediments. When further induration of the rock requires the use of a diamond saw for splitting the cores, the sediment is considered to be lithified, and, in the case of carbonate sediment, is called a limestone.

At Site 288 the boundaries between ooze and chalk and between chalk and limestone are reasonably well defined. There is a thin transition zone between ooze and chalk (at about 130 m). In the chalk interval there is a reversal horizon (to ooze) in the Maestrichtian (between 573 and 580 m). A larger transition zone exists between chalk and limestone. The drilling disturbance drops at about the ooze-chalk boundary. Another one occurs within the chalk interval.

At Site 289 the boundary between ooze and chalk is not well defined. Below the level where the sediment is first called a chalk (at about 243 m), ooze and chalk intervals alternate. It is not always easy to decide whether the sediment is still an ooze, or should be called a chalk. In such cases the term "chalk/ooze" has been used in the core logs. There is a gradual downward increase in the amount of chalk. However, ooze horizons persist up to the level where the sediment becomes a limestone (at about 1007 m). A reversal in induration (to ooze) occurs in the limestone interval at about 1185 meters. This reversal horizon has the same age as the reversal in the chalk interval at Site 288. Major drops in drilling disturbance occur at the ooze and ooze (-chalk) boundary, and at about the level where chalk becomes the predominant rock type.

Siliceous Fossil and Calcium Carbonate Percentages

Core averages of percentages of siliceous fossils, as determined from smear slides onboard the ship, have been plotted in the next column. Percentage estimates in smear slides have a limited accuracy, and only general trends should be considered.

At Site 288, siliceous fossils are present in small quantities only, up to a core average of 6%. Between 450 and 750 meters no siliceous fossils were seen. Chert is present in the lower part of the sequence up to the early Miocene, suggesting that larger amounts of siliceous fossils must have been present originally.

At Site 289 the amount of siliceous fossils slightly increases downwards, from an average of 2% at the top to 6% at 960 meters. Below this there is a sudden increase, with wild fluctuations. This increase coincides with the Eocene-Oligocene boundary. The fluctuations are probably due to local dissolution of siliceous fossils and reprecipitation of the silica as chert nodules. Below 1050 meters no siliceous fossils have been observed. They either have been completely dissolved, or have been transformed to silicospheres (see description of SEM samples).

Calcium carbonate percentages were determined in the DSDP shore laboratory. The values, plotted in the same column as the siliceous fossil percentages, are not core averages, but represent individual spot samples, taken onboard the ship. Also compare with bulk X-ray analyses (Table 1).

At Site 288 the calcium carbonate percentages fluctuate between 90% and 100%, between 100 and 700 meters subbottom depth. The lower values above 100 meters probably reflect the presence of volcanic ash, while the lower percentages below 700 meters reflect the presence of both ash and silica.

At Site 289 calcium carbonate percentages fluctuate between 90% and 100% over most of the sequence. Only one lower value of 51% was recorded at 1037 meters, in the interval with larger amounts of siliceous fossils. The almost pure carbonate samples below 1100 meters are probably from sediments in which all siliceous fossils have been dissolved and the silica reprecipitated elsewhere as chert nodules. Nonsystematic sampling onboard the ship has produced results not entirely representative for all lithologic changes. Analyses of bulk samples would no doubt have given a truer picture.

Seismic Reflectors

The estimated depths of seismic reflectors are given in Tables 2 and 3 and plotted on Figures 1 and 2. The velocities used to determine the reflector depths are derived from shipboard Hamilton frame measurements. It should be pointed out that the coring program at Site 289 was continuous and that at Site 288 was discontinuous, and hence depth estimates at Site 289 are the more reliable.

The pattern of reflectors at the two sites is different. At Site 288, there are stronger reflectors in the upper part of the section than at Site 289. The strongest reflectors at Site 288 are at 0.22, 0.50, and 0.74 sec (two-way travel time) at estimated subbottom depths of 172, 415, and 617 meters, respectively, while those at Site 289 are at 0.84, 0.94, and 1.00 sec at 711, 814, and 873 meters, respectively. The greater energy return to the surface from higher levels at Site 288 is almost certainly attributable to the greater abundance of chert in the higher levels of that stratigraphic column.

Porosity

The plotted values represent core averages.

At Site 288 porosity decreases from just under 70% at the top to about 60% at 210 meters. At that level the porosity drops to 50% and remains fairly constant to about 650 meters. Below 650 meters values decrease rapidly, with large fluctuations, to about 20%. The latter decrease coincides with the change from chalk to limestone.

At Site 289 the porosity decreases gradually from about 70% at the top of the sequence to about 40% at 1000 meters. Below this level, values decrease sharply, fluctuating between 10% and 40%, increasing somewhat again below 1150 meters. These lower values are all in the chert-bearing lithified limestone interval. The large number of data points reflects the continuous coring at this site.

Bulk Density

The plotted values represent core averages. The bulk density trends compare closely with those of porosity.

At Site 288 the bulk density increases from 1.6 at the top to just under 1.7 at 210 meters. At this level there is a

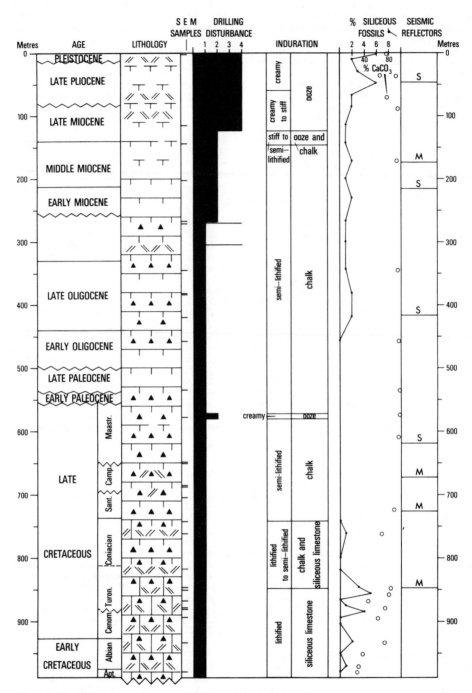

Figure 1. *Site 288, stratigraphic column and plots of physical properties.*

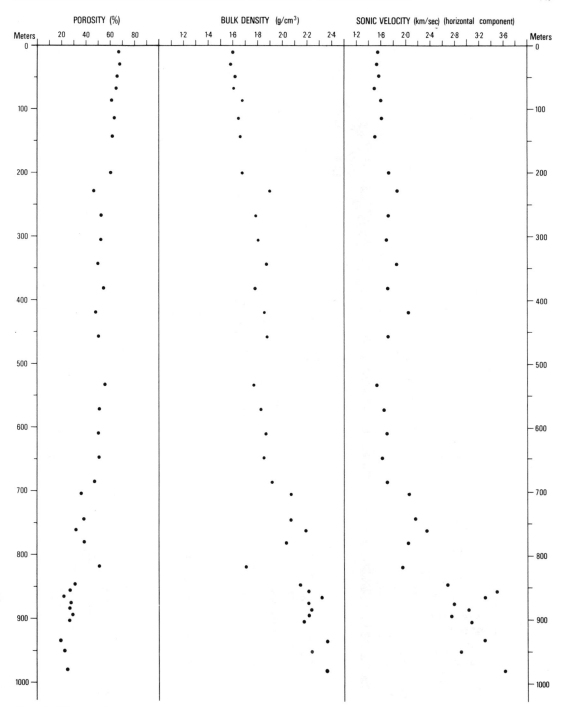

Figure 1. *(Continued).*

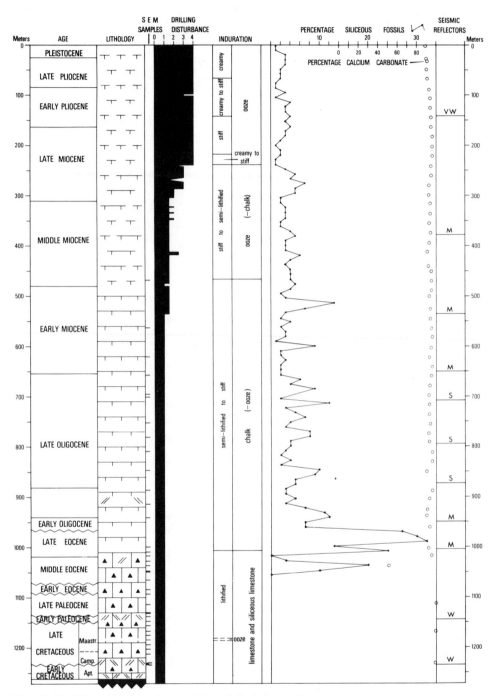

Figure 2. *Site 289, stratigraphic column and plots of physical properties.*

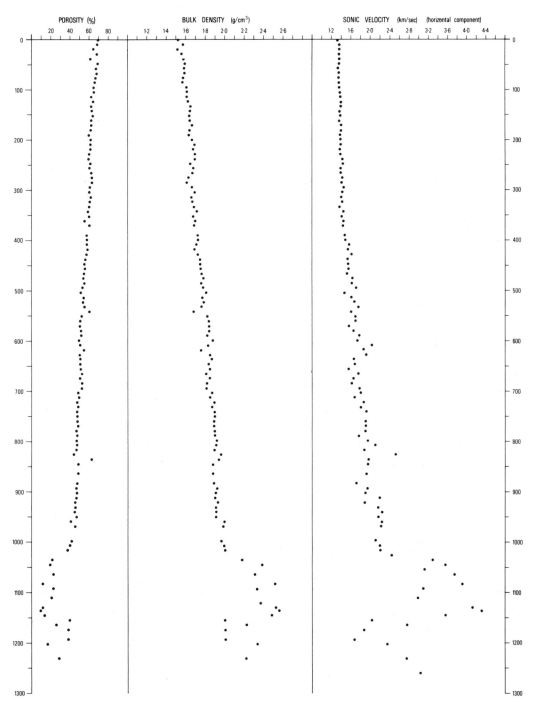

Figure 2. *(Continued).*

TABLE 1
Bulk X-Ray Data, Sites 288, and 289

Sample Depth Below Sea Floor (m)	Diff.	Amor.	Calc.	Dolo.	Quar.	Cris.	K-Fe.	Plag.	Mica	Chlo.	Mont.	Paly.	Trid.	Clin.	Pyri.	Bari.	Sepi.	Augi.
Hole 288																		
2.70	57.6	31.5	97.5		0.8			1.7										–
16.40	77.9	63.6	83.7		0.5			12.0										3.7
72.20	50.0	19.2	97.9		0.5			1.6										–
88.70	43.2	8.4	100.0		–			–										–
Hole 288A																		
457.80	42.6	7.4	100.0	–	–	–	–	–	–		–	–	–	–		–		
535.10	41.9	6.3	100.0	–	–	–	–	–	–		–	–	–	–		–		
535.90	44.3	10.2	100.0	–	–	–	–	–	–		–	–	–	–		–		
578.10	40.3	3.7	100.0	–	–	–	–	–	–		–	–	–	–		–		
579.30	39.3	2.1	100.0	–	–	–	–	–	–		–	–	–	–		–		
609.90	38.8	1.3	100.0	–	–	–	–	–	–		–	–	–	–		–		
649.40	81.4	48.0	22.3	2.3	5.9	–	–	–	5.5		2.6	60.1	–	1.2				
762.00	73.8	53.2	55.7	–	2.9	–	–	10.3	–		4.0	4.3	–	19.5		3 3		
762.00	48.7	18.1	88.4	–	10.0	–	–	–	–		–	–	–	–		1.6		
762.40	79.8	55.6	2.7	–	4.5	–	–	16.7	3.9		15.0	13.7	–	35.0		8.5		
850.80	79.4	65.0	80.2	–	11.2	5.8	–	–	–		–	–	1.5	–		1.3		
858.10	63.7	37.6	69.9	–	0.2	–	–	4.1	–		3.9	–	–	17.2		4.7		
876.70	56.7	9.4	54.2	–	13.9	22.8	–	–	–		1.8	–	4.8	–		2.5		
884.80	46.4	16.1	72.3	–	25.6	–	–	–	–		–	–	–	–		2.1		
895.00	47.8	16.8	79.0	–	17.0	–	–	–	–		1.4	–	–	–		2.6		
913.60	56.3	24.4	53.8	–	31.8	10.3	–	1.3	–		–	–	2.1	0.8		–		
934.20	44.9	14.2	67.5	–	30.1	–	–	–	–		–	–	–	–		2.4		
952.30	57.5	12.8	49.2	–	21.3	22.5	–	–	–		–	–	5.5	–		1.6		
971.10	57.0	13.9	32.2	–	37.5	28.0	–	1.1	–		–	–	–	1.2		–		
980.50	57.6	27.0	38.2	–	41.7	10.4	3.7	1.8	0.9		2.2	–	–	1.0		–		
Site 289																		
0.70	49.1	18.0	99.1		0.9	–	–	–	–	–	–		–	–		–		–
67.10	45.8	12.6	99.7		0.3	–	–	–	–	–	–		–	–		–		–
144.40	44.3	10.2	100.0		–	–	–	–	–	–	–		–	–		–		–
182.60	42.1	6.6	100.0		–	–	–	–	–	–	–		–	–		–		–
260.20	43.3	8.5	100.0		–	–	–	–	–	–	–		–	–		–		–
336.20	42.6	7.4	100.0		–	–	–	–	–	–	–		–	–		–		–
374.10	44.2	10.0	100.0		–	–	–	–	–	–	–		–	–		–		–
450.00	42.1	6.6	100.0		–	–	–	–	–	–	–		–	–		–		–
488.90	43.3	8.5	100.0		–	–	–	–	–	–	–		–	–		–		–
564.30	43.5	8.9	100.0		–	–	–	–	–	–	–		–	–		–		–
602.00	42.6	7.4	100.0		–	–	–	–	–	–	–		–	–		–		–
678.30	44.7	10.8	100.0		–	–	–	–	–	–	–		–	–		–		–
716.30	43.4	8.7	100.0		–	–	–	–	–	–	–		–	–		–		–
754.40	44.7	10.8	100.0		–	–	–	–	–	–	–		–	–		–		–
790.20	46.6	13.9	100.0		–	–	–	–	–	–	–		–	–		–		–
829.90	41.8	6.1	100.0		–	–	–	–	–	–	–		–	–		–		–
887.30	57.5	30.8	93.0		–	–	6.3	–	–	–	–		–	–	0.6	–		–
915.80	60.6	35.9	95.8		–	–	2.9	–	–	–	1.3		–	–		–		–
925.80	46.4	13.5	100.0		–	–	–	–	–	–	–		–	–		–		–
950.80	59.1	34.0	100.0		–	–	–	–	–	–	–		–	–		–		–
960.40	44.1	9.8	100.0		–	–	–	–	–	–	–		–	–		–		–
1001.60	43.7	9.2	100.0		–	–	–	–	–	–	–		–	–		–		–
1036.80	61.3	23.2	70.8		5.3	16.3	–	–	–	–	–		–	6.3		–	1.4	–
1065.20	39.6	2.6	100.0		–	–	–	–	–	–	–		–	–		–	–	–
1112.30	39.8	2.9	100.0		–	–	–	–	–	–	–		–	–		–	–	–
1138.30	70.6	48.2	55.8		1.2	–	1.6	–	–	1.6	3.7		–	15.0		–		21.2
1194.40	39.2	1.9	100.0		–	–	–	–	–	–	–		–	–		–		–
1230.50	72.1	39.3	47.6		6.5	–	3.6	–	5.4	0.6	1.3	35.1	–	–		–		–
1231.60	80.4	33.5	–		11.3	–	5.7	–	4.8	0.6	2.3	75.3	–	–		–		–
1233.60	49.1	14.5	89.7		–	–	–	–	–	–	1.6	3.9	–	1.5		–	3.2	–
1259.50	39.3	2.1	100.0		–	–	–	–	–	–	–		–	–		–	–	–
1261.50	79.1	52.4	–		8.7	–	59.6	–	3.1	–	28.7		–	–		–	–	–
1261.80	87.4	69.7	–		7.2	–	41.7	–	8.6	–	42.5		–	–		–	–	–
1262.30	38.6	1.0	94.9		4.1	–	–	–	1.0		–		–	–		–	–	–

TABLE 2
Estimated Depths to Seismic Reflectors at Site 288

Reflector Intensity[a]	Reflector Depth (2 way Travel Time Below Sea Floor)	Interval Velocity to Next Reflector (m/sec)	Estimated Depth to Reflector (m)
S	0.0	1540	0
M	0.06	1580	46
S	0.22	1730	172
M	0.27	1740	215
S	0.50	1690	415
S	0.74	1630	617
M	0.81	2090	674
M	0.86	2290	726
M	0.97	2290	847

[a]S = strong, M = moderate.

TABLE 3
Estimated Depths to Seismic Reflectors at Site 289

Reflector Intensity[a]	Reflector Depth (2-way Travel Time Below Sea Floor)	Interval Velocity to Next Reflector (m/sec)	Estimated Depth to Reflector (m)
S	0.0	1550	0
VW	0.18	1580	140
M	0.48	1750	377
M	0.66	1950	535
M	0.78	1970	652
S	0.84	2050	711
S	0.96	1970	814
S	1.00	2200	873
M	1.07	2150	951
M	1.12	4000	1004
W	1.19	2500	1144
W	1.26	(?Basalt)	1232

[a]S = strong, M = moderate, VW = very weak.

jump to 1.9. Between 210 and 600 meters, values fluctuate between 1.9 and somewhat less than 1.8. Below 650 meters values increase, with fluctuations, to almost 2.4.

At Site 289 values increase gradually from just over 1.5 at the top to 2.0 at about 1000 meters. Below that level values increase rapidly, with large fluctuations, to almost 2.6, decreasing somewhat again below 1150 meters.

Sonic Velocities

Onboard the ship horizontal and vertical sonic velocity components were measured on sediment cubes, cut from the core. Differences between the two components were detected, especially in the more indurated parts of the sediment sequence. Investigations into the causes of this anisotropy did not form part of the present study, not because of lack of interest, but because of time involved. This matter will be pursued in the future. For the present study only the core averages of the horizontal velocity component have been plotted, as this is the component normally recorded during shipboard measurements. At both sites there is a close parallelism between the sonic velocity trends and those of porosity and bulk density.

At Site 288 the sonic velocity is between 1.5 and 1.6 km/sec near the surface. It remains fairly constant, fluctuating between these two values, to about 175 meters.

Between 150 and 200 meters there is a stop in the velocity, and from 200 to about 500 meters, values fluctuate between 1.7 and 2.1. The value of 2.1 at 420 meters is almost rigid. Below 500 meters values drop to under 1.6, after which they gradually increase to 1.7 at about 700 meters. Below that level there is a jump in sonic velocity to over 2.0. From 700 to about 830 meters values fluctuate between 1.9 and 2.4. Below 830 meters there is another jump, to over 2.6. Between 830 meters and the bottom of the hole values fluctuate widely between 2.0 and 3.6.

At Site 289 values decrease regularly from just under 1.6 at the top to about 2.4 at 1000 meters. Below this level values increase rapidly with large fluctuations, to about 4.5. Below 1150 meters there is a slight decrease, again with large fluctuation.

SCANNING ELECTRON MICROSCOPIC FEATURES OF THE SEDIMENTS

Introduction

At Site 288, the highest sample studied comes from only 0.9 meters below the sediment-water interface. At Site 289, the highest sample studied was taken at 468.9 meters, at about the level where ooze (-chalk) changes to chalk (-ooze). Apart from the unindurated ooze samples, freshly broken surfaces only were studied.

Site 288

The sediments near the top of the hole consist of whole tests and broken fragments of calcareous and siliceous organisms (Figure 3). Fragmentation can take place in the water column during settling (see Lisitzin, 1971, fig. 127), and, near the sediment-water interface, by infaunal activity.

During the first stages of lithification water expulsion is the main process. The sediment becomes stiff and crumbly when a grain-supporting framework is established (Packham and van der Lingen, 1973). Further diagenesis takes place through a complicated process of dissolution and reprecipitation (see earlier review of progressive carbonate diagenesis studies).

Through dissolution of foraminifera and easily dissolvable coccoliths, micritic, anhedral particles are produced. Such particles are called "micarb" by DSDP. These micritic particles and the remaining fossil tests form the substrate ("nucleation sites," "seed crystals") for calcite precipitation. Dissolution and reprecipitation are taking place simultaneously (Figures 4 and 5, 202.6 meters). Precipitation starts in the central parts of coccolith placoliths and on micarb particles. Welding of individual particles start at the boundary ooze-chalk (Figure 4).

Discoasters are particularly susceptible to calcite overgrowth (Figures 4 and 6). This overgrowth must have started before or soon after burial. Seldom have discoasters without overgrowth been observed in DSDP cores. This phenomenon has been explained by the fact that discoasters are single crystals, and as such are ideal "seed crystals" (Wise and Kelts, 1972; Matter, 1974). Adelseck et al. (1973) reproduced discoaster overgrowth experimentally. Figure 6 shows two stages of discoaster overgrowth. The new calcite envelops some coccoliths, thus contributing to the lithification (welding) process.

The first chert nodule occurs at 267.3 meters. This nodule consists of a fine-grained matrix in which irregular patches of coarser silica occur (Figure 7). The fine-grained matrix consists of chalcedony crystals (Figure 8), while the coarser patches are made up of interlocking quartz crystals (Figure 9).

The next step in recrystallization is the bridging of the space between the proximal and distal shields of coccolith placoliths (Figure 10, 271.4 m).

Dissolution and reprecipitation increases downwards. Euhedral calcite crystals, most of them formed by overgrowth of fossil fragments, increase in size. Some of these crystals show pits, resembling negative crystals (Figure 11, 382.0 m). It may well be that these pits are typical for calcite crystals enclosing fossil fragments. Compaction by crushing of fossil tests seems to play a minor role only. Most foraminifera seem to remain intact (Figure 12, 420.1 m).

At 457.6 meters, most micarb particles have become subhedral, showing some crystal faces (Figures 13 and 14). The central part of some coccoliths become filled with new calcite, while others develop angular euhedral overgrowths along their circumference (Figure 14). Pore space decreases, but average porosity in this interval is still 50%. Figure 14 shows some clusters of radiating subhedral calcite crystals. A similar cluster is also visible in Figure 11. They are the (partly overgrown?) nan-

nofossil *Sphenolithus moriformis* (A.R. Edwards, personal communication).

The outline of coccoliths becomes more angular downwards (Figure 15, 535.3 m). The size of euhedral calcite and the degree of welding also increases.

At about the Cretaceous-Tertiary boundary (550 m) both the bulk density and the sonic velocity decrease somewhat. The corresponding increase in porosity is less pronounced. Not far below this boundary there is a temporary reversal in induration (to ooze). Figures 16 and 17 are SEM photographs of this ooze horizon. The particles, however, do not show a much lower diagenetic stage. For instance, the space between the distal and proximal shields of coccolith placoliths is filled in with secondary calcite. Micritic micarb particles and the surfaces of foraminifera have euhedral calcite overgrowths. There is, however, no sign of welding. It is difficult to explain this phenomenon. Maybe at this level the balance between solution and reprecipitation was more in favor of solution than in the horizons directly above and below. On the other hand, it may be a reflection of a change in paleooceanographic chemistry. Supporting this possibility is the fact that a similar reversal is observed at Site 289, at about the same age level, though at a much greater depth.

At 650 meters diagenesis has increased again. Particles are welded together by overgrowth calcite (Figures 18 and 19). The matrix locally exists of a fairly tight packing of subhedral to euhedral micarb particles (Figure 19). Coccoliths consist of angular "keystone plates" (Figure 18). This angularity seems to be characteristic of Mesozoic coccolith forms (A.R. Edwards, personal communication). It is not always easy to distinguish between primary (taxonomic) angularity and secondary overgrowth. Only features like irregular size of the plates and interpenetration of calcite crystals can establish secondary changes. Disintegration of angular coccoliths produces abundant rhombic particles which might be mistaken for authigenic dolomite rhombs (Figure 18). This is all the more relevant as at this very level real dolomite rhombs do occur, large ones (Figure 20) and small ones (Figure 21). The presence of dolomite is confirmed by the bulk X-ray data (Table 1).

There is a substantial change in the overall composition of the sediment at this level. Above 650 meters calcium carbonate percentages are between 90% and 100%. The bulk X-ray diffraction data indicate 100% calcite for the crystalline fraction. Below this level, the calcium carbonate percentage decreases, and the bulk X-ray data indicate the presence of dolomite, quartz, cristobalite, plagioclase, mica, montmorillonite, palygorskite, tridymite, clinoptilolite, and barite, in varying quantities. This clearly indicates mixing of biogenic sediment with volcanic debris, most of which has subsequently been altered. It is quite likely that the magnesium for the dolomite has been derived from the volcanic material.

The presence of a substantial nonbiogenic component locally gives the sediment a "dirty" look (Figure 22, 705.5 m, and Figure 23, 818.3 m) and diagenetic changes are more difficult to observe.

Below 700 meters there is a further decrease in porosity and increase in bulk density and sonic velocity. The sediment gradually changes from a chalk to limestone.

Figure 3. *Site 288, 0.91 meters, Pleistocene, foram-nanno ooze, porosity (15.94 m) 69%, scale bar 12μ; unconsolidated sediment near the sediment-water interface, consisting of nannofossils and abundant broken fragments of nannofossils and foraminifera.*

Figure 4. *Site 288, 202.6 meters, middle Miocene. nanno chalk, porosity (202.2 m) 60%, scale bar 16μ; discoasters have completely been overgrown by secondary calcite, showing euhedral crystal faces; the secondary calcite has welded fossil fragments together; welding can also be seen in other parts of the picture; note corroded coccoliths.*

The limestone is for the greater part silicified. The matrix can become quite dense, and the sediment shows imprints of coccoliths (Figure 24, 743.2 m).

Figure 23 (818.3 m) is from a sample with a substantial nonbiogenic component. Silica spherules are present (Figures 25 and 26). Both cristobalite and tridymite were recorded in a sample not far below this one (Table 1). This seems to support the theory of cristobalite formation by Lancelot (1973), mentioned earlier.

The next two samples studied are close together (847.0 and 850.8 m). The coccoliths are very angular (Figure 27, 847.0 m). The growth of euhedral calcite crystals inside foram chambers has advanced further (Figure 28). Figure 29 shows a foram chamber completely filled with a single crystal. It is very difficult to determine the nature of a crystal from its shape only. Analysis by EDAX would be required to establish its identity (equipment not available to the authors). Figure 27 shows a peculiar amorphous material, enveloping carbonate particles. The same feature can be seen on Figure 52 (Site 289, 886.2 m). Figure 27 also shows some fine threads. Such threads have been observed in some samples, and are probably fungi, grown in the sample after collection. The enveloping material might also be a fungus growth. If not, it could be a form of amorphous silica.

Silica diagenesis is well shown in Figures 30 and 31 (850.8 m). The first shows a ghost of either a radiolarian or a foraminiferal chamber, filled with silica. The groundmass is also silicified, embedding coccoliths.

Figure 31 shows silica spherules, together with a clinoptilolite crystal. This kind of paragenesis has been discussed earlier. The silica spherules are of the "sugary surface" type. The bulk X-ray data indicate the presence of cristobalite and tridymite, but no clinoptilolite. However, clinoptilolite is recorded in the 2 to 20 μm fraction.

Figure 32 (867.8 m) shows advanced recrystallization of coccolith cement. Welding of particles is well advanced. Chert nodules from the same level show a smooth breakage surface (Figure 33).

Because of the substantial amount of silicification in this lowermost interval, progressive changes in carbonate diagenesis are difficult to observe. Porosity has by now decreased to between 20% and 30%, while bulk density and sonic velocity have increased substantially.

Figure 34 (879.5 m) shows the boundary between a chert nodule and (silicified) chalk. The chalk shows faint and smoothed outlines of coccoliths (Figure 35). The smoothness is probably due to silica coating and not to dissolution. It is quite common in samples from silicified carbonate rock to observe a coating of carbonate particles with a substance looking like icing sugar. The fracture surface of the chert is smooth, showing some fracture lines (Figure 36). The only possible organic remains (or ghosts) are spherical objects (Figure 36). These may represent recrystallized radiolarians.

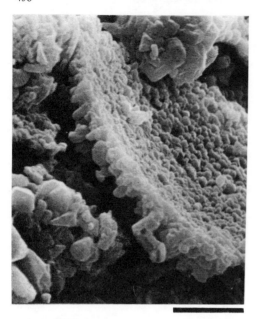

Figure 5. *Site 288, 202.6 meters, middle Miocene, nanno chalk, porosity (202.2 m) 60%, scale bar 16μ; fragment of a foraminifera, showing granular surface due to dissolution.*

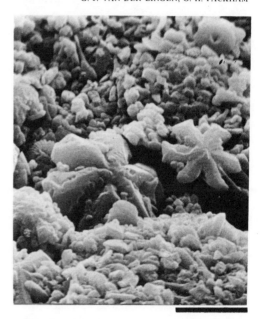

Figure 6. *Site 288, 202.6 meters, middle Miocene, nanno chalk, porosity (202.2 m) 60%, scale bar 16μ; discoasters in the center of the picture show two stages of calcite overgrowth, the right-hand side one being least overgrown; note the abundance of micarb particles, partly overgrown with new calcite.*

Thin sections of silicified limestones often show spherical objects, filled with chalcedony or microquartz (e.g., Lancelot, 1973). Such spheres are called "silicospheres" (not to be confused with the silica spherules, which are of an entirely different order of magnitude). They generally are silicified radiolarian molds or single foraminiferal chambers. Figure 37 (894.5 m) shows a roughly spherical body consisting of platy silica, probably chalcedony. This most likely is a silicosphere. The sediment around this silicosphere shows the usual angular coccoliths.

Figures 38 to 40 are from the lowest sample studied (979.9 m). Figure 38 is a good example of a silicosphere, while Figure 39 shows the faint outline of a silicified foraminifera. Figure 40 shows a remarkably well preserved coccolith in a silica matrix.

Site 289

At Site 289, the overall sediment thickness is much greater than at Site 288. The Tertiary is about twice as thick, while the Cretaceous is about four times thinner than at Site 288. Another important difference between the two sites is the absence of chert and volcanic debris from the late Eocene onwards at Site 289. For further discussion, see Site 289 Site Report. This has the advantage that carbonate diagenesis in the top thousand meters at Site 289 can be studied without the interference of silica or volcanic debris.

Where appropriate, the scanning electron photomicrographs of Site 289 will be compared with those from Site 288.

Figure 7. *Site 288, 267.3 meters, late Oligocene, chert nodule in nanno chalk, scale bar 300μ; dense chert with irregular coarser crystalline patches.*

Figure 8. *Site 288, detail of Figure 7, scale bar 6μ; dense part of chert, showing chalcedony crystallites.*

Figure 10. *Site 288, 271.4 meters, early Miocene, nanno chalk, porosity (270.7 m) 53%, scale bar 8μ; foraminiferal fragments, nannofossils, and calcareous spicules (foraminiferal spicules?); many particles are welded together; the space between proximal and distal shields of coccolith placoliths is partly bridged by secondary calcite (arrow); new calcite is growing in the central parts of coccoliths (lower right-hand corner).*

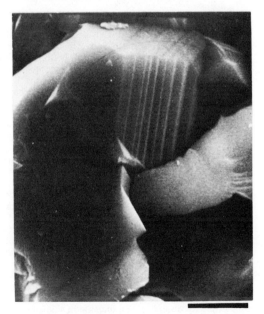

Figure 9. *Site 288, detail of Figure 7, scale bar 6μ; interlocking quartz crystals in coarser crystalline patch of chert.*

The first sample studied (468.9 m) shows partly dissolved coccoliths, overgrown discoasters, and abundant anhedral micarb particles (Figure 41). These micarb particles clearly are derived from disintegrated nannofossils and foraminifera. The secondary calcite on discoasters has incorporated (and welded) some coccolith particles. Minor recrystallization has taken place in the central parts of coccoliths. A similar diagenetic stage at Site 288 is shown in Figures 4 to 6 (202.6 m).

More pronounced overgrowth and welding is shown in Figure 42 (562.4 m). At 620.4 meters, micarb particles start to show euhedral crystal faces due to overgrowth (Figures 43 and 44). Secondary calcite bridges the space between proximal and distal placolith shields (Figure 44). This phenomenon is even better shown in Figure 45 (657.5 m). Calcite growth in the central parts of coccolith placoliths has become more pronounced. A comparable diagenetic level at Site 288 is shown in Figure 13 (457.6 m).

Figures 46 and 47 show calcite needles. Similar needles also exist at Site 288 (Figure 10). Euhedral crystal faces and an imprint of a coccolith clearly indicate secondary overgrowth. This has obscured the

Figure 11. *Site 288, 382.0 meters, late Oligocene, foram nanno chalk, porosity (382.7 m) 55%, scale bar 8μ; coccoliths show corrosion and calcite precipitation; particles are welded together; large euhedral calcite crystals in the center (some having "negative crystals") are probably all overgrown fossil particles; radiating cluster of calcite crystals is the nannofossil* Sphenolithus moriformis.

Figure 12. *Site 288, 420.1 meters, late Oligocene, nanno chalk, porosity (420.2 m) 48%, scale bar 160μ; breakage surface showing uncrushed foraminifera.*

morphology of the needles. They may be either calcareous sponge spicules or foraminiferal spicules.

Figure 48 (704.8 m) shows pronounced dissolution of coccoliths. One placolith shows secondary calcite growth in between shields. It seems that part of this secondary calcite has subsequently been dissolved as well. This has interesting consequences. We stated earlier that dissolution and reprecipitation take place simultaneously. The balance between the two processes apparently can change in time, and already precipitated calcite can get dissolved again.

Figure 49 (752.1 m) shows increased calcite overgrowth on coccoliths and micarb particles. At about 847.0 meters the coccoliths have developed an angular outline through the development of calcite crystal faces. If such coccoliths disintegrate, they produce "pseudodolomite" rhombs (Figure 50). A similar phenomenon was observed in Figure 18 of Site 288 (650.0 m). Figure 50 shows a "euhedral" discoaster, the center of which is strongly pitted. It is difficult to establish whether this pitting is primary (no secondary calcite deposited), or whether it is due to dissolution or to breakage. Figure 51, from the same sample, shows euhedral calcite growth inside a foraminiferal chamber, and in the surrounding matrix. A comparable stage is shown in Figure 28 of Site 288 (847.0 m).

Figure 52 (886.2 m) shows partly dissolved coccoliths as well as secondary calcite overgrowth on coccoliths. Abundant micritic subhedral grains can be seen. Larger grained euhedral calcite cement welds particles together. A radial calcite crystal aggregate, similar to those described from Figure 14 (Site 288, 457.6 m), probably being the nannofossil *Sphenolithus moriformis,* is also present.

Figure 53 (914.3 m) can be directly compared with Figure 43 (620.4 m). Both show part of a radiolarian, partly covered with micritic particles. However, in Figure 53 the radiolarian test has been corroded more, and the micritic grains have obtained euhedral outlines through overgrowth.

At 941.2 meters, welding seems further advanced (Figure 54). Some coccoliths have obtained a smooth surface (either through solution or by coating with "icing sugar" silica). Some wisp-like bridging structures can be seen (see Figures 54 and 55) similar to those described by Packham and van der Lingen (1973, Plate 4, fig. 4 and 5). The authors think that such bridging structures consist of silica. Though the first chert was observed somewhat lower in the sequence (at 1009 m), some silica may already be present at this level. Smearslide estimates do indicate an increase in siliceous fossil content at about this level (Figure 2).

The next photomicrograph (Figure 56, 1009.1 m) is from a sample near the top of the limestone interval. At this level the first chert was observed. Euhedral overgrowth on foraminiferal fragments, coccoliths, and micarb particles is well advanced.

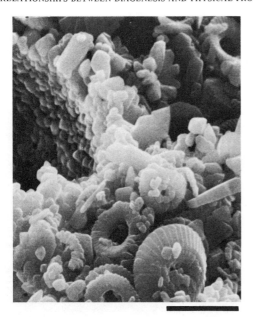

Figure 13. *Site 288, 457.6 meters, early Oligocene, nanno chalk, porosity (457.8 m) 52%, scale bar 12u; coccoliths show extensive calcite recrystallization in their central parts and in between proximal and distal shields; most micarb particles have obtained euhedral outlines; the granular inner surface of the foraminiferal fragment starts to develop new calcite overgrowth.*

Figure 14. *Site 288, 457.6 meters, early Oligocene, nanno chalk, porosity (457.8 m) 52%, scale bar 8μ; radiating calcite clusters are the nannofossil* Sphenolithus moriformis; *these and many micarb particles show euhedral crystal faces; some coccoliths have developed extensive calcite growth along their edges (arrow).*

Figure 57 (1017.2 m) shows a fairly tight packing of coccoliths, coccolith fragments, and micarb. The diagenetic stage seems somewhat less than at higher levels.

The next sample was taken just below the sudden decrease in porosity and increase in bulk density and sonic velocity. Welding has become very significant (Figure 58, 1027.1 m). The central area of some coccoliths has become filled "to overflowing" with euhedral calcite crystals. Overgrowth calcite crystals interpenetrate each other. The calcite crystals inside foraminiferal chambers have reached relatively large sizes (Figure 59).

Comparisons with Site 288 are virtually impossible, because of the presence of volcanic material and silica at that site. Figure 32 (Site 288, 867.8 m) probably depicts about the same diagenetic level.

Volcanic and altered volcanic material is present in the lower 200 meters of Site 289, though in far lower quantities than at Site 288. The bulk X-ray data indicate the presence of quartz, cristobalite, potash feldspar, mica, chlorite, montmorillonite, palygorskite, tridymite, clinoptilolite, barite, and sepiolite (Table 1). These minerals, in limited combinations, are irregularly distributed. In contrast to the lower part of Site 288, pure

carbonate horizons are intercalated with impure carbonate ones. A minor amount of volcanic and altered volcanic material is also present at a higher level (in samples from 887.3 and 915.8 m).

The next sample illustrated comes from 1036.8 meters. The bulk X-ray data at this level indicate cristobalite, tridymite, and quartz. This relatively large amount of silica is clearly shown in the scanning electron micrographs. Figure 60 shows discoasters and coccoliths, overgrown with secondary calcite. The dense groundmass contains irregular subspherical clusters, resembling (cristobalite-tridymite) silica spherules. The discoasters in Figure 61 are coated with an amorphous "icing sugar" substance. It was suggested earlier that such "icing sugar" may be silica. This suggestion is supported in this case by the X-ray data. Note that the silica coating gives the discoaster the appearance of being strongly corroded. Figure 62 is another example of nannofossils embedded in a dense silica matrix (compare with Figure 35, Site 288). Porosity at this level is about 22%.

The sample from 1047.2 is a fairly pure carbonate. Figure 63 shows recrystallized foraminiferal chambers, some of them almost entirely filled with large, freely grown calcite crystals. Figure 64 shows advanced recrystallization of cement and coccoliths. Some

Figure 15. *Site 288, 535.3 meters, late Paleocene, nanno-foram chalk, porosity (535.1 m) 56%, scale bar 12μ; coccoliths and coccolith fragments showing advanced calcite overgrowth and welding. Note zeolite crystal left of center.*

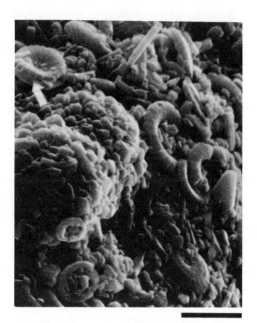

Figure 16. *Site 288, 578.2 meters, middle Maestrichtian, foram-nanno ooze, porosity (572.4 m) 52%, scale bar 8μ; unwelded mixture of coccoliths, coccolith fragments, and foraminifera; some coccoliths well preserved, others showing secondary calcite filling in the space between proximal and distal shields (arrow); note recrystallized surface of foraminifera.*

pressure solution can be seen. Impressions of coccoliths can be seen in some of the larger calcite crystals. Small cracks in the limestone have become filled with sparry calcite (Figure 65). The first stylolites were observed in the cores at about this level.

The X-ray data of the next sample (1065.2 m) indicate pure carbonate. Both dissolution and recrystallization of calcite can be seen (Figure 66).

Figure 67 (1084.2 m) shows very fine fibrous material covering a calcite substrate. This fibrous material looks like cristobalite (-tridymite). No X-ray data are available from this level.

The next photomicrograph is from a pure carbonate limestone (Figure 68, 1093.8 m). It shows the common diagenetic features of this limestone interval; overgrown nannofossils, pressure solution features, welding, a matrix of coccolith fragments, and relatively large secondary calcite crystals. It has, however, some anhedral to subhedral micritic particles, common to lower diagenetic levels.

Figure 69 (1130.5 m) shows densely packed and welded coccoliths in a granular calcite matrix. Some smooth surfaces are visible ("icing sugar" silica?). Porosity is about 11%.

The next sample is just below this one (1131.6 m). Figure 70 shows the inside of a foraminiferal chamber. The inside surface has a very peculiar texture of interlocking small rods. This structure has not been seen by the authors before. Closer examination shows these

rods to be the ends of elements constructing the wall of the chamber and therefore probably consist of calcite. Foraminiferal wall structures consisting of calcite needles in a three-dimensional random array are known from the benthonic *Quinqueloculina*, and are considered a taxonomic characteristic of that genus (Cheriff and Flick, 1974). Though that structure does not look exactly like the one shown in Figure 70, diagenetic modification could convert it to that. However, the foraminifera in Figure 70 clearly is a planktonic species. Normal free-growing calcite crystals are attached to the inner surface. Completely filled foraminiferal chambers are shown in Figure 71.

At 1138.1 meters the sediment contains some silica again. "Icing sugar" coating is common (Figure 72). Secondary silica has filled most of the pore space (Figure 73), and the porosity is 7% only.

As mentioned in the description of the physical properties, there is a reversal in the general trends below 1150 meters. This level coincides with the Cretaceous-Tertiary boundary. A similar phenomenon was observed at Site 288. Induration is generally lower than above. At both sites there is an ooze horizon in the Maestrichtian. The photomicrographs of Site 289 (Figures 74 to 76) compare closely with those of sediments of about the same age at Site 288 (Figures 16

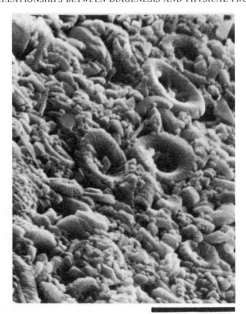

Figure 17. *Site 288, 578.2 meters, middle Maestrichtian, foram-nanno ooze, porosity (572.4 m) 52%, scale bar 8μ; unwelded mixture of coccoliths and micarb particles; note well-preserved coccolith in upper right-hand corner; the coccolith left of center has the intershield space filled up with secondary calcite.*

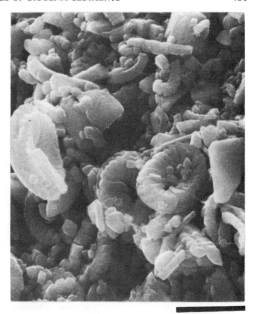

Figure 18. *Site 288, 650.0 meters, Campanian, nanno chalk, porosity (648.4 m) 51%, scale bar 12μ; coccoliths and coccolith fragments welded together; note angular "keystone" plates, typical of Mesozoic forms; extensive secondary calcite growth; note wisp-like bridging structures.*

and 17). Particles are loosely packed and hardly any welding is present (Figure 74, 1155.8 m). Some of the coccoliths show a lower degree of overgrowth (Figure 76, 1175.3 m). Foraminifera are less filled with new calcite (Figure 75, 1167.4 m). Porosity has increased to 40%.

However, as at Site 288, the diagenetic stage is not very much lower than directly above. It certainly never returns to the level of the ooze at the top of the section.

At 1203.9 meters, diagenesis seems to have "picked up" again. Porosity has decreased to about 17%. Figure 77 shows a densely packed, and somewhat welded, mass of angular fossil fragments and crystals. Figure 78 (1231.0 m) shows a similar picture. Pressure solution can be observed. Most micarb particles have perfect euhedral shapes. Some coccoliths have smooth surfaces, probably due to solution (no sign of silica coating). Others show well-developed crystal faces.

Figure 79 (1231.6 m) is from the surface of a slickensided fault. This surface is coated with clay minerals. A similar phenomenon was described by Packham and van der Lingen (1973, Plate 7, fig. 5), and was explained by dissolution and migrating away of the calcite from the fault surface.

The last sample photographed is a silicified, impure limestone (Figure 80, 1232.7 m). It shows a dense matrix with imprints of nannofossils (porosity about 18%). Similar imprints were also described from Site 288 (Figure 23, 742-743 m).

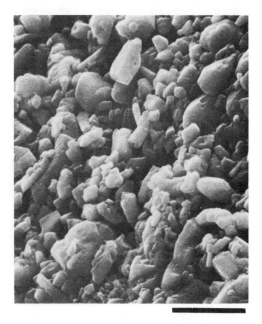

Figure 19. *Site 288, 650.0 meters, Campanian, nanno chalk, porosity (648.4 m) 51%, scale bar 8μ; densely packed coccoliths and subhedral granular calcite.*

70460

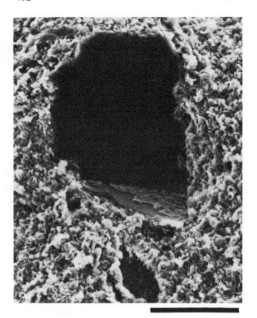

Figure 20. *Site 288, 650.0 meters, Campanian, nanno chalk, porosity (648.4 m) 51%, scale bar 80µ; hollow, formerly containing a dolomite rhomb.*

Figure 22. *Site 288, 705.5 meters, Santonian, nanno chalk, porosity (705.5 m) 36%, scale bar 20µ; coccoliths in a dense matrix of largely nonbiogenic (volcanic) material.*

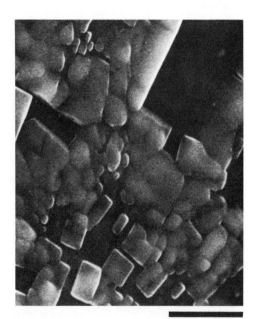

Figure 21. *Site 288, detail of Figure 20, scale bar 8µ; bottom of hollow covered with small dolomite rhombs.*

Figure 23. *Site 288, 818.3 meters, Coniacian-Turonian, zeolite-rich claystone, porosity (818.4 m) 51%, scale bar 20µ; sporadic nannofossils in a volcanogenic clay matrix.*

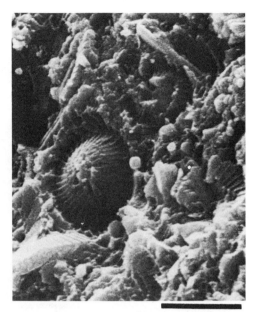

Figure 24. *Site 288, 743.2 meters, late Coniacian, nanno chalk, porosity (743.3 m) 43%, scale bar 6μ; coccoliths and coccolith impressions in a dense, largely nonbiogenic (volcanic) matrix.*

Figure 26. *Site 288, detail of Figure 25, scale bar 8μ; silica spherules with "sugary" surface.*

Figure 25. *Site 288, 818.3 meters, Coniacian-Turonian, zeolite-rich claystone, porosity (818.4 m) 51%, scale bar 16μ; imperfectly formed silica spherules.*

Figure 27. *Site 288, 847.0 meters, ?Turonian, foram-nanno chalk, porosity (847.0 m) 35%, scale bar 8μ; angular coccoliths (Mesozoic forms) with minor secondary calcite overgrowth. Smooth shapes are probably artifacts (fungi?).*

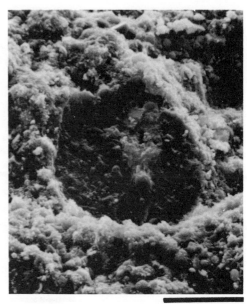

Figure 28. *Site 288, 847.0 meters, ?Turonian, foram-nanno chalk, porosity (847.0 m) 35%, scale bar 8μ; euhedral calcite crystals inside a foraminiferal chamber.*

Figure 30. *Site 288, 850.8 meters, ?Turonian, silicified limestone, porosity (850.7 m) 35%, scale bar 40μ; silicosphere, a spherical object (radiolarian?) filled with secondary silica; the matrix surrounding the silicosphere for the greater part also consists of silica.*

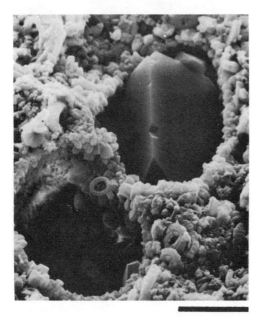

Figure 29. *Site 288, 847.0 meters, ?Turonian, foram-nanno chalk, porosity (847.0 m) 35%, scale bar 20μ; foraminiferal chambers; one filled with single diagenetic crystal; walls recrystallized.*

Figure 31. *Site 288, 850.8 meters, ?Turonian, silici-fied limestone, porosity (850.7 m) 35%, scale bar 20μ; silica spherules with sugary surface, and euhedral clinoptilolite crystal.*

Figure 32. *Site 288, 867.8 meters, ?Turonian, chert-bearing nanno limestone, porosity (867.8 m) 22%, scale bar 2μ; extensive development of secondary calcite, welding particles together; note relatively large euhedral calcite cement.*

Figure 34. *Site 288, 879.5 meters, early Turonian, chert-bearing nanno limestone, porosity (875.8 m) 27%, scale bar 800μ; boundary between chert (upper half of picture) and silicified limestone (lower half).*

Figure 33. *Site 288, 867.8 meters, ?Turonian, chert in nanno limestone, scale bar 8μ; smooth breakage surface of chert nodule.*

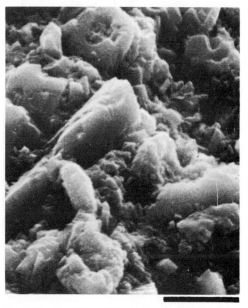

Figure 35. *Site 288, 879.5 meters, detail of Figure 34, scale bar 6μ; silicified limestone part, showing coccoliths embedded in a silica matrix.*

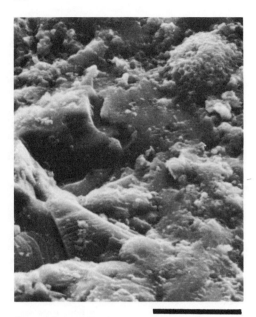

Figure 36. *Site 288, 879.5 meters, detail of Figure 34, scale bar 20μ; chert part, showing smooth silica surface with fracture marks; spherical object in upper right corner may be recrystallized radiolarian.*

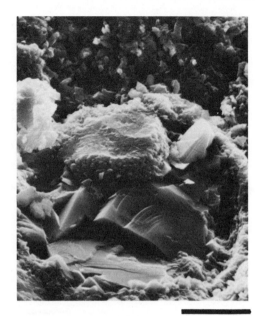

Figure 38. *Site 288, 979.9 meters, Aptian, silicified limestone, porosity (980.4 m) 25%, scale bar 20μ; silicosphere, a spherical fossil (foraminifera?) filled with chalcedony.*

Figure 37. *Site 288, 894.5 meters, Cenomanian, limestone, porosity (894.9 m) 29%, scale bar 10μ; subspherical object ("silicosphere"), consisting of radiating sheet-like silica (probably chalcedony); matrix contains abundant coccoliths in a silica groundmass.*

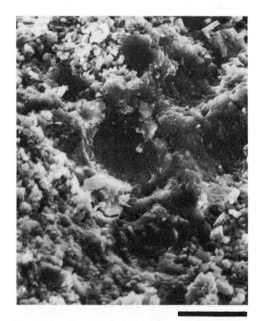

Figure 39. *Site 288, 979.9 meters, Aptian, silicified limestone, porosity (980.4 m) 25%, scale bar 20μ; faint outline of a silicified foraminifera (4 chambers showing); groundmass is a dense mixture of nannofossils and secondary silica.*

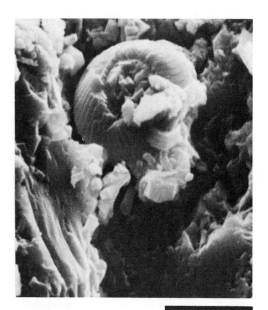

Figure 40. *Site 288, 979.9 meters, Aptian, silicified limestone, porosity (980.4 m) 25%, scale bar 20μ; fairly well preserved coccolith in a dense, silicified matrix; some calcite overgrowth in the central part of the coccolith.*

Figure 41. *Site 289, 468.9 meters, middle Miocene, nanno-foram chalk/ooze, porosity (467.3 m) 55%, scale bar 5μ; overgrown discoasters; coccoliths and coccolith fragments, showing both dissolution and re-precipitation; some welding by secondary calcite; note abundant anhedral micarb particles.*

CORRELATION WITH SEISMIC REFLECTORS

Earlier in this paper mention was made of correlations of major seismic reflectors with sediment physical properties and scanning electron microscopic features for drill sites in the Southwest Pacific by Packham and van der Lingen (1973). Their findings can be further tested on the data from Sites 288 and 289. Because of the more complete sampling at Site 289 and the absence of chert in the post middle Eocene, the thick ooze section there is ideal for this purpose. At Site 288, as indicated above, the acoustic behavior of the sequence differs considerably from that at Site 289, this is in part due to the presence of more chert in the column.

Site 289

The shallowest reflector (very weak) is at about 140 meters in the upper Pliocene in the vicinity of the change of the ooze from creamy to stiff, but lies above the level where drilling disturbance drops significantly (250-300 m). The first reflector corresponds to a small step in the acoustic impedance (see site report) resulting mainly from a slight change in the density curve.

The 250-300 meter zone in which the degree of drilling disturbance drops does not slow any significant excursion in the acoustic impedance, nor is there a major reflection event.

A moderately strong reflection corresponds to a depth of 376 meters in middle Miocene stiff to semilithified ooze. There is no significant peak in the acoustic impedance curve, but this is the level at which there is a change in gradient in the depth velocity curve.

The third reflector at an estimated depth of 535 meters is in lower Miocene semilithified to stiff chalk. From this level onwards down the hole, the amplitude of the excursions acoustic velocity curve becomes significantly greater and the anisotropy between vertical and horizontal velocities increase. This clearly marks a change in the degree of lithification of the sediments.

Although SEM examination shows an increase in the amount of welding of grains and the development of overgrowths, the most significant change at this level is probably a change in the distribution of grain contacts (? and orientation) indicated by the increased anisotropy. The wide variation in velocities indicates that some layers and more susceptible to modification than others. In terms of macroscopic features, this reflector marks the lower limit of significant drilling disturbance.

The fourth reflector corresponds to an estimated depth of 652 meters in semilithified chalk at the Miocene-Oligocene boundary. There is a broad low in the acoustic velocity at this depth between maxima at 600 and 700 meters, but no corresponding low in the

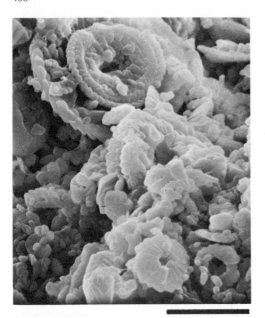

Figure 42. *Site 289, 562.4 meters, early Miocene, nanno-foram chalk, porosity (562.7 m) 53%, scale bar 5μ; pronounced calcite overgrowth on fossil particles, and substantial welding.*

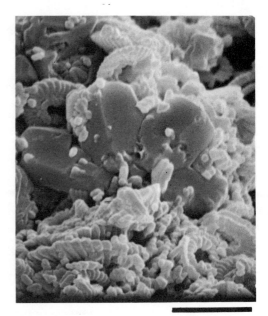

Figure 44. *Site 289, 620.4 meters, early Miocene, nanno-foram chalk, porosity (620.3 m) 55%, scale bar 5μ; discoaster with secondary calcite overgrowth, showing some pitting; substantial overgrowth on coccoliths.*

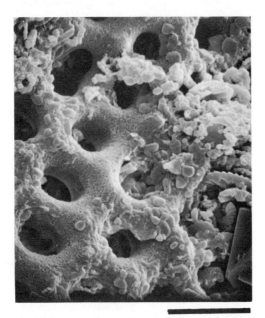

Figure 43. *Site 289, 620.4 meters, early Miocene, nanno-foram chalk, porosity (620.3 m) 55%, scale bar 10μ; slightly corroded radiolarian, covered with clusters of subhedral micarb particles; some overgrowth on cocco-liths.*

density or increase in porosity. The explanation of this low probably lies in the nature and frequency of grain contacts. Velocity anisotropy is at a minimum also. No clear explanation can be obtained from examination of the SEM photographs, but there is a strong suggestion of grain dissolution.

The first of the strong reflection events is at an estimated depth of 711 meters in upper Oligocene chalk. At this depth there is another velocity low immediately below the 700-meter maximum indicated above, thus resulting in a significant reflection coefficient. An SEM examination of a sample from 704.8 meters (Figure 48) shows pronounced dissolution of coccoliths as well as precipitation of secondary calcite. The solution may be responsible for decreasing grain contacts and impeding sound transmission.

The next strong reflector is at an estimated depth of 814 meters, again in upper Oligocene chalk. Sediment velocity measurements indicate a low immediately above this level and a high corresponding to it. The evidence available suggests that this reflector is of similar origin to the two strong reflectors immediately above. The next reflector at an estimated 873 meters is also a strong one and may correspond to a further velocity low at 885 meters where grain solution has again been observed.

The next reflector has been placed near the base of the lower Oligocene chalk at 950 meters and does not correspond to any significant excursion in the acoustic impedance curve. At this level the welding of grains is further advanced.

Figure 45. *Site 289, 657.5 meters, late Oligocene, nanno-foram chalk/ooze, porosity (657.7 m) 54%, scale bar 5μ; secondary calcite overgrowth on discoasters has incorporated other fossil fragments (see also coccolith impressions in calcite); secondary calcite fills the central parts of coccoliths and bridges the space between placolith shields; note corrosion of some coccoliths.*

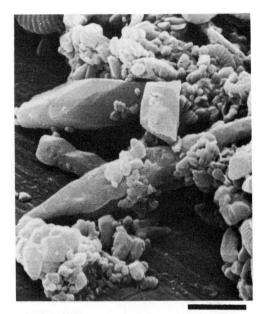

Figure 46. *Site 289, 686.2 meters, late Oligocene, nanno-foram chalk/ooze, porosity (686.8 m) 51%, scale bar 8μ; calcareous spicules, overgrown with secondary calcite (crystal faces); identity unknown, may be calcareous sponge spicules or foraminiferal spicules.*

The velocity data indicate that the next reflector is located at 1004 meters in upper Eocene chalk. However, it is most likely that this is some 20 meters too high. At this lower level in the middle Eocene there is a sudden downwards increase in sonic velocity, bulk density, and a decrease in porosity. SEM photographs indicate a sudden increase in the quantity of diagenetic overgrowths. Minor quantities of chert are also present.

In general, below this level, the acoustic impedance increases steadily to 1138 meters, it then decreases sharply at about 1150 meters. Although the grains in this lower velocity sediment display considerable diagenesis, they are loosely packed. The weak reflector with an estimated depth of 1144 meters probably correspond to this change.

The deepest reflector identified is at 1.26 sec two-way travel time subbottom and is tentatively placed at 1232 meters at the base of the chalk horizon. It may alternatively represent the limestone-basalt contact at 1262 meters.

Site 288

Because coring was not continuous at this site, it is not possible to define the depth velocity curve with the same degree of accuracy as at Site 289. Comparisons

with Site 289 will be made where appropriate in this discussion.

The highest reflector is at a depth of 46 meters subbottom in upper Pliocene sediments. This is above the level where any diagenetic changes might be expected, and its existence might be explained by volcanic ash layers present at that depth.

The second reflector is at an estimated depth of 172 meters is middle Miocene. This is below the level at which the sediment passes from ooze to chalk, the shallowest level of this transition compared with Site 289 could be the result of erosion (the lower Pliocene is missing) or slumping (disturbance is visible on the seismic record). In the latter case dewatering of the sediment may have followed mass movement. Curiously, no reflector associated with the ooze-chalk transition is seen on the record. Although it is not lithologically explainable, there is an increase in sonic velocity at about the level of this reflector. A further increase takes place at the level of the next reflector (215 m) together with an increase in bulk density. The formation of diagenetic calcite was noted at 202 meters. This reflector may correspond to that at 376 meters at Site 289.

A sonic velocity peak occurs in the vicinity of the next reflector (estimated depth, 415 m) in upper Oligocene chalk. The degree of dissolution and reprecipitation of calcite increases downwards at about this level, and

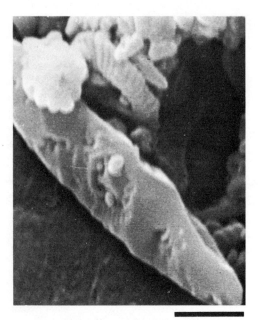

Figure 47. *Site 289, 686.2 meters, late Oligocene, nanno-foram chalk/ooze porosity (686.8 m) 51%, scale bar 6μ; calcareous spicules as in Figure 46; note impression of coccolith in secondary calcite.*

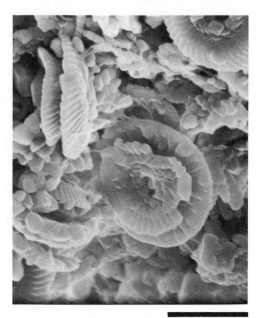

Figure 49. *Site 289, 752.1 meters, late Oligocene, foram-nanno chalk/ooze, porosity (752.1 m) 49%; scale bar 5μ; extensive calcite overgrowth on coccoliths; particles are welded; note wisp-like bridging of particles (silica?).*

Figure 48. *Site 289, 704.8 meters, late Oligocene, foram-nanno chalk/ooze, porosity (704.5 m) 47%, scale bar 5μ; strongly corroded coccoliths; at the same time secondary calcite growth in central areas, in between placolith shields, and on top of placoliths.*

cherts make a sporadic appearance. No simple explanation for this velocity peak can be offered.

The next three reflectors occurring in the chalk section have estimated depths of 617, 674, and 726 meters. The first falls within Maestrichtian beds, the second in Campanian, and the third in Santonian. The origin of these reflectors may lie in alternations in the degree of lithification as observed to be associated with reflectors at depths between 652 and 951 meters at Site 289. Depth estimates at Site 288 may not be accurate because of the paucity of data. It appears that the two lower reflectors under discussion have a calculated depth some 15 to 20 meters too shallow. If they were that much deeper, they would correspond to the top and bottom of a significant velocity excursion. The presence of chert beds in the section is extremely likely to mask the effects of carbonate diagenesis in that the reflections may be composite events.

The deepest reflection has been computed to correspond to a depth of 847 meters close to the junction between Turonian chalk and limestone. The conversion of chalk to limestone is associated at this site with substantial amounts of silicification. This reflector, however, corresponds in origin to that at 1004 meters in the upper Eocene at Site 289.

CONCLUSIONS AND DISCUSSION

From the studies of progressive diagenesis, published so far, a few clear concepts have emerged. On the other hand, several fundamental questions still remain unanswered. Some of these concepts and questions will be

Figure 50. *Site 289, 847.0 meters, late Oligocene, nanno chalk/ooze, porosity (848.0 m) 48%, scale bar 5μ; overgrown discoaster, showing strong pitting; extensive calcite overgrowth on coccoliths and micarb particles; corroded radiolarian fragment at bottom of picture.*

Figure 51. *Site 289, 847.0 meters, late Oligocene, nanno chalk/ooze, porosity (848.0 m) 48%, scale bar 10μ; foraminiferal chamber showing advanced stage of calcite recrystallization of walls and inner space; surrounding sediment also shows abundant new calcite formation.*

discussed briefly, referring to results from Sites 288 and 289 where appropriate.

1) For the transition of a carbonate ooze into a limestone there is no need for calcium carbonate to be introduced from an outside source. Schlanger et al. (1973) first suggested that the progressive diagenesis of carbonate sediments in the central Pacific was calcite conservative. Matter (1974) came to the same conclusion for sediments in the Arabian Sea.

2) As reported earlier in this paper, Schlanger et al. (1973) proposed a model for carbonate diagenesis. Their model suggests that 3-cc carbonate ooze (density 1.35; porosity 80) first changes to 1.7-cc chalk (density 1.60; porosity 65), which finally gets transformed into 1-cc limestone (density 2.0; porosity 40). This model can be compared with data from Sites 288 and 289.

At Site 288 the uppermost carbonate sediments have a bulk density of about 1.6, and a porosity of between 65 and 70. At Site 289 these figures are between 1.5 and 1.6, and between 65 and 70, respectively. Both sites thus have a higher density and lower porosity than the ooze in the Schlanger et al. model. As intraparticle porosity (especially from foraminifera) can contribute substantially to overall porosity and density, this difference might be explained by differences in foraminifera percentages in the sediments. The model of Schlanger et al. was based on data from Site 167. At that site

foraminifera are present in abundance (numerical estimates of sediment components became routine practice in the DSDP only from Leg 21 onwards). However, in the top parts of the sequences at Sites 288 and 289, foraminifera are also present in abundance. Water depths cannot explain the differences either (3176 m at Site 167, 3000 m at Site 288, and 2206 at Site 289).

The equivalent stage to the Schlanger et al. intermediate chalk stage was reached at a depth of about 300 meters at Site 167. At Site 288 this stage (as far as density and porosity are concerned) is already reached at 50 meters, and at Site 289 at 100 meters. Both levels are still in the ooze interval.

Things become more complicated in the limestone intervals at Sites 288 and 289, due to the presence of silica and volcanic material. The siliceous limestone at Site 288 (lowermost 150 m) has density values fluctuating between 2.2 and 2.4. Porosity values vary between 20% and 30%. The last relatively pure carbonate sediment occurs at about 600 meters, which is still in the chalk interval. At 609.9 meters, the calcium carbonate percentage is 96%, the density 1.86, and the porosity 51%. In the limestone interval, the highest calcium carbonate percentage (80%) was recorded at 858.6 meters. The density at that level is 2.20 and the porosity 27. At Site 289 the limestone interval has density values fluctuating between 2.0 and 2.4, and porosity values between 15%

Figure 52. *Site 289, 886.2 meters, early Oligocene, foram-nanno chalk, porosity (886.2 m) 48%, scale bar 5μ; coccoliths show dissolution and reprecipitation of calcite; radiating clusters of subhedral calcite crystals are the nannofossil* Sphenolithus moriformis; *note welding of particles; amorphous substance at lower left corner is probably fungus growth.*

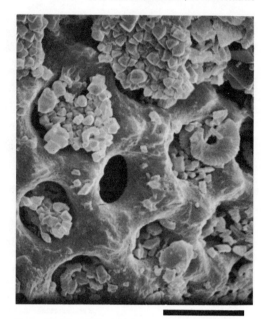

Figure 53. *Site 289, 914.3 meters, early Oligocene, nanno-foram chalk, porosity (914.8 m) 47%, scale bar 10μ; compare with Figure 43; strongly corroded radiolarian, covered with euhedral micarb particles.*

and 40%. The lowermost part of this interval has relatively pure carbonate horizons. For instance, at 1167 meters, the calcium carbonate percentage is 99%. The density at this level is 2.22, and the porosity is 26.

In the Schlanger et al. model, the sediment is compacted to a third of its original volume during the transition from ooze to limestone. A similar type of calculation that forms the basis of their model can be applied to, for instance, Site 289. Provided no calcite has been introduced from outside, 10 cc of ooze (density 1.55, porosity 70%) has changed to about 4.3 cc of limestone (density 2.22, porosity 26%). This is a compaction to less than half the original volume. Keeping in mind that the ooze at the top of Site 289 is already more compacted than the ooze in the model, numerically the compaction compares closely with the model.

There are, however, several pitfalls in this type of approach. Firstly, it is not certain that the original ooze which is now compacted to limestone, had the same density and porosity at the time it was close to the sediment-water interface, as the present-day ooze. Secondly, the proviso of the system being calcite conservative as a whole does not mean that within the system calcite cannot move from one horizon to another. That this takes place is suggested by the presence of stylolites and clay-coated slickensides. The formation of chert nodules,

through the exchange of calcite for silica, must also have resulted in zones richer in calcium carbonate.

However, keeping in mind these problems, it can fairly be stated that considerable compaction takes place during diagenesis. This compaction does take place not so much by crushing of microfossil tests (such as foraminifera) as by disintegration of larger microfossils into smaller particles (micarb) through dissolution; by closer packing of particles; and by interpenetration of particles through pressure solution. This might in part provide an answer to the problem, stated by Bathurst (1969), that no obvious signs of compaction (e.g., crushed foraminiferal tests) can be detected in many lithified carbonate muds (micrites) and that consequently the large amount of calcite cement, filling the pore space, must have come from an outside source.

3) Dissolution and reprecipitation within the sediment takes place simultaneously. Whether calcite gets dissolved or acts as seed crystals depends on the form in which the calcite occurs. Physicochemical conditions of the interstitial water can change in time, shifting the balance between dissolution and reprecipitation. This may explain, for instance, the changing induration levels (chalk and ooze) at Site 289.

4) No doubt paleooceanographic water chemistry further complicates the progressive diagenetic processes. A good example is an Oligocene chalk in the Atlantic Ocean, drilled during Leg 3 of the DSDP (Wise and Kelts, 1972). This chalk has very little overburden, and is underlain by hundreds of meters of unindurated ooze (up to Cretaceous

Figure 54. *Site 289, 941.2 meters, early Oligocene, nanno-foram chalk/ooze, porosity (941.0 m) 45%, scale bar 5μ; coccoliths show corrosion and secondary calcite overgrowth; particles are welded together with subhedral to euhedral calcite cement; note wisp-like bridging structures and smooth "icing-sugar" coating of coccoliths in top part of picture (silica?).*

Figure 55. *Site 289, 999.4 meters, late Eocene, foram-nanno chalk, porosity (999.4 m) 43%, scale bar 5μ; pronounced overgrowth with secondary calcite of coccoliths, discoasters, and micarb particles; note interpenetration of calcite crystals and fossil fragments (welding); some wisp-like bridging structures (silica?) are also present.*

in age). Wise and Kelts (1972) argue that induration took place soon after burial, while the sediment was still close to the sediment-water interface, and suggest that cementation was the result of paleooceanographic conditions. An interesting complicating factor is the presence of silica spherules, which Wise and Kelts think were formed *after* the carbonate cementation.

The ooze horizons in the Maestrichtian at Sites 288 and 289 can also best be explained by different paleooceanographic conditions. Many authors think that the worldwide existence of chert in the Eocene is also due to environmental conditions (e.g., Lancelot, 1973). Combining these two events, it would be another example of worldwide (catastrophic?) changes in Late Cretaceous-early Tertiary time. In the Australasian region the initial breakup of Gondwanaland took place during this period.

5) Lancelot (1973) suggested that chert formation is an early-diagenetic process. Wise and Kelts (1972) concluded that silica spherules in an Oligocene chalk in the Atlantic were formed after calcite cementation, thus being a late-diagenetic process. These two aspects of silica diagenesis underline the necessity, as expressed earlier by the present authors, to treat the formation of quartz-ose chert nodules and the formation of (cristobalite-tridymite) silica spherules as two separate processes.

It is difficult to find unambiguous criteria to decide whether (quartz-) silica diagenesis is an early process or not. The present authors have little to add to the excellent discussion by Lancelot (1973).

6) Another problem is the influence of silica diagenesis on carbonate diagenesis. The replacement of calcite by silica must result in calcium carbonate enrichment outside the chert nodules. This extra calcium carbonate is available for precipitation on the existing calcite substrate, or for crystallization in inter- and intraparticle pore space. If chert nodules are formed soon after burial, substantial carbonate diagenesis should then take place simultaneously.

In most deep-sea biogenic sediments chert is not present in sediments younger than Eocene. This is the case, for instance, at Site 289. At this site the carbonate sediment above the chert interval changes very gradually only. Notwithstanding an overburden of 1000 meters and a time interval of about 40 m.y., ooze horizons still exist until just above the chert-bearing interval. The various sediment-physical properties change gradually with depth, without any jumps. The situation is quite different at Site 288, where chert is present in sediments as young as early Miocene. Carbonate-diagenetic stages, equivalent to those at Site 289, are at much shallower levels. The change from ooze to chalk is more "positive" and the physical properties graphs

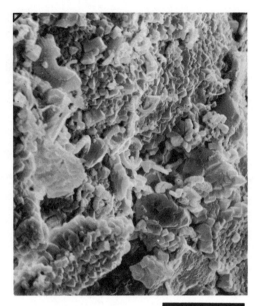

Figure 56. *Site 289, 1009.0 meters, late Eocene, nanno-foram limestone, porosity (1008.8 m) 39%, scale bar 10μ; foraminiferal fragments, coccoliths and micarb particles welded together by extensive euhedral calcite precipitation.*

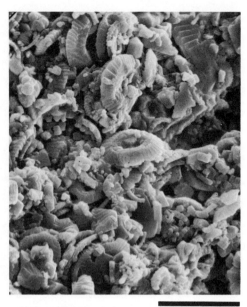

Figure 57. *Site 289, 1017.2 meters, middle Eocene, nanno-foram limestone, porosity (1017.2 m) 38%, scale bar 10μ; tightly packed nannofossils and micarb particles; note "icing sugar" coating (silica?) in upper right-hand corner.*

show several jumps. The conclusion seems inescapable that silica diagenesis affects carbonate diagenesis.

If the above is the case, then the fact that chert is present in older sediments, or, when present in younger sediments, increases in abundance with time, could create a false impression of progressive diagenesis, dependent on time and depth of burial.

7) Heat flow is another parameter influencing diagenesis, which could create a false impression of progressive diagenesis dependent on time and depth of burial. Plate tectonics theory states that there is a relationship between the age of the oceanic crust, its depth below sea-level, and heat-flow (Sclater, 1972). This means that the first sediment deposited on newly created oceanic crust is subject to relatively high temperatures. In time, the sediment pile increases in thickness, while at the same time the heat flow decreases. Consequently, the influence of heat flow on diagenesis decreases gradually as well.

8) Notwithstanding these complicating factors, time and depth of burial play their part in diagenesis as well.

9) Another problem is the difficulty in determining how much dissolution of microorganisms has taken place in the water column during settling, at the sediment-water interface (further influenced by infaunal activity), and after burial (diagenesis proper). Pressure solution is the only type of solution which clearly has taken place after burial. Pressure solution and reprecipitation are therefore the only unambiguous indicators of "burial diagenesis."

10) Keeping these many complicating factors in mind, it is nevertheless possible to compose a broad picture of progressive carbonate diagenesis.

Discoasters are the first organisms to develop secondary calcite overgrowth. This starts very soon after settling. Calcite precipitation on other microfossil particles only begins after compaction and water expulsion has advanced to the stage that a grain supporting framework is established. This level is reached in the lower part of the stiff-ooze interval, where the sediment becomes crumbly. Calcite is then first precipitated along the edges and in the central areas of coccolith placoliths, and on micarb particles. This process continues and the next observable stage is the connection by secondary calcite of the proximal and distal shields of placoliths. Free-growing euhedral calcite crystals start to form inside foraminiferal chambers. Secondary calcite overgrowth continues to increase in size, developing euhedral crystal faces. Secondary calcite fills up more and more of pore space, at the same time welding particles together. In the most advanced diagenetic stage observed in deep-sea sediments, practically all fossil particles are covered with subhedral to euhedral calcite overgrowth, the central areas of coccoliths are filled "to overflowing" with granular calcite, and most remaining foraminiferal chambers are filled with secondary calcite. A large part of the pore space is filled with granular calcite cement. Porosity may have decreased from between 70% and 80% to about 25% (e.g., Site 289).

Figure 58. *Site 289, 1027.1 meters, middle Eocene, foram-nanno limestone, porosity (1027.4 m) 30%, scale bar 5μ; advanced welding of nannofossils through calcite overgrowth and coating with "icing sugar" (silica?); note central area of coccolith filled "to overflowing" with secondary granular calcite, in lower right-hand corner.*

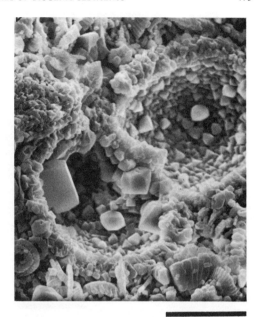

Figure 59. *Site 289, 1027.1 meters, middle Eocene, foram-nanno limestone, porosity (1027.4 m) 30%, scale bar 5μ; relatively large euhedral calcite crystals inside foraminiferal chambers.*

11) Because of the highly complicated picture of silica and carbonate diagenesis, a simple set of causative diagenetic factors for seismic reflectors cannot easily be delineated. From deep-sea drilling sites in the Southwest Pacific (Packham and van der Lingen, 1973; This paper), the following list of possible correlations between sediment properties and seismic reflectors can be compiled: (a) the creamy-stiff ooze boundary, (b) the lower part of the stiff-ooze interval, where the sediment becomes crumbly, (c) the stiff-ooze-semilithified chalk boundary, where grain welding commences, (d) alternating ooze-chalk intervals, (e) the level where the bulk density starts to increase more rapidly with depth as crystal growth and solution reduce porosity, (f) the chalk-limestone boundary, (g) levels below which non-biogenic (e.g., volcanic) material is mixed with biogenic sediment, (h) chert horizons, (i) the top boundaries of silicified sediment intervals.

ACKNOWLEDGMENTS

We wish to thank Miss R.J. Burr, University of Canterbury, Christchurch, and Mr. Allan Terrill, University of Sydney, for their assistance in taking the scanning electron photomicrographs, and printing the final photographs. We are grateful to Mr. E.T.H. Annear, New Zealand Geological Survey, Christchurch, for drafting Figures 1 and 2, and to Miss C.M. Johnstone, New Zealand Geological Survey, Christchurch, for typing the final manuscript.

Figure 60. *Site 289, 1036.8 meters, middle Eocene, nanno limestone, porosity (1036.8 m) 22%, scale bar 5μ; slightly pitted overgrown discoasters, embedded in a silica groundmass; some of the silica has subspherical shapes (cristobalite?).*

Figure 61. *Site 289, 1036.8 meters, middle Eocene, nanno limestone, porosity (1036.8 m) 22%, scale bar 5μ; discoasters embedded in "icing sugar" silica.*

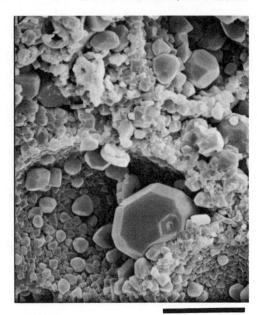

Figure 63. *Site 289, 1047.2 meters, middle Eocene, nanno limestone, porosity (1049.7 m) 19%, scale bar 10μ; foraminiferal chambers, partly filled with secondary euhedral calcite crystals; note "negative crystal" in largest calcite crystal.*

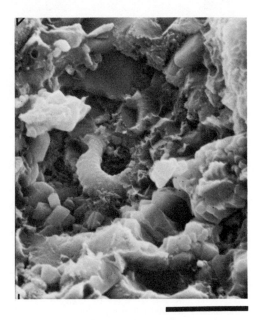

Figure 62. *Site 289, 1036.8 meters, middle Eocene, nanno limestone, porosity (1036.8 m) 22%, scale bar 5μ; coccoliths embedded in a dense silica matrix.*

Figure 64. *Site 289, 1047.2 meters, middle Eocene, nanno limestone, porosity (1049.7 m) 19%, scale bar 10μ; advanced recrystallization of nannofossils; discoaster at left center shows impressions of coccoliths in the secondary calcite; note pressure solution (arrow).*

Figure 65. *Site 289, 1047.2 meters, middle Eocene, nanno limestone, porosity (1049.7 m) 19%, scale bar 10μ; cracks in limestone are filled with sparry calcite; note euhedral calcite overgrowth in groundmass.*

Figure 67. *Site 289, 1084.2 meters, lower Eocene, siliceous limestone, porosity (1083.9 m) 11%, scale bar 30μ; calcite substrate covered with a fine fibrous mineral, probably cristobalite; the same mineral is also present in the dense groundmass.*

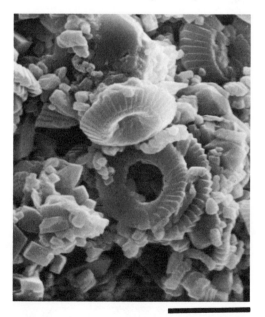

Figure 66. *Site 289, 1065.2 meters, middle Eocene, nanno-foram limestone, porosity (1065.5 m) 23%, scale bar 5μ; coccoliths show dissolution and recrystallization; note euhedral crystal faces on granular calcite cement.*

Figure 68. *Site 289, 1093.8 meters, late Paleocene, limestone, porosity (1093.6 m) 22%, scale bar 5μ; welded mass of nannofossils and micarb particles; advanced calcite overgrowth.*

Figure 69. *Site 289, 1130.5 meters, late Paleocene, foram-nanno limestone, porosity (1131.6 m) 11%, scale bar 5μ; densely packed coccoliths, welded together by secondary calcite; some smooth areas (silica coating?) are visible just right of center.*

Figure 71. *Site 289, 1131.6 meters, late Paleocene, limestone, porosity (1131.6 m) 11%, scale bar 5μ; foraminiferal chamber, completely filled with interlocking calcite (?) crystals; note recrystallized chamber wall.*

Figure 70. *Site 289, 1131.6 meters, late Paleocene, limestone, porosity (1131.6 m) 11%, scale bar 10μ; inner surface of foraminiferal chamber covered with rod-like crystals (calcite?); also free-growing euhedral calcite crystals.*

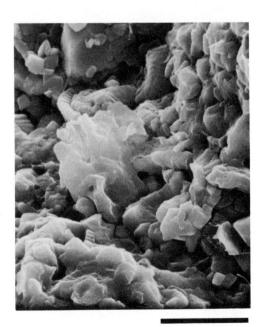

Figure 72. *Site 289, 1138.1 meters, early Paleocene, limestone, porosity (1138.2 m) 7%, scale bar 5μ; nannofossils embedded in dense ("icing sugar") silica.*

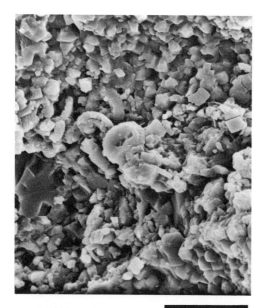

Figure 73. *Site 289, 1138.1 meters, early Paleocene, limestone, porosity (1138.2 m) 7%, scale bar 10μ; carbonate fossil fragments embedded in a silica matrix.*

Figure 75. *Site 289, 1167.4 meters, middle Maestrichtian, siliceous limestone, porosity (1167.1 m) 26%, scale bar 10μ; porous packing of abundant micarb particles, surrounding a foraminifera only partly filled with secondary calcite.*

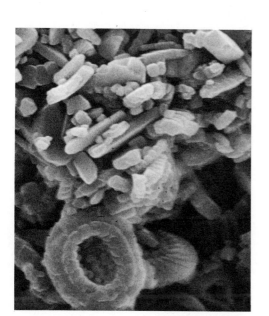

Figure 74. *Site 289, 1155.8 meters, middle Maestrichtian, limestone, porosity (1155.9 m) 40%, scale bar 5μ; sample shows higher porosity and less welding than samples above; still substantial secondary calcite overgrowth.*

Figure 76. *Site 289, 1175.3 meters, middle Maestrichtian, siliceous limestone, porosity (1175.3 m) 38%, scale bar 10μ; micarb particles and relatively well preserved coccoliths; limited welding.*

Figure 77. *Site 289, 1203.9 meters, early Maestrichtian, limestone, porosity (1203.9 m) 28%, scale bar 10μ; densely packed and welded angular coccoliths and micarb particles; probably silica in matrix.*

Figure 78. *Site 289, 1231.0 meters, Campanian, limestone, porosity (1231.3 m) 26%, scale bar 5μ: angular coccoliths showing dissolution and calcite overgrowth; relatively large subhedral granular calcite cement; a fair amount of welding has taken place; note the many rhombic micarb particles (coccolith fragments).*

REFERENCES

Adelseck, C.G., Geehan, G.W., and Roth, P.H., 1973. Experimental evidence for the selective dissolution and overgrowth of calcareous nannofossils during diagenesis: Geol. Soc. Am. Bull., v. 84, p. 2755-2762.

Bathurst, R.G.C., 1969. Problems of lithification in carbonate muds: Geol. Assoc. Proc., v. 81, p. 429-440.

Berger, W.H. and von Rad, U., 1972. Cretaceous and Cenozoic sediments from the Atlantic Ocean. *In* Hayes, D.E., Pimm, A.C., et al., Initial Reports of the Deep Sea Drilling Project, Volume 14: Washington (U.S. Government Printing Office), p. 787-954.

Buurman, P. and van der Plas, L., 1971. The genesis of Belgian and Dutch flints and cherts: Geol Mijnbouw, v. 50, p. 9-27.

Cheriff, O.M. and Flick, H., 1974. On the taxonomic value of the wall structure of *Quinqueloculina*. Micropaleontology, v. 20, p. 236-244.

Deer, W.A., Howie, R.A., and Zussman, M.A., 1963. Rock-forming minerals. v. 4—Framework silicates: London (Longmans, Green).

Greenwood, R., 1973. Cristobalite: its relationship to chert formation in selected samples from the Deep Sea Drilling Project: J. Sediment. Petrol., v. 43, p. 700-708.

Heath, G.R., 1973. Cherts from the eastern Pacific, Leg 16, Deep Sea Drilling Project. *In* van Andel, T.H., Heath, G.R., et al., Initial Reports of the Deep Sea Drilling Project, Volume 16: Washington (U.S. Government Printing Office), p. 609-613.

Heath, G.R. and Moberly, R., 1971. Cherts from the western Pacific, Leg 7, Deep Sea Drilling Project. *In* Winterer, E.L., Riedel, W.R., et al., Initial Reports of the Deep Sea Drilling Project, Volume 7: Washington (U.S. Government Printing Office), p. 991-1007.

Hsü, K.J., 1974. In Zürich: pelagic sediments discussed: Geotimes, v. 19, p. 24.

Kastner, M. and Siever, R., 1973. Diagenesis in Mediterranean Sea core samples from Site 124—Balearic Rise, and Site 125A—Ionian Basin. *In* Ryan, W.B., Hsü, K.J., et al., Initial Reports of the Deep Sea Drilling Project, Volume 13: Washington (U.S. Government Printing Office), p. 721-726.

Lancelot, Y., 1973. Chert and silica diagenesis in sediments from the central Pacific. *In* Winterer, E.L., Ewing, J.I., et al., Initial Reports of the Deep Sea Drilling Project, Volume 17: Washington (U.S. Government Printing Office), p. 377-405.

Lancelot, Y. and Ewing, J.I., 1972. Correlation of natural gas zonation and carbonate diagenesis in Tertiary sediments from the North-West Atlantic. *In* Hollister, C.D., Ewing, J.I., et al., Initial Reports of the Deep Sea Drilling Project, Volume 11: Washington, (U.S. Government Printing Office), p. 791-799.

Lisitzin, A.P., 1971. Sedimentation in the world ocean: SEPM Spec. Publ. No. 17.

Matter, A., 1974. Burial diagenesis of pelitic and carbonate deep-sea sediments from the Arabian Sea. *In* Whitmarsh, R.B., Weser, O.E., Ross, D.A., et al., Initial Reports of the Deep Sea Drilling Project, Volume 23: Washington (U.S. Government Printing Office), p. 421-469.

Moberly, R. and Heath, G.R., 1971. Carbonate sedimentary rocks from the Western Pacific: Leg 7, Deep Sea Drilling Project. *In* Winterer, E.L., Riedel, W.R., et al., Initial Reports of the Deep Sea Drilling Project, Volume 7:

Figure 79. *Site 289, 1231.6 meters, Campanian, limestone, scale bar 5μ; clay-covered surface of slickensided fault.*

Figure 80. *Site 289, 1232.7 meters, Aptian, ash-bearing nanno limestone, porosity (1233.0 m) 18%, scale bar 5μ; impressions of coccoliths in dense, ash-bearing matrix.*

Washington (U.S. Government Printing Office), p. 977-985.

Oehler, J.H., 1973. Tridymite-like crystals in cristobalitic "cherts": Nature Physical Science, v. 241, p. 64-65.

Packham, G.H. and van der Lingen, G.J., 1973. Progressive carbonate diagenesis at deep sea drilling sites 206, 207, 208, and 210 in the Southwest Pacific, and its relationship to sediment physical properties and seismic reflectors. *In* Burns, R.E., Andrews, J.E., et al., Initial Reports of the Deep Sea Drilling Project, Volume 21: Washington (U.S. Government Printing Office), p. 495-521.

Pimm, A.C., Garrison, R.E., and Boyce, R.E., 1971. Sedimentology synthesis: lithology, chemistry, and physical properties of sediments in the Northwestern Pacific Ocean. *In* Fischer, A.C., Heezen, B.C., et al., Initial Reports of the Deep Sea Drilling Project, Volume 6: Washington (U.S. Government Printing Office), p. 1131-1252.

Schlanger, S.O., Douglas, R.G., Lancelot, Y., Moore, T.C., and Roth, P.H., 1973. Fossil preservation and diagenesis of pelagic carbonates from the Magellan Rise, central North Pacific Ocean. *In* Winterer, E.L., Ewing, J.I., et al., Initial Reports of the Deep Sea Drilling Project, Volume 17: Washington (U.S. Government Printing Office), p. 407-427.

Sclater, J.G., 1972. Heat flow and elevation of the marginal basins of the Western Pacific: J. Geophys. Res., v. 77, p. 5705-5719.

van der Lingen, G.J., Andrews, J.E., Burns, R.E., Churkin, M., Davies, T.A., Dumitrica, P., Edwards, A.R., Galehouse, J.S., Kennett, J.P., and Packham, G.H., 1973. Lithostratigraphy of eight drill sites in the Southwest Pacific—preliminary results of Leg 21 of the Deep Sea Drilling Project. *In* Fraser, R. (Ed.), Oceanography of the South Pacific 1972: New Zealand Commission for UNESCO, Wellington, p. 299-313.

von der Borch, C.C., Galehouse, J. and Nesteroff, W.D., 1971. Silicified limestone-chert sequences cored during Leg 8 of the Deep Sea Drilling Project: a petrologic study. *In* Tracey, J.I., Jr,, Sutton, G.H., et al., Initial Reports of the Deep Sea Drilling Project, Volume 8: Washington (U.S. Government Printing Office), p. 819-827.

von Rad, U. and Rösch, H., 1972. Mineralogy and origin of clay minerals, silica, and authigenic silicates in Leg 14 sediments. *In* Hayes, D.E., Pimm, A.C., et al., Initial Reports of the Deep Sea Drilling Project, Volume 14: Washington (U.S. Government Printing Office), p. 727-751.

Wise, S.W. and Hsü, K.J., 1971. Genesis and lithification of a deep sea chalk: Ecolog. Geol. Helv., v. 64, p. 273-278.

Wise, S.W. and Kelts, K.R., 1972. Inferred diagenetic history of a weakly silicified deep sea chalk: Gulf Coast Assoc. Geol. Soc. Trans., p. 177-203.

Wise, S.W., Buie, B.F., and Weaver, F.M., 1972. Chemically precipitated cristobalite and the origin of chert: Ecolog. Geol. Helv., v. 65, p. 157-163.

G. J. van der Lingen and G. H. Packham

ERRATA

Page 443, line 22 should read: "... The latter chalk-ooze ..."

Page 443, line 23 should read: "... absence of chert at this interval at ..."

Page 444, line 21 of the abstract should read: "... diagenesis influenced by heat-flow ..."

Page 445, line 30 in the 1st column should read: "... and Packham and van der Lingen (1973) ..."

Page 445, line 54 in the 1st column should read: "... South Atlantic (Wise ..."

Part III

SILICA DIAGENESIS

Editor's Comments
on Papers 6 Through 9

The next four papers deal exclusively with silica diagenesis. Paper 6 by Heath and Moberly is one of the first detailed accounts of silica diagenesis as encountered in deep-sea sediments. It is based mainly on data from leg 7 of the DSDP in the central Pacific. The paper is well illustrated with photographs of thin sections and one SEM photograph. The paper discusses both nodular and bedded cherts. The authors emphasize the unique stratigraphic position (Eocene) of most deep-sea chert horizons, in both the Atlantic and Pacific Oceans (see Figure 1 in the Introduction), and they suggest that this position might be explained by an abrupt change in paleo-oceanographic conditions. They discuss the possible silica sources for the nodular and bedded cherts, and for the former they suggest a biogenic, rather than a volcanic, source. The source for the bedded chert is more difficult to establish.

The authors consider the formation of chert to be a two-stage process. Dissolved biogenic silica is first redeposited as finely crystalline cristobalite, which is subsequently transformed to quartz. Their analysis of this process is the outline of what other authors would later call the "maturation process" (Papers 7 and 16). They also show foresight in suggesting that a simple maturation process could be complicated by geochemical and thermodynamic factors.

Lancelot (Paper 7) studied the silica diagenesis in cores from DSDP leg 17 (central Pacific). The carbonate diagenesis of one of the leg 17 drill sites (167) is discussed by Schlanger and Douglas (Paper 2). Lancelot distinguishes two types of chert, "porcelanitic chert" (predominantly disordered cristobalite) and "quartzose chert" (composed of quartz and/or chalcedony). He noticed that the type of chert depends on the host lithology rather than on the age or depth of burial. Cherts in carbonate sediments seem to be predominantly quartzose, while cherts in clayey sediments are mainly porcelanitic. From these data, he concludes that the maturation theory is probably less important in explaining the distribution of the various silica phases.

To explain his observations, the author develops the theory that certain foreign (metallic) cations and permeability play controlling roles in the formation of chert. In clayey sediments, for instance, which have a high metallic cation content, disordered cristobalite is formed. The disordered structure of cristobalite is open enough to accommodate large cations. In carbonate sediments with little or no cations (apart from Ca and Na) and with a higher permeability, quartz can precipitate directly from solution. Pertinent to Lancelot's theory is his argument that chert formation is an early-diagenetic process.

Lancelot's paper was a real milestone. It stimulated much discussion, more research, and new experiments. The controversy between Lancelot's theory and the maturation theory was a major topic discussed at the 9th International Sedimentological Congress (von Rad and Rösch 1975; Kastner and Keene 1975).

The papers so far have dealt with silica diagenesis in the Atlantic and Pacific Oceans. Froehlich's paper (Paper 8) describes silica diagenesis in Eocene sediments from the Indian Ocean near Madagascar (leg 25) and compares these with silica diagenesis in Cretaceous epicontinental pelagic chalks from the Paris Basin. For further comparison he adds observations from DSDP samples collected in the Atlantic Ocean. Similarities between oceanic and epicontinental biogenic sediments have been noted by many authors (Papers 14 and 20).

Froehlich describes in great detail the various silica phases. What are now generally called "lepispheres," he calls "cristobalite-tridymite spherules." He observed that cristobalite-tridymite is dominant in strongly silicified samples, while quartz is dominant only in weakly silicified samples. Moreover, he states that samples with quartz contain more alumino silicates, which, incidentally, is in direct contrast with Lancelot's observations and theory.

Froehlich distinguishes four main types of silicification in chalk, the first of which is weakly-silicified chalk, in which quartz in the form of whiskers is the main silica phase, with little or no cristobalite-tridymite. The samples contain an important amount of clay and/or clinoptilolite. The second type is moderately-silicified chalk, in which quartz and cristobalite-tridymite occur in approximately equal amounts. The samples contain little clay but often have large quantities of clinoptilolite. The third type is strongly-silicified chalk, in which cristobalite-tridymite is dominant with little quartz. The third type of samples have little or no alumino-silicates. The fourth type is chalcedonic chert, which consists of equal amounts of quartz (or chalcedony) and cristobalite-tridymite with little or no alumino-silicates.

Using these four types, the author describes samples from the DSDP and the Paris Basin. His general conclusions are very interesting because they differ from those of most other workers. He states that silicification is not only independant of the degree of lithification of the chalk but independant of age and depth of burial. He also argues that the amount of silica in biogenic sediments is a direct reflection of the paleo-oceanography, that is, of fluctuations in the supply of silica to the oceans. He sees a hydrolyzing climate as especially conducive to abundant silica supply. Leclaire (Paper 18) expresses a similar opinion.

According to Froehlich, cristobalite-tridymite spherules and clinoptilolite are formed before the quartz whiskers. Chalcedony is the last silica phase to form and is restricted to massive siliceous horizons where it replaces (pseudomorphs) carbonate. Unlike the quartz whiskers, chalcedony is not formed at the expense of the cristobalite-tridymite spherules. Tentatively, the author suggests that there is a close relationship between quartz whiskers and fibrous chalcedony.

The last paper in this group, by Flörke et al. (Paper 9), links earlier mineralogical studies of opaline minerals with the results from the DSDP. By looking at SEM photographs, X-ray diffraction patterns, and experimental results, it tries to clear up some confusions regarding nomenclature. The authors stress that secondary silica in deep-sea sediments, which is called "cristobalite," "α-cristobalite," etc., is similar to earlier-defined "opal-CT," and therefore should be so called. ("CT" stands for "cristobalite-tridymite".) They reject the term "porcellanite" as being ill defined. They also emphasize that no environment of deposition can be inferred from the presence of opal-CT.

REFERENCES

Kastner, M.; and Keene, J. B. (1975) Diagenesis of pelagic siliceous ooze. 9th Internat. Sedimentol. Cong., Nice. Thème VII: 8–9.

von Rad, U.; and Rösch, H. (1975) Progressive chertification of siliceous sediments in the Cretaceous and Tertiary North Atlantic. 9th Internat. Sedimentol. Cong., Nice. Thème I: 18.

6

Reprinted from pp. 991–1007 of *Initial Reports of the Deep Sea Drilling Project,*
Volume 7, Pt. 2, E. L. Winterer *et al.,* Washington, D.C.: U.S. Government Printing
Office, 1971, 1757 pp.

CHERTS FROM THE WESTERN PACIFIC, LEG 7, DEEP SEA DRILLING PROJECT

G. R. Heath, Department of Oceanography, Oregon State University, Corvallis, Oregon
and
Ralph Moberly, Jr., Hawaii Institute of Geophysics, University of Hawaii, Honolulu, Hawaii

ABSTRACT

Nodular and bedded cherts are present at all sites drilled on
Leg 7. Nodules are usually associated with carbonates, but
also occur in siliceous sequences. Bedded cherts, usually
porcelaneous, are restricted to noncarbonate deposits. Textural and mineralogical characteristics suggest that deep-sea
cherts form in two stages. In the first stage, biogenous opal is
dissolved and reprecipitated as finely crystalline cristobalite
to produce porous porcelanites. The cristobalite is either
deposited as interstitial matrix, or replaces pre-existing
calcite or montmorillonite. In the second stage, the cristobalite inverts to quartz, and the remaining porosity is lost.
The end product is a classic dense vitreous chert. The second
stage inversion may be primarily a solid-solid zero-order
reaction of the type described by Ernst and Calvert. Neither
the mineralogy nor texture of cherts is related to the age of
enclosing sediments in a simple way. However, the occurrence
of quartz-rich cherts seems to be favored by higher-than-
average temperatures in the sediment.

INTRODUCTION

Chert or porcelanite was sampled at all seven sites
drilled on Leg 7 (Figure 1, Table 1). The ages,
associated lithotypes, textural characteristics and mineral compositions of these rocks are varied enough to
make them ideal subjects for a preliminary evaluation
of the nature and geological significance of deep-sea
cherts.

Many characteristics of the Leg 7 siliceous deposits are
strongly reminiscent of the Monterey Formation of
California (Bramlette, 1946; Ernst and Calvert, 1969).
Such features as the association of biogenous silica,
cristobalitic porcelanite, and quartz-rich chert, as well
as the irregular distribution of silicification phenomena,
are shared by the outcropping and deep-sea deposits.
Thus, it is possible to reinforce arguments based on
sparse drill-hole data with evidence from exposed
sections where the spatial relationship of the various
forms of silica are more clearcut.

TABLE 1. Cherts and Porcelanites Recovered on Leg 7, Deep Sea Drilling Project

61-0-1-CC	Silicified laminated siltstone, light brownish-gray (5YR6-4/1). Radio-larian. Age: Santonian—early Campanian.

TABLE 1 – *Continued*

	White silty chert. Radiolarian.
61-1-1-1	Grayish-yellow brown (10YR5/1) porcelanite. Radiolarian.
61-1-1-CC	Chert nodule. Olive gray (5Y4/1) and light yellowish-gray (25Y7/1). Dense vitreous core, porous porce-laneous radiolarian rind.
62-0-6-1	Light olive gray (5Y6-7/1) nodular chert intergrown with chalky nan-nofossil limestone. No siliceous microfossils. Age: Late late Oligocene (N3).
62-0-6-CC	As 62-0-6-1.
63-0-7-2	Dark yellow brown (10YR3/2) vitreous chert. No siliceous micro-fossils. Age: Middle Oligocene (early N2).
64-1-11-CC	Light gray (N6-N7) to olive gray (5Y5/1) nodular chert. No sili-ceous microfossils. Age: Late Middle Eocene (P12-P14).

TABLE 1 – *Continued*

65-0-14-CC	Porcelanite or porcelaneous mudstone. Dark brownish-gray (5YR-3/1). Radiolarian. 125.5 meters. Age: Earliest Oligocene.
65-0-17-CC	As 65-0-14-CC.
	Silicified laminated arenite (turbidite). Pale yellowish-brown (10-YR6/3). Radiolarian. 137.5 meters. Age: Middle-late Eocene.
65-1-1-CC	Dark brownish-gray porcelanite and plae yellowish-brown silicified arenite as 65-0-14-CC and 65-0-17-CC.
65-1-2-CC	As 65-1-1-CC.
65-1-3-CC	As 65-1-1-CC.
65-1-4 and 65-1-5	Contain cavings of silicified rock described above.
66-0-4-CC	Porcelanite grading to cristobalitic mudstone. Brownish-black (10YR-2/1) to dark yellowish-brown (10YR3/2). Radiolarian. Age: Late Oligocene (extrapolating sedimentation rates).
66-0-5-CC	As 66-0-4-CC.
	Laminated silicified siliceous turbidite or turbidite porcelanite. Moderate yellowish-brown (10YR-5/4) and grayish-orange pink (5YR7/2). Radiolarian. Age: Late Oligocene (extrapolating sedimentation rates).
67-1-2-CC	Porcelanite. Dusky yellowish-brown (10YR7/2). ? Radiolarian. Age: Possibly late Early Eocene.
	Laminated porcelanite. Pale yellowish-brown (10YR7/3). ? Radiolarian.

FIELD CHARACTERISTICS

The cherts and porcelanites collected on Leg 7 fall into two major classes, corresponding in a general way to the nodular and bedded cherts of the geological literature.

The nodules are the typical form encountered in carbonate sequences (Sites 62, 63, 64), but, as shown by the occurrences at Sites 61 and 67, they are not restricted to such sequences. Nodules small enough to enter the 5 centimeter coring bit have been sampled complete. More commonly, only fragments (usually milled by grinding at the bottom of the hole) or cored sections are recovered. Some idea of the maximum size

of deep-sea nodules is revealed by dredged samples. The largest of these have horizontal dimensions in excess of 2 by 1 meters, and are as much as 10 centimeters thick. Such nodules are comparable in size and shape to those recorded in exposed carbonate sequences of many geological ages and from numerous localities throughout the world.

The deep-sea nodules range from simple triaxial ellipsoids to irregular anastomosing bodies of silica, which are often rich in chalk inclusions, and grade unpredictably to ordinary chalk. The suggestion that the geometrical complexity of chert nodules decreases with the proportion of impurities in the enclosing limestone (Sujkowski, 1958) is not borne out by the Leg 7 samples. The complexity bears no simple relation to the composition of the associated carbonate, nor to any other obvious parameters such as depth of burial or age of enclosing sediment.

The mode of formation of the nodular chert is discussed in more detail in a subsequent section. However, in general, the silica appears to replace individual carbonate particles, and often pseudomorphs fine details of the former calcite tests.

The second group of cherts, the bedded type, is found in both siliceous oozes and pelagic clays, but never in the pelagic carbonates. Because bedded chert has not been dredged and since the Leg 7 occurrences are not quite like bedded cherts on land, little of their lateral dimensions or of the nature of their lateral aboundaries can be inferred. J. Ewing (personal communication) has observed that the upper surface of the chert sequence in the Central Basin consists of numerous discontinuous reflectors (evidenced by overlapping hyperbolae on the 3.5-kHz reflection records). Thus, individual silicified layers may cover as little as a few square meters, and probably rarely exceed a few thousand square meters in extent.

The thickness of silicified beds is usually in the 2 to 5 centimeter range. Few layers thicker than 10 centimeters were encountered on Leg 7. Individual beds at Sites 65 and 66 are separated by one to several meters of apparently unaltered siliceous ooze. Unfortunately, the alternation of extremely hard and soft unconsolidated material is virtually impossible to sample adequately using existing coring techniques. Thus, the nature of the ooze-porcelanite contacts is poorly known. A few of the porcelanite fragments retain adjacent softer sediment. In these cases, the boundary is gradational over a few millimeters.

The bedded cherts encountered during Leg 7 appear to represent an early stage in the formation of the classic bedded sequences observed on land. The presence of abundant reactive silica (opal) suggests that only time is

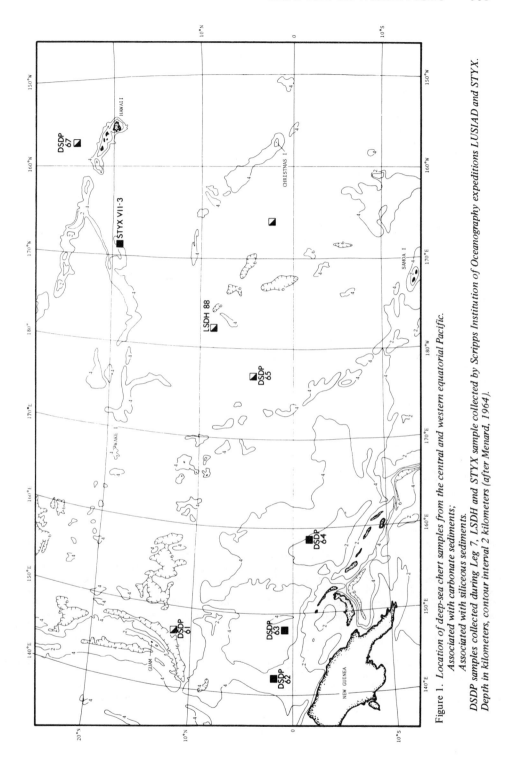

Figure 1. Location of deep-sea chert samples from the central and western equatorial Pacific.
◧ *Associated with carbonate sediments;*
◪ *Associated with siliceous sediments.*
DSDP samples collected during Leg 7, LSDH and STYX sample collected by Scripps Institution of Oceanography expeditions LUSIAD and STYX.
Depth in kilometers, contour interval 2 kilometers (after Menard, 1964).

required to produce much greater quantities of chert. Whether or not the ultimate product will be rhythmically bedded is still open to question. Hopefully, later legs of the Deep Sea Drilling Project will be able to penetrate thicker and older chert sequences, and thereby provide an answer to this question.

AGE OF CHERTS

One of the striking early discoveries of the Deep Sea Drilling Project was the Eocene, and more particularly middle Eocene, age of most Atlantic deep-sea cherts. Although the Leg 7 cherts range in age from late Cretaceous to late Oligocene (Table 1), most of the bedded occurrences are of Eocene age, particularly those in the Central Basin. The previously mentioned problem of coring interbedded ooze and chert makes accurate determinations of the age of chert beds very difficult. Probably the best date for the top of the opaque layer in the Central Basin comes from Scripps core LSDH 88P. Here, slightly porcelaneous cristobalitic claystone is directly overlain by radiolarian ooze belonging to the upper part of the *Podocyrtis mitra* Zone, that is from the upper third of the middle Eocene (T. C. Moore, personal communication). The other deep-sea chert from the equatorial Pacific collected by Scripps Institution prior to the Deep Sea Drilling Project, the large nodule from STYX 7, Station 3, contains poorly preserved nannofossils of probably late Lutetian age, again late middle Eocene (T. C. Moore, personal communication).

The widespread occurrence of these Eocene cherts points to a catastrophic, or at least abrupt change in geologic or oceanographic conditions about 45 to 50 million years ago. There is little evidence or necessity for a marked change in the silica budget to supply the material for the cherts. A relatively minor shift in the locus of biogenous silica deposition from present high to lower latitudes would provide vastly more reactive amorphous silica than is found in the Eocene cherts. The problem is rather one of identifying factors which would result in mobilization and concentration of silica in beds of a characteristic age. The Central Basin porcelanites are silicified cristobalite-(tridymite)-montmorillonite-(clinoptilolite) claystones, rather than silicified siliceous oozes, which are abundant above, between and below the cherts. This suggests that some chemical peculiarity of altering volcanic debris may be responsible for the accumulation of chemical silica. Alternatively, the volcanic debris may be no more than a by-product of some event which also modified oceanographic conditions near the sea floor. Temporary stagnation of the bottom water, due either to tectonic "silling" of large areas of the Pacific and North Atlantic Oceans (connected during the early Tertiary) or influx of dense hypersaline water (as perhaps suggested by the unusual mineralogy at Site 12 in the eastern North Atlantic,

Peterson *et al.*, 1970) could lead to changes in *p*H or dissolved silica of sufficient magnitude to produce silicification of surface sediments.

The evidence for such a dramatic oceanographic event in the middle Eocene is far from conclusive. In part, this reflects the minimal amount of geochemical work done on Deep Sea Drilling Project cores so far, and in part, a lack of basic knowledge of the oceanographic parameters responsible for many of the geochemical and mineralogical peculiarities observable in deep-sea sediments.

Perhaps the only valid conclusions at present are that cherts in thick carbonate sequences are not isochronous, although they do occur most abundantly in the lower Tertiary, whereas cherts in noncarbonate sequences are commonly but not invariably first encountered in middle Eocene deposits. The catastrophists may be considered in the ascendency, but the battle is not yet over!

MINERALOGY

All three common polymorphs of silica are represented in the Leg 7 cherts. Cristobalite and quartz (α − forms in both cases) are the dominant minerals. Each ranges in abundance from less than 1 per cent to almost 100 per cent of chert samples. Tridymite is present in many cristobalite-rich porcelanites, and in associated claystone (? altered volcanic debris).

The quartz is generally chalcedonic in thin section. It yields sharp X-ray diffraction patterns, indicating that crystallites are considerably coarser than 1000 Å. This high degree of crystallinity is found regardless of the abundance of the quartz or age of the sample.

The crystobalite, in contrast, is always poorly crystallized. The crystallite size normal to 101, determined from the Scherrer equation (Klug and Alexander, 1954), ranges from about 50 to 130 Å. As is obvious from Figure 2, this crystallite size bears no simple relation to the age of sediments enclosing the chert. This does not prove, however, that the size is unrelated to the age of chertification. There simply is no means of estimating such an age. For comparison, the crystallite sizes of typical Quaternary, Lower Miocene and Upper Eocene radiolarian opals are also shown on Figure 2. It is clear that chert cristobalite, despite its poor crystallinity, consists of much more ordered silica than does opal. Conclusions based on the few data of Figure 2 should be viewed with some caution. Nevertheless, it appears that there is a clear gap separating the chert and opal data. This implies that chert does not arise by the crystallization of cristobalite on aging opal nuclei.

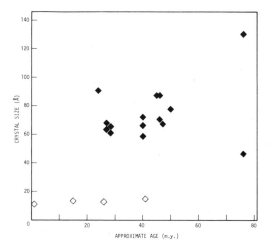

Figure 2. *Size of cristobalite crystals normal to (101) as a function of geologic age. Deep-sea cherts—solid diamonds, radiolarian opal—open diamonds.*

The ratio of cristobalite to quartz (based on the relative heights of their 101 reflections), like the crystal size, bears no simple relation to the age of enclosing sediment (Figure 3). Again, the age of nodule-formation is probably a more logical abscissa, but is indeterminate at present.

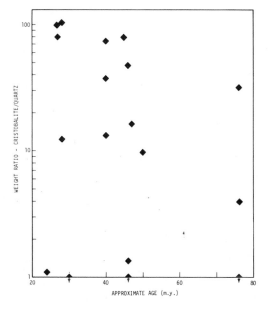

Figure 3. *Cristobalite/quartz ratio in deep-sea cherts as a function of geologic age. Peak-height ratios converted to weight ratios using a factor of 9.5 (Rex, 1969).*

Because cristobalite and tridymite are metastable under the conditions at which the nodules are found, it

might be expected that they would be least abundant where the temperature is highest (providing it remains within the quartz stability field). This assumption is borne out by the mineralogy of silica adjacent to the intrusive basalt at Site 62. This silica is all quartz. Even 50 meters above the basalt, quartz is still abundant in chert nodules.

Unfortunately, *in situ* temperatures were not measured during the drilling of Leg 7 holes. However, the bulk density and water content of a sediment sample can be used to make a semi-quantitative estimate of its thermal conductivity (Langseth, 1965). Such values, together with the heat flow value and bottom water temperature at each site, plus the depth of the cherts below the sea floor, can be used to calculate crude *in situ* temperatures at the level of the cherts. The possible errors inherent in such a procedure are obvious, yet a plot of the relative abundances of cristobalite and quartz against such a temperature scale yields an intuitively reasonable pattern (Figure 4).

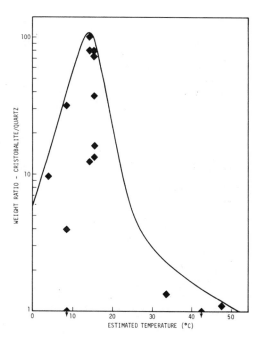

Figure 4. *Variation of cristobalite to quartz ratio in Leg 7 cherts and porcelanites as a function of estimated* in situ *temperature. See text for method of deriving temperatures. Peak-height ratios converted to weight ratios using a factor of 9.5 (Rex, 1969).*

At the lowest temperatures there are few cherts, and these tend to be fairly quartz-rich. Since these include the oldest (Cretaceous) samples, it is possible that, given enough time, quartz will form directly from dissolved silica. Alternatively, the chert may have formed at much higher temperatures than prevail at the site today. As will be discussed later, the age of thermal history of a sample almost certainly influence its mineralogy.

At temperatures in the vicinity of 15°C, the ratio of cristobalite to quartz is at a maximum. This could mean that the conversion of opal to cristobalite is relatively rapid at this temperature, whereas the inversion of cristobalite to quartz is relatively slow. At higher temperatures, the second reaction is accelerated to the extent that quartz becomes the dominant or even the sole silica mineral (Figure 4), as is the case in samples close to intrusive basement.

Tridymite is present in many of the porcelanites from Sites 65 and 66. Although its 4.30 Å peak reflection is difficult to resolve from the 4.26 Å (100) reflection of quartz, tridymite can be identified by comparing the composite 4.3 Å peak with the 3.343 Å (101) reflection of quartz. When tridymite is present, the spacing of the peaks exceeds the quartz value of 0.91 Å, and their intensity-ratio exceeds the normal range of 0.2 to 0.3. In some of the Site 65 porcelanites, the 4.3 Å reflection is twice as intense as the 3.34 Å peak. Tridymite is most abundant in slightly indurated claystone from the core catcher of Core 17, Hole 65.0, and from the core catcher of Scripps piston core LSDH 88P. In both cases, the clay is rich in cristobalite, clinoptilolite and montmorillonite, and is probably altered volcanic debris. The presence of aluminum and other metal cation impurities appears to stabilize the tridymite structure (Buerger, 1954), but the parameters controlling its formation are not known. However, the unusually high concentrations of nonsilica minerals in the tridymite-bearing porcelanites points to the influence of readily available metal cations which may be necessary to initiate the formation of this mineral (Frondel, 1962).

TEXTURE

The nodular and bedded cherts show rather different textural characteristics. The nodules from carbonate sections will be considered first, since their textural features are most readily interpreted.

As mentioned in the chapter on "Siliceous Sediments", the first sign of silica cementation is the tangential "welding" of radiolarian tests. In extreme cases (such as, the chalk at the top of Core 8, Hole 64.0), this "welding" creates a spongy network of silica which can be freed by treatment of the rock with dilute acid, and then handled without destroying the original form of the rock fragments.

Such material, however, does not appear to be a precursor of nodular chert. In all the calcareous sequences sampled on Leg 7, the chert nodules are surrounded by a region containing severely corroded opaline skeletons or none at all. To the extent that this phenomenon negates the requirement for an exotic source of silica, it simplifies the interpretation of the chert occurences. However, it also eliminates the most promising source of evidence as to the factors controlling the locations at which nodules accrete. The same problem remains unsolved for nodular cherts in carbonate sequences on land. It is easy to present plausible arguments for subtle variations in pH or increased permeability at nodule locations, but in reality, it is still not known why nodules and networks of accreted silica favor certain stratigraphic levels. In any case, it appears the nodules in both shallow and deep-water carbonates derive their silica from opaline skeletal material deposited with the carbonate, rather than from external sources like volcanic exhalations.

The first stage in the formation of a nodule is the filling of empty foraminiferal chambers with silica (Figures 5 and 6). This silica is usually chalcedonic quartz, often as one or two generations of radiating bladed crystals. Cristobalite is rare, but occasionally lines filled the chambers.

The second stage is the replacement of the micritic groundmass of the carbonate rock (mostly nannofossils) by extremely fine-crystalline cristobalite (Figures 7, 8 and 9). The transition zone from virtually pure calcite to virtually pure silica is usually quite narrow, rarely more than 1 millimeter and occasionally as little as 50 microns. The anastomosing network of silica which characterizes the more irregular nodules results from the preservation of "islands" of carbonate, rather than from abnormally diffuse zones of silicification.

The third stage in the formation of a nodule is the replacement of foraminiferal tests by silica. This replacement is apparently a slow process, which occurs through crystal by crystal replacement of the radiating calcite prisms of the test walls (Figures 10 and 11). The resulting silica is in the form of radiating blades of chalcedonic quartz, which preserve both the shape and crystallographic orientation of the antecedent calcite.

The final infilling of pore spaces, which converts the somewhat porcelaneous replaced-chalk to "classic" chert apparently occurs later, and perhaps much later in the history of the rock. The textural evidence allows the silicification stages to be ordered in time, but cannot reveal the actual times involved. The rarity of dense vitreous cherts and the inverse relation of quartz content to porosity suggest that the final stage of silicification does not occur until the matrix cristobalite begins to invert to quartz.

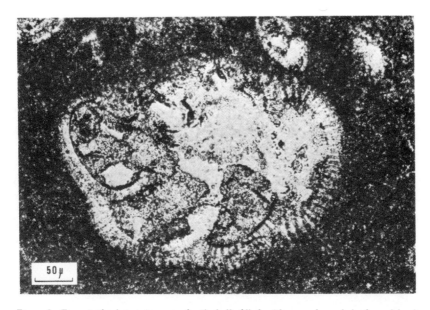

Figure 5. *Foraminiferal test in nannofossil chalk filled with secondary chalcedony (clear). Bar is 50 microns long. Plane polarized light. 7-64-1-11-CC.*

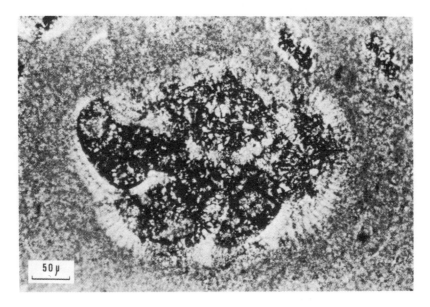

Figure 6. *As Figure 5, but under crossed nicols.*

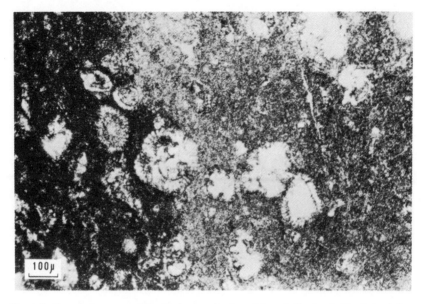

Figure 7. *Contact between calcite (dark-nannofossils and foraminifera) and silica (cristobalite and lesser chalcedony) at the edge of a chert nodule in chalk. Bar is 100 microns long. Plane polarized light. 7-64-1-11-CC.*

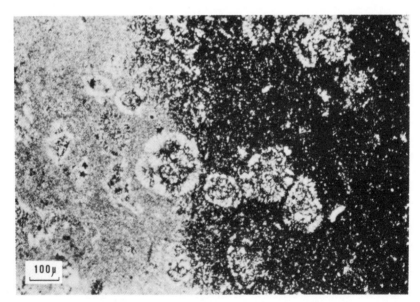

Figure 8. *As Figure 7, but under crossed nicols.*

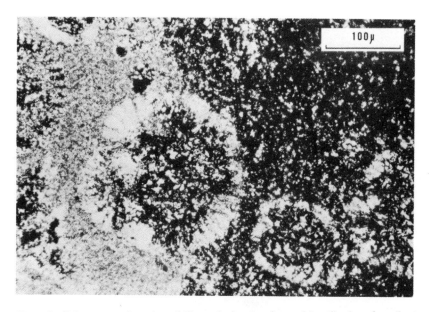

Figure 9. *Enlargement of portion of Figure 8, showing sharp calcite-silica boundary. Bar is 100 microns long. Crossed nicols. 7-64-1-11-CC.*

The bedded cherts are much more difficult to interpret because of their fine grain size.

The first stage in the formation of many of these cherts is the development of a cristobalitic and often tridymitic claystone, apparently by alteration of montmorillonite and perhaps amorphous volcanic debris. None of the claystones examined in thin section contains recognizable biogenous remains, so it is not clear whether the silica for the cristobalite is derived from opaline debris within the rock, from alteration of montmorillonite (or associated volcanic material), or from nearby opal-bearing beds.

The cristobalite-montmorillonite claystones (often containing clinoptilolite) are more indurated than associated pelagic clays, but are still quite friable, and often shrink markedly or disintegrate on drying.

The second stage of chert formation involves the crystallization of sufficient additional cristobalite to create a well-indurated porcelanite (Figure 12). The porosity and permeability of such a porcelanite is much less than the associated mudstone, but the two rock types are virtually indistinguishable in thin sections. Porcelanites of this type contain up to 100 times as much cristobalite as quartz — the highest proportion of cristobalite of any of the siliceous rocks sampled on Leg 7.

The radiolarian-bearing cherts (for example, at Site 61) are the most difficult to interpret. The preservation of siliceous tests is often excellent (Figures 13 and 14), and the tests are close packed, indicating that secondary silica was not derived from the solution of skeletons within the silicified bed. However, no examples of partly silicified ooze were sampled on Leg 7, so the circumstances surrounding this type of silicification cannot be evaluated in this report.

The final stage in the formation of bedded cherts, as of nodules, is the inversion of cristobalite to quartz and the loss of virtually all remaining porosity. The inversion appears to spread from centers throughout the rock (Figure 15), creating clear masses of chalcedonic (and sometimes unstrained) quartz in place of the fine cristobalite mosaic. Where present, the interiors of radiolarian tests and the tests themselves are usually the first portions of the rock to invert. The recrystallization obliterates virtually all the siliceous skeletons preserved in the cristobalitic porcelanites.

DEVELOPMENT OF CHERT

The evidence from thin section and X-ray diffraction studies of the deep-sea samples, in conjunction with field observations and experimental work on exposed siliceous sediments, allows scientists to deduce something of the geochemical life history of a marine chert.

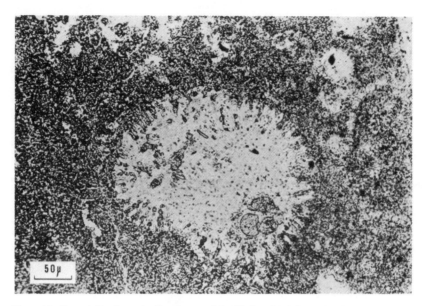

Figure 10. *Foraminiferal test inside chert nodule filled and largely replaced by chalcedonic silica. A few calcite crystals remain. Bar is 50 microns long. Plane polarized light. 7-64-11-CC.*

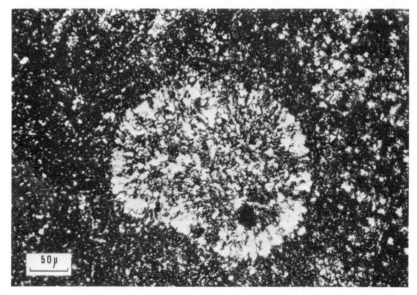

Figure 11. *As Figure 10, but under crossed nicols.*

Figure 12. *Electron micrograph of typical Central Basin porcelanite. Flakes are remnants of cristobalite-montmorillonite-clinoptilolite claystone which has been impregnated with cristobalite. Bar is 2 microns long. 7-66-0-4-CC.*

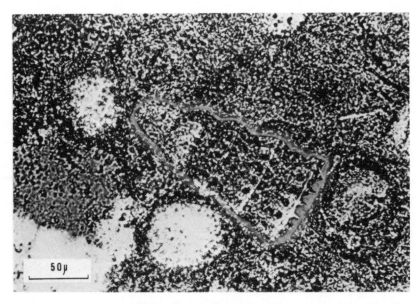

Figure 13. *Costate dictyomitrid (radiolarian) in Cretaceous chert nodule, showing preservation of surface ornamentation (margin retouched). Bar is 50 microns long. Crossed nicols 25° from extinction 7-61-1-1-CC.*

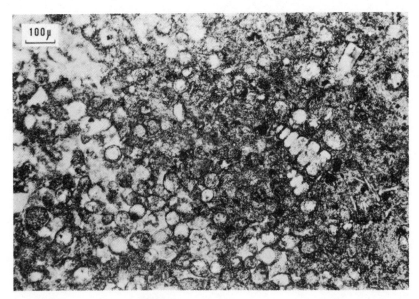

Figure 14. *Cretaceous radiolarian chert, showing underformed closepacked siliceous tests in cristobalite (dark) and chalcedony (light) matrix. Bar is 100 microns long. Plane polarized light. 7-61-1-1-CC.*

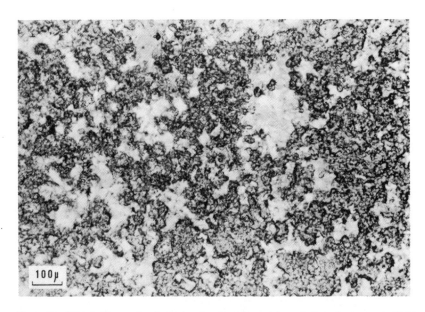

Figure 15. *Cristobalite-quartz chert, showing irregular patches of chalcedonic quartz (light, low relief), developing by inversion of finely cristalline cristobalite. Bar is 100 microns long. Crossed nicols 15° from extinction. 6-67-1-2-CC.*

Both textural and mineralogical data indicate that the formation of chert is usually in a two stage process. Initially, the biogenous opal is converted to finely crystalline crystobalite, usually, but not always, in the form of porcelanite. Later, this cristobalite inverts to quartz, which is the characteristic constituent of all pre-Cenozoic cherts. What can be said about the chemical reactions involved in the two stages of this process?

The bulk of evidence from Leg 7 cherts indicates that the first stage involves dissolution of the silica. The absence of siliceous microfossils in most cherts, the presence of severely corroded siliceous skeletons or barren areas around cherts, and the volume of silica in the cherts indicate that the cristobalite silica must migrate distances of centimeters to meters from its point of origin to a growing chert mass.

The inferred solution step, while facilitating the migration of silica, complicates the interpretation of the mineralogy of the porcelanites. If the opal-porcelanite conversion involved a solid-solid transition, it would be easy to attribute the formation of cristobalite to the influence of "inherited" opal structure. However, since cristobalite is far from its stability field at sea-floor temperatures and pressures (Figure 16), its crystallization from solution is decidedly anomalous from a thermodynamic point of view. Two possible explanations

could account for this anomaly. Firstly, by analogy with the carbonate system, it is possible that the direct precipitation of quartz, the thermodynamically stable form of silica, is inhibited by some other component of the system. Thus, magnesium ions in seawater inhibit the nucleation of calcite, the thermodynamically stable form of calcium carbonate under oceanic T-P conditions, and lead to the precipitation of an unstable form, aragonite. In the same way, cristobalite may be the only form of silica which can nucleate in interstitial solutions. The alternative explanation is that the dissolved opal consists of "domains" of sufficient size that they retain their original cristobalite structure. Such a hypothesis does not accord with existing knowledge of the state of dissolved amorphous silica, but cannot be dismissed until more is known of the chemical conditions prevailing in the sediment at sites of chert formation.

Experimental work on the alteration of opal to cristobalite is not particularly useful as a guide to reactions in deep-sea sediments. Mizutani (1966) states that the reaction is first-order with respect to silica, but his arguments include considerable circular reasoning, and his experimental conditions (particularly the use of 0.08N KOH as the interstitial medium) are so unlike those of the real world that little confidence can be placed in his results. Other investigations of the reaction are subject to similar criticism, and this problem appears unusually ripe for study at the present time.

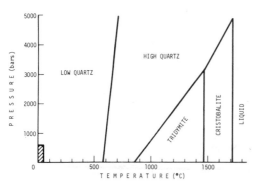

Figure 16. *Pressure-temperature diagram for SiO₂ (after Tuttle and Bowen, 1958). Shaded area is P-T field enclosing* in situ *conditions of Leg 7 cherts.*

Field data for Cenozoic opal-porcelanite occurrences are open to the same range of interpretations as the deep-sea samples. Thus, the Monterey Formation includes porcelanite which contains fragmented diatom frustules in a fine-grained cristobalitic matrix (Bramlette, 1946; Ernst and Calvert, 1969). Lateral facies changes from diatomite to porcelanite are not accompanied by significant changes in the thickness of mapped beds, suggesting that the cristobalite has not developed by partial solution of opal within the beds. However, the field evidence is equivocal, and neither the precise source of the matrix silica nor the distance it has migrated have been satisfactorily established.

The second reaction, the inversion of cristobalite to quartz, is somewhat easier to interpret, both because of some unusually unequivocal textural evidence and because recent experimental work by Ernst and Calvert (1969) is directly applicable to deep-sea chert.

Quartz is thermodynamically stable in deep-sea sediments, whereas cristobalite is not. Therefore, the persistence of the latter mineral for tens of millions of years indicates that the activation energy for the inversion is large. In fact, Ernst and Calvert (1969) calculate a value of 23 kcal/mole for this reaction. Since the free energy of formation of both quartz and cristobalite is about 197 kcal/mole (Fournier and Rowe, 1962), the large activation energy indicates that the inversion requires a drastic reorganization of bonds within the crystal lattice. The necessity for such reorganization becomes apparent when one compares the structures of the two minerals (Figure 17). It is not surprising that cristobalite is quite stable at low temperatures, and requires either a great deal of time or a source of thermal energy to invert to quartz.

Ernst and Calvert (1969), on the basis of an extensive series of hydrothermal experiments on porcelanite from the Monterey Formation, conclude that the inversion is zero-order. They picture an inversion front migrating through each cristobalite particle at a velocity which depends only on the temperature. Such a concept is strikingly supported by textural relations in a cristobalite-quartz vein in a porcelanite from Hole 65.0 (Figures 18 and 19). The saw-tooth contacts between the high-relief cristobalite and low-relief chalcedonic quartz are classic solid-solid inversion fronts. The low porosity of the two mineral masses in this occurrence virtually excludes the possibility of a solution phase, except perhaps in a thin film at the inversion fronts. It also eliminates the requirement for additional silicification at this stage in the history of the rock.

Unfortunately, such clear-cut examples are very rare, and most quartz-bearing cherts show evidence of a more complex recrystallization history. In general, quartz appears to nucleate at many points in a given porcelanite, and increases in abundance by an increase in the

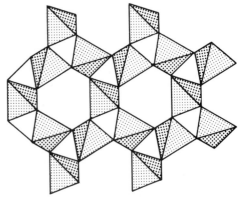

(a)   (b)

Figure 17. *Crystal structure of (a) α - cristobalite projected on (001) and (b) α - quartz projected on (001) to illustrate bond reorganization which must accompany inversion. SiO₄ tetrahedra are shaded. Adapted from Deer et al. (1963).*

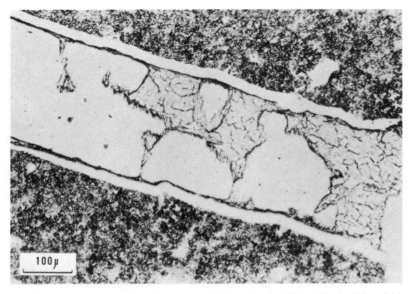

Figure 18. *Vein of silica in cristobalitic porcelanite showing "frozen" cristobalite (high-relief) — quartz (low relief) inversion fronts. Bar is 100 microns long. Plane polarized light. 7-66-0-5-CC.*

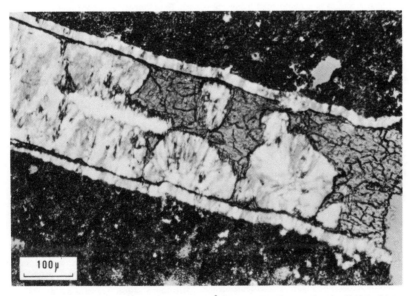

Figure 19. *As Figure 18, but crossed nicols 10° from extinction, to show radiating sheaves of chalcedony.*

139

Figure 20. *Masses of chalcedonic quartz growing by inversion of finely crystalline cristobalite. Bar is 100 microns long. Crossed nicols 10° from extinction. 7-67-1-2-CC.*

size of nuclei as well as the appearance of new nuclei (Figure 20). Because the quartz masses are considerably less porous than the original cristobalite, and because the growth of quartz masses is accompanied by severe degradation or destruction of textural features within the cristobalite (such as, molds and tests of siliceous organisms), at least partial solution of silica must accompany the inversion. To what extent this is an incidental phenomenon, rather than evidence for a distinctly different mechanism from that proposed by Ernst and Calvert, remains to be established.

One of the most important reasons for understanding the chemistry of chert formation is related to its potential use as a paleotemperature indicator. Clearly, if the Ernst-Calvert model is correct, young cherts rich in quartz provide strong evidence of abnormally high temperatures within the enclosing sediments. Hopefully, further experimental work guided by observations of textural features and mineral associations within the deep-sea cherts will allow such evidence to be quantified.

ACKNOWLEDGEMENTS

Sincere thanks are due to T. C. Moore, Jr., W. R. Riedel and J. Lipps for stratigraphic advice and for the electron photomicrograph (Figure 12). Discussions with E. L. Winterer, R. W. Rex, T. C. Moore and J. I. Ewing have been most helpful.

REFERENCES

Bramlette, M. N., 1946. The Monterey Formation of California and the origin of its siliceous rocks. *U. S. Geol. Survey Profess. Paper 212.* 57 pp.

Buerger, M. J., 1954. The stuffed derivatives of the silica structures. *Am. Mineralogist.* **39,** *(7-8), 600.*

Deer, W. A., Howie, R. A. and Zussman, J., 1963. Framework Silicates. *Rock-Forming Minerals, Volume 4.* London (Longmans), 435 pp.

Ernst, W. G. and Calvert, S. E., 1969. An experimental study of the recrystallization of porcelanite and its bearing on the origin of some bedded cherts. *Am. J. Sci.* **267-A,** 114.

Frondel, C., 1962. Silica Minerals. *The System of Mineralogy.* 7th edition. **3,** New York (John Wiley), 334 pp.

Klug, H. P. and Alexander, L. E., 1954. *X-Ray Diffraction Procedures.* New York (John Wiley), 716 pp.

Langseth, M. G., 1965. Techniques of measuring heat flow through the ocean floor. **In** Terrestrial heat flow. Lee, W. H. K., (Ed.), *Am. Geophys. Union, Geophys. Monog. No. 8.* 58-77.

Menard, H. W., 1964. Marine Geology of the Pacific. New York (McGraw-Hill), 271 pp.

Mizutani S., 1966. Transformation of silica under hydrothermal conditions. Nagoya Univ., *J. Earth Sci.,* **14,** 56.

Peterson, M. N. A., Edgar, N. T., von der Borch, C., Cita, M. B., Gartner, S., Goll, R. and Nigrini, C., 1970. Site 12. **In** Peterson, M. N. A. *et al., 1970. Initial Reports of the Deep Sea Drilling Project, Volume 2.* Washington (U. S. Government Printing Office), 249.

Rex, R. W., 1969. X-ray mineralogy studies — Leg 1. **In** Ewing, M. *et al., 1969. Initial Reports of the*

Deep Sea Drilling Project, Volume 1. Washington (U. S. Government Printing Office), 354.

Sujkowski, Z. L., 1958. Diagenesis. *Bull. Am. Assoc. Petrol. Geologists.* **42**, (11), 2692.

Tuttle, O. F. and Bowen, N. L., 1958. Origin of granite in the light of experimental studies in the system $NaAlSi_3O_8$-$KAlSi_3O_8$-SiO_2-H_2O. *Geol. Soc. Am. Mem. 74.*

7

Reprinted from pp. 377–405 of *Initial Reports of the Deep Sea Drilling Project*,
Volume 17, E. L. Winterer *et al.*, Washington, D.C.: U.S. Government Printing
Office, 1973, 930 pp.

CHERT AND SILICA DIAGENESIS IN SEDIMENTS FROM THE CENTRAL PACIFIC[1]

Yves Lancelot, Lamont-Doherty Geological Observatory of Columbia University, Palisades, New York

ABSTRACT

Chert recovered during Leg 17 of the Deep Sea Drilling Project in the central Pacific consists of porcelanitic and quartzose nodules and beds. Porcelanitic chert made of disordered cristobalite is found in clay-rich sediments, while quartzose nodules are restricted to carbonate environments. It is suggested that the mineralogical nature of the chert is directly influenced by the composition of the sediments in which it develops, and the roles of foreign cations and permeability are emphasized.

Recrystallization of silica appears to result from differences in solubility between amorphous and crystalline silica. The transfer from amorphous to crystalline forms might be controlled by the nature of the silicates present in the sediments.

INTRODUCTION

Since scientists from the first legs of the Deep Sea Drilling Project recognized that chert was widely spread in ocean basins, in the lowermost Tertiary and older sediments, much interest has been shown in trying to understand the processes involved in its formation. Although progress has been made in determining the nature of the chert, no satisfactory explanation has been given to account for its worldwide distribution as well as for its preferential occurrence in some sections of the stratigraphic column. The purpose of this study is to present some new observations and use the diversity of chert-bearing sediments recovered in the central Pacific to try a new approach to the question.

During Leg 17, chert was encountered in very different sedimentary environments and the mineralogical nature of the chert appears to be strongly influenced by the composition of the host sediments. This observation is used as a guide to investigate the processes involved in the diagenetic recrystallization of silica.

Although the chert studies here is exclusively that recovered from central Pacific sediments, the author believes that the conclusions reported in this paper could apply to the deep-sea cherts from most oceanic basins.

NATURE OF THE CHERT

Occurrence

Chert was found at every Leg 17 site (Figure 1). It lies predominantly in lower Tertiary sediments but is very common also in Mesozoic layers (Table 1). It occurs in biogenic sediments (calcareous and siliceous) as well as in deep-sea brown zeolitic clays.

---

[1]Lamont-Doherty Geological Observatory Contribution No. 1975.

Stratigraphic Setting

Most of the chert encountered during earlier drilling is reportedly of lowermost Tertiary age and older, with very rare younger occurrences in Oligocene and even more recent sediments. A detailed investigation of the stratigraphic occurrences of chert recovered by drilling in the Atlantic and Pacific shows that there are, in fact, considerable variations in the age of the youngest chert reported, not only in the post-Eocene sediments but also within the middle and lower Eocene stratigraphy. However, on a regional basis the correlation of the chert distribution in the stratigraphic column is usually rather good. For example, such a correlation can be observed at Sites 33 and 34 in the eastern Pacific for some middle Miocene chert (McManus et al., 1970); at Sites 62 and 63 in the western Pacific where the youngest chert is of late Oligocene age (Winterer et al., 1971); and is especially striking at Sites 70, 71, and 72 in the equatorial Pacific (Tracey et al., 1971) where upper Oligocene chert is reported in sediments of exactly the same age (*Theocyrtis annosa* radiolarian zone). This distribution implies that the occurrence of chert is not related to a simple overburden or late diagenesis effect independent of the stratigraphy. If this were the case, the occurrence of the younger chert would be more randomly distributed in the stratigraphic scale.

At all of the Leg 17 sites the youngest occurrence of chert constitutes an abrupt change in the lithology. Generally it occurs in middle Eocene sediments except at Site 164 where chert is present in Oligocene layers. Whether or not such a lithologic change can be correlated on the seismic profiler with the top of the opaque layer is not completely established (see Ewing, this volume). As a rule the entire section cored in the central Pacific can be divided in three units (from bottom to top):

(1) Mesozoic sediments with variable amounts of chert: The variations in the abundance of chert are apparently not correlatable from one site to another although there seems to be a consistently smaller amount of chert occurrence

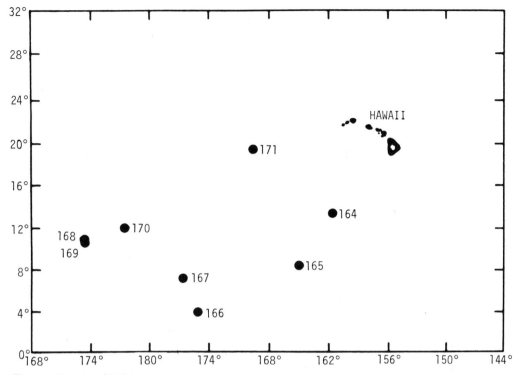

Figure 1. *Location of drill sites.*

near the base of the Maastrichtian and upper part of the Campanian.

(2) Lowermost Tertiary sediments with abundant chert: Most of the chert is restricted to the middle Eocene, but within this epoch regional correlations are possible. At Sites 164 and 165 in the Line Islands Chain a correlative occurrence can be observed in the lower-middle Eocene sediments (*Thyrsocyrtis triacantha* and *Podocyrtis ampla* radiolarian zones). Another regional correlation can be observed at Sites 166 and 167 where the youngest chert is of middle to late Eocene age (*Podocyrtis chalara* and *Podocyrtis goetheana* radiolarian zones).

(3) Post-Eocene sediments with very rare chert: Except for Site 164, the post-Eocene sediments recovered during Leg 17 do not contain chert. This observation apparently applies to the great majority of the deep-sea sediments in the world oceans.

Lithology of Host Sediments

The fact that chert was found in different kinds of pelagic sediments (biogenic and nonbiogenic) makes the analysis particularly interesting. Table 1 indicates the lithologic facies in which chert was found in Leg 17 cores.

It is especially interesting to note that sediments apparently devoid of any siliceous organisms, such as most of the upper Cretaceous brown clays of Site 164, contain relatively abundant chert.

It should also be noted that the volcanogenic sediments such as tuff and volcanic siltstones, very frequent at Leg 17 sites, are conspicuously devoid of chert. A detailed investigation of the mineralogical composition of the sediments shows that the chert-bearing sediments have some common characteristics. The most significant trends observed in the chert-bearing layers are: high K-feldspar/plagioclase ratio, high palygorskite/montmorillonite ratio, high clinoptilolite/phillipsite ratio. Obviously further detailed analysis is required to establish more firmly these correlations as the X-ray analyses were only performed on a rather random sampling of the sedimentary column at each site (see X-ray mineralogy chapter in this volume). Site 164 provides the most useful data because the relatively small amount of biogenic components leaves a clear picture of the silicate distribution, even in the composition of bulk samples.

A general survey of the mineralogical composition of all the chert-bearing sediments reported in previous Deep Sea Drilling Reports has been attempted; and although correlations are rather imprecise for reasons mentioned above, the same general trends can be observed grossly; and no major contradictions have been found. It is also worth mentioning that the association of chert with silica-rich silicates such as palygorskite, sepiolite, and clinoptilolite on land is abundantly reported in the literature. For example, many of the lower Tertiary silicifications observed in sedimentary basins from the African continent are accompanied by important

143

TABLE 1
Summary of Chert Occurrence

Core Iden-tification	Nature of Chert	Nature of Sediments	Age of Sediments
164-3-2	Porcelanitic Chert	Radiolarian-rich zeolitic clay	Early Oligocene
164-4-1	Porcelanitic chert	Clayey radiolarian ooze	Middle Eocene
164-6-1	Porcelanitic chert	Clayey radiolarian ooze	Middle to ? early Miocene
164-7-1 164-7-5	Porcelanitic chert	Zeolitic brown clay	? Late Cretaceous
164-8-1	Porcelanitic chert	Zeolitic brown clay	Late Cretaceous
164-10-1 164-10-4	Porcelanitic chert; fragments and beds or nodules	Zeolitic brown clay	? Santonian to ? Campanian
164-11-1 164-11-2	Porcelanitic chert	Zeolitic brown clay	? Santonian to ? Campanian
164-12-1 164-12-CC	Porcelanitic chert	Zeolitic brown clay	? Santonian to ? Campanian
164-13-1 164-13-CC	Porcelanitic chert	Zeolitic brown clay	? Santonian to ? Campanian
164-14-1 164-14-CC	Porcelanitic chert	Zeolitic brown clay	? Santonian to ? Campanian
164-15-1 164-15-CC	Porcelanitic chert	Zeolitic brown clay	? Santonian to ? Campanian
164-17-1 164-17-CC	Porcelanitic chert	Zeolitic brown clay	Late Cretaceous
164-19-1 164-19-2 164-19-3 164-19-CC	Porcelanitic chert	Zeolitic brown clay and claystone	Late Cretaceous
164-22-CC	Porcelanitic mudstone	Mudstone	Late Cretaceous
164-23-CC	Sand-size chert fragments (mainly recrystal-ized radiolarians)	No recovery	? Cenomanian-? Turonian
164-14-1	Small quartzose to chalcedonic chert fragments	Some coarsely crystalline calcite	?
164-25-1	Small quartzose to chalcedonic chert fragments	Mud containing common cocco-lites and some coarsely crystal-line calcite	Albian
164-26-CC	Small quartzose to chalcedonic chert fragments	Mud containing rare coccolites	Albian
164-27-1	Quartzose chert nodules	No recovery	?
165A-14-CC	Small porcelantic chert fragments	Radiolarian ooze	Middle Eocene
165A-15-CC	Small quartzose chert fragments	No recovery	Early-middle Eocene

TABLE 1 – *Continued*

Core Iden-tification	Nature of Chert	Nature of Sediments	Age of Sediments
165A-16-CC	Quartzose chert	Limestone contain-ing shallow-water carbonates ((turbidite)	Late Maastrichtian
165A-17-1 165A-17-CC	Quartzose chert	Limestone	Late Cam Campanian-early Maastrichtian
165A-23-CC	Recrystallized radiolarians	Volcanic siltstone	? Campanian-? Santonian
166-12-1	Porcelanitic chert nodule	Radiolarian ooze	Late Eocene
166-16-5	Porcelanitic chert	Brown clay and radiolarian ooze	Middle-late
166-17-CC	Porcelanitic chert	Ashy radiolarian ooze	Middle Eocene
166-18-2 166-18-CC	Porcelanitic chert	Brown clay	?
166-19-3 166-19-CC	Porcelanitic chert	Zeolitic siltstone	?
166-22-CC	Porcelanitic chert	Zeolitic brown clay	Early Cretaceous
166-23-2	Porcelanitic chert	Zeolitic ashy mudstone	Early Creta Cretaceous
166-27-2	Quartzose chert breccia with drusy quartz veinlet	Zeolitic nanno-fossil marl	Hauterivian
166-28-1 166-28-2	Quartzose chert brecciated	Zeolitic nanno-fossil marl	Hauterivian
167-33-1 167-33-CC	Quartzose chert	Nannofossil chalk	Middle Eocene
167-34-CC	Quartzose chert	Nannofossil chalk	Middle Eocene
167-35-CC	Quartzose chert	Nannofossil chalk	Middle Eocene
167-36-CC	Quartzose chert	Partly silicified nannofossil limestone	Middle Eocene
167-37-CC	Quartzose chert	Partly silicified nannofossil lime limestone	Middle Eocene
167-38-CC	Quartzose chert	Partly silicified nannofossil limestone	Paleocene
167-39-1 167-39-CC	Quartzose chert	Partly silicified nannofossil limestone	Paleocene
167-40-CC	Quartzose chert	Partly silicified nannofossil limestone	Paleocene (Danian)
167-41-CC	Quartzose chert	Partly silicified nannofossil lime-stone and nannofossil chalk	Paleocene (Danian)
167-42-1 to 167-42-5	Quartzose chert	Chalky limestone	Middle Maastrichtian

TABLE 1 – *Continued*

Core Identification	Nature of Chert	Nature of Sediments	Age of Sediments
167-43-2 167-43-CC	Quartzose chert fragments (drill cuttings)	No recovery	Middle Maastrichtian
167-44-1	Quartzose chert	Nannofossil chalk and limestone	Middle Maastrichtian
167-48-CC	Quartzose chert	Nannofossil chalk	? Maastrichtian ? Campanian
167-49-CC	Quartzose chert	Nannofossil chalk	? Campanian
167-50-CC	Quartzose chert	Nannofossil chalk	? Campanian
167-51-CC	Quartzose chert	Foraminiferal-nannofossil chalk	Campanian
167-52-1 167-52-CC	Quartzose chert	Nannofossil chalk	Campanian
167-53-1 167-53-CC	Quartzose chert	Nannofossil chalk	Campanian
167-54-CC	Quartzose chert	Nannofossil chalk	Campanian
167-55-CC	Quartzose chert	Nannofossil chalk	Campanian
167-56-CC	Quartzose chert	Nannofossil chalk	? Campanian-? Santonian
167-59-3	Quartzose chert slightly porcelanitic	Marly limestone	? Turonian
167-60-2	Quartzose to procelanitic chert	Marly limestone and tuffaceous limestone	Cenomanian
167-61-1	Quartzose to porcelanitic chert	Marly limestone	Cenomanian
167-62-2 167-62-3 167-62-4	Quartzose to porcelanitic chert	Marly limestone	Cenomanian
167-63-1 167-63-2 167-63-3	Quartzose chert, sometimes brecciated, and porcelanitic chert	Limestone, brecciated limestone, pebbly marly mudstone, and marly limestone	Cenomanian
167-64-1 to 167-64-5	Quartzose to porcelanitic chert	Limestone and marly limestone	Cenomanian
167-65-1 to 167-65-3	Quartzose chert	Limestone and tuffaceous limestone	Cenomanian
167-67-1 to 167-67-3	Quartzose chert	Limestone and tuff	Cenomanian
167-68-2 167-68-3	Quartzose and porcelanitic chert	Limestone and shale	Early Cretaceous
167-69-1 to 167-69-3	Quartzose and porcelanitic chert	Limestone and shale	Early Cretaceous
167-70-1 to 167-70-4	Quartzose chert	Limestone and shale	Early Cretaceous
167-71-1 167-71-2	Quartzose chert	Limestone and shale	Early Cretaceous
167-72-1	Quartzose chert	Limestone	Early Cretaceous
167-73-1 167-73-2	Quartzose chert	Limestone	Early Cretaceous (?Hauterivian)

TABLE 1 – *Continued*

Core Identification	Nature of Chert	Nature of Sediments	Age of Sediments
167-74-1 167-74-2	Quartzose chert	Limestone	Early Cretaceous Cretaceous
167-75-1	Quartzose chert	Limestone	Early Cretaceous Cretaceous
167-76-1 167-76-2	Quartzose chert	Limestone	Early Cretaceous
167-77-1	Quartzose chert	Limestone	Early Cretaceous
167-78-1	Quartzose chert	Limestone	Early Cretaceous
167-81-CC	Quartzose chert	Limestone	Early Cretaceous
167-82-CC	Quartzose chert	Silicified limestone	Early Cretaceous
167-84-CC	Quartzose chert	Limestone	Early Cretaceous
167-85-CC	Quartzose chert	No recovery	? Early Cretaceous to ? Late Jurassic
167-86-CC	Quartzose chert	No recovery	? Early Cretaceous to ?Late Jurassic
167-88-1	Quartzose chert	Limestone	? Early Cretaceous to ? Late Jurassic
167-92-1	Quartzose chert	Limestone	Berriasian-Late Tithonian
167-93-1 167-93-2	Quartzose chert	Limestone	Berriasian-Late Tithonian
167-94-1	Quartzose chert	Limestone	Tithonian
168-4-1 to 168-4-5	Small porcelanitic chert fragments	Zeolitic radiolarian bearing brown clay	Late Eocene
169-i-1	Porcelanitic chert	Zeolitic claystone	Late Maastrichtian
169-2-CC	Porcelanitic chert	Zeolitic mudstone	Late Maastrichtian
169-3-CC	Porcelanitic chert	Zeolitic clay with rare nannofossils	Campanian
169-8-CC	Quartzose chert	Nannofossil chalk	Cenomanian
169-9-CC	Quartzose chert	Nannofossil chalk	Early Cenomanian
169-10-CC	Quartzose to porcelanitic chert	Nannofossil-bearing zeolitic claystone	Late Albian
170-3-CC	Quartzose chert	Nannofossil chalk	Early Maastrichtian
170-4-CC	Quartzose chert	Limestone	Early Maastrichtian-Late Campanian

145

TABLE 1 – *Continued*

Core Iden-tification	Nature of Chert	Nature of Sediments	Age of Sediments
170-5-CC	Quartzose chert	Limestone	Late Campanian
170-12-CC	Pieces of ?quartzose chert (cavings)	Zeolitic brown clay and nanno-fossil chalk	Late Turonian
170-13-CC	Quartzose chert	Foraminiferal-nannofossil chalk	Early Cenomanian
170-14-CC	Quartzose chert	Nannofossil chalk	Early Cenomanian
171-9-1 to 171-9-6 171-9-CC	Small fragments of quartzose chert	Nannofossil-foraminiferal ooze	Middle Eocene to Maastrichtian
171-10-4 171-10-5 171-10-CC	Quartzose chert	Nannofossil-foraminiferal ooze	Maastrichtian
171-11-4 171-11-5 171-11-CC	Quartzose chert	Foraminiferal-nannofossil ooze	Maastrichtian
171-12-CC	Quartzose chert	Foraminiferal-nannofossil ooze	Maastrichtian
171-13-4 171-13-5 171-13-CC	Quartzose chert	Foraminiferal-nannofossil ooze	Maastrichtian
171-14-CC	Quartzose chert	Foraminiferal-nannofossil ooze	Maastrichtian
171-15-1 to 171-15-6 171-15-CC	Quartzose chert	Foraminiferal-nannofossil chalk	Maastrichtian
171-16-3 to 171-16-6 171-16-CC	Quartzose chert	Foraminiferal-nannofossil chalk	Maastrichtian
171-17-2 to 171-17-6 171-17-CC	Quartzose chert	Foraminiferal-nannofossil chalk	Maastrichtian
171-18-CC	Quartzose chert	Foraminiferal-nannofossil chalk	Maastrichtian
171-21-2 171-21-CC	Quartzose chert	Foraminiferal-nannofossil chalk	Campanian

accumulations of palygorskite (Millot, 1964). Similar observations have been mentioned in previous Deep Sea Drilling Reports (Rex, 1970; von Rad and Rösch, 1972).

Composition

The composition of the chert has been studied using three different methods: thin sections examination, X-ray diffraction analysis, and scanning electron microscope observations. In most cases the three methods have been used on the same sample or portion of sample so that they provide complementary observations.

Terminology

Before describing the results, it is necessary to point out some problems of terminology and to define the meaning of the terms used in this paper.

Various terms are used in the literature to designate diagenetic concentrations of silica in sedimentary environments. Chert and flint usually refer to hard nodules, and sometimes beds, that are largely quartz and/or chalcedony and show a conchoidal fracture and a vitrous luster. These terms are often opposed to porcelanite (or porcellanite) which designates low density, more or less porous and dull-lustered varieties of chert made of opaline silica or cristobalite (Bramlette, 1946; Ernst and Calvert, 1969). Although the term porcelanite is confusing, as it is used quite extensively, at least in Europe, to designate contact metamorphosed marls, it is wisely accepted and shall be used in this paper.

The term chert shall be used in a very broad sense to designate any recrystallized form of silica and the adjectives "quartzose" and "porcelanitic" shall precise its mineralogical nature.

Still more important problems arise from the variety of terms used to designate the different polymorphs of silica that enter the composition of chert (see Table 2).

Nonterrigenous silica appears in deep-sea sediments under different states of crystallization, ranging from amorphous to well-crystallized.

Amorphous silica is present in the tests of radiolarian and diatoms and in siliceous sponge spicules it is characteristic of biologically precipitated silica. A typical X-ray diagram of radiolarian ooze shows only a very broad and flat bulge of amorphous scattering, extending from about 2.9 to 4.8 Å, without any identifiable diffraction peak (Figure 2). Very often in the literature such amorphous silica is referred to as opal or opaline silica, although it has been long recognized that opal is a poorly but definitely crystallized form of silica (Levin and Ott, 1932, 1933). Flörke (1955) defined its structure and showed that opal was made of disordered cristobalite in which the stacking of "low" cristobalite layers is periodically interstratified with "low" tridymite layers. This disordered cristobalite is also sometime called opal-cristobalite. Typical X-ray diffraction traces from disordered cristobalite appear in Figure 3.

The term chalcedony is widely used for designating both a fibrous impure variety of quartz and cryptocrystalline quartz. In the latter case it is often used in the adjective form: chalcedonic quartz. This appears logical if one considers the identical X-ray signature of these varieties. However, as there is a clear and easy distinction between these two varieties under the microscope, in this report the name chalcedony shall be restricted to the fibrous variety of quartz, and microcrystalline to cryptocrystalline quartz shall be referred to as quartz.

Mineralogy of the Chert

Previous reports, and particularly the most detailed ones (Heath and Moberly, 1971; von Rad and Rösch, 1972; Berger and von Rad, 1972), have shown that deep-sea chert can be divided into two categories, namely, (1) porcelanitic cherts, predominantly disordered cristobalite, and (2) quartzose cherts composed of quartz and/or chalcedony. The same two types have been observed at Leg 17 sites. Furthermore, there appears to be a consistent relationship between the mineralogical nature of the chert and the composition of the sediments in which it is found. As a rule, chert occurring in clayey layers (zeolitic clays,

TABLE 2
Different Terminologies Used in Chert Mineralogy

Terms Used in This Report	Equivalent Terms Used in the Literature	Mineralogical Structure	Occurrence in Sediments (nonterrigenous)
Amorphous Silica	Opal, opaline silica	SiO_4 tetrahedrous randomly oriented, high water content[a]	Tests of siliceous organism (radiolarians, diatoms, sponge spicules)
Disordered Cristobalite	Opal (Berger and von Rad, 1972)[b] Lussatite[c] (Berger and von Rad, 1972) Opal-cristobalite (Flörke, 1955) Cristobalite (DSDP X-ray reports)	Disordered stacking of "low"-cristobalite and "low"-tridymite layers[a]	porcelanitic chert; peripheral parts of quartzose chert nodules; partially silicified carbonate siliceous or shaly rocks
Chalcedony	Chalcedony	Optically fibrous quartz (slightly disordered quartz structure giving relatively broad quartz peaks on X-ray diagrams)	void filling and fracture filling in both quartzose and porcelanitic cherts
Quartz	Chalcedony[d] Quartz	Cryptocrystalline, microcrystalline, and coarsely crystalline quartz (broad to sharp quartz peaks on X-ray diagrams, depending on the size of the crystallites)	quartzose cherts, fracture filling in quartzose chert

[a]See X-ray diagrams in Figures 2 and 3.

[b]Opal described by these authors appears intermediate between amorphous silica and disordered cristobalite (highly disordered cristobalite).

[c]Lussatite was described by Mallard (1890) as a fibrous variety of cristobalite.

[d]See text for explanation.

clayey radiolarian oozes, marls, and marly limestones) is predominantly disordered cristobalite, while chert in carbonate sediments is predominantly quartz. Quartzose chert in carbonates shows a tendency to form nodules, while in clays and marls porcelanitic cherts often occur in thin beds or as a mere impregnation of the sediment.

Cherts in Clays

Site 164 provides an exceptional opportunity to study the nature of chert in clay deposits, as the entire section consists of deep-sea brown zeolitic clays with only variable amounts of Radiolarians in the upper part and very rare carbonates near the base.

X-ray analysis shows that, except for the extreme base of the column, chert at Site 164 has a remarkably constant composition in the entire section ranging in age from middle Tertiary to early Cretaceous. It consists of abundant disordered cristobalite and variable amounts of quartz. It is clearly seen in thin sections (Figure 4) that quartz and/or chalcedony are concentrated only inside former voids that are probably radiolarian molds and that the matrix is disordered cristobalite. Whenever the voids appear unfilled or absent or filled with disordered cristobalite in thin sections, no quartz peak is observed on the X-ray diagrams (Figure 5). In fact it has been possible to verify under the microscope that each increase in the quartz/disordered cristobalite ratio observed on the diffractograms corresponds to an increase in the abundance of chalcedony-filled radiolarian molds. At the base of the section two samples show a different composition. They are made exclusively of quartz and chalcedony (164-24-CC and 164-25-CC). Their origin is not clear as the samples come apparently from drill cuttings. It is remarkable however, that their occurrence seems to correspond to a general change in the lithology as carbonate was observed in smear slides made from the same

sample. The carbonate components consist of abundant calcareous nannofossil and some coarsely crystalline calcite (fracture filling?).

The boundary between chert and nonsilicified sediment is rather transitional in nature. In thin sections the transition is marked by the disappearance of chalcedony fillings inside the radiolarian molds and a gradual outward increase in clay minerals and zeolite contents. X-ray diagrams confirm these observations (Figure 5). It is generally in such transitional zones that abundant small silica spherules made of blade-shaped radiating crystals are observed on scanning electron micrographs (Figures 6 and 7). They occur as a lining on the inner parts of voids (radiolarian molds). Similar spherules have been described by Wise and Hsü (1971), Wise et al. (1972), von Rad and Rösch (1972), and Berger and von Rad (1972). Wise et al. (1972) identified these spherules as disordered cristobalite by running X-ray diffraction analyses on insoluble residues from chalks and chalk inclusions in cherty radiolarian oozes. Some very small spherules have a strongly different shape, and it is not known at present if they also consist of reprecipitated silica (Figure 8).

The observations from Site 164 have been confirmed by analysis of samples from clay-rich layers at other Leg 17 sites. There seems to be a consistent correlation between the porcelanitic nature of the chert and the presence of clay minerals in the host sediments. A survey of the nature of the chert described in previous DSDP reports leads to similar conclusions.

Chert in Carbonates

The composition of the chert recovered from carbonate sediments is remarkably different from that described above. Macroscopically, chert nodules have a massive, hard, and dense aspect. They show a vitrous luster and con-

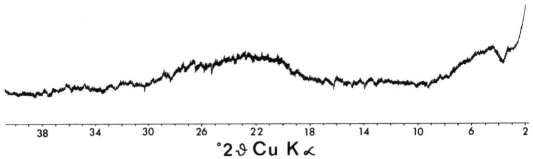

Figure 2. *X-ray diffraction pattern of a radiolarian ooze (116-16-2, 150 cm).*

choidal fracturing. Although the nodular form appears most common, thin layers and stringers were observed in the lower part of Hole 167.

X-ray analyses and thin section examination show that these cherts are composed almost exclusively of chalcedony and quartz. Disordered cristobalite is regularly observed only at the periphery of chert nodules and seems to be characteristic of the zone of transition from chert to the sedimentary matrix. X-ray diagrams indicate that quartz is predominantly "chalcedonic" in nature (either chalcedony or cryptocrystalline quartz) although very well-crystallized quartz is often found associated with fractures or geodes.

A typical aspect of these cherts in thin sections is that of a fine-grained mosaic of quartz with variable amounts of calcareous organism remains. These remains are usually well-preserved foraminifera that have had their chambers filled with chalcedony. In other cases they show evidence of dissolution and may even be completely replaced by silica (Figures 9 and 10). In one sample, small clusters of dark-brown needles are found associated with these chalcedony-filled foraminifera. Their definite mineralogical nature is unknown, and they are tentatively identified as rutile needles (Figure 11). In the lower part of Hole 167 (from Core 62 down) the carbonate lithification processes reach such an extent that sparry calcite has precipitated in some of the foraminifera chambers while others are filled with silica. Many of the foraminifera show cooccurence of sparry calcite and silica (Figures 12 and 13). As calcite filling occurs only at subbottom depths (860 meters in this case) where the sediment overburden is sufficient to produce extensive lithification of the limestone, while silica-filled foraminifera are found much higher in the sediment column, sparry calcite can be considered a late diagenesis product. It probably precipitated preferentially in foraminifera chambers that had been left unfilled during previous silica precipitation. In some cases of cooccurrence of silica and calcite, it seems that calcite has been growing from the foraminifera tests and replacing part of the silica (Figure 13), although the order of crystallization cannot be clearly established.

The contact between chert and carbonate sediment is characterized by a transitional zone in which disordered cristobalite is the main component of the matrix, while chalcedony and quartz are restricted to the foraminifera and/or radiolarian chambers. The boundary between quart-

zose chert and disordered cristobalite is generally sharp and outlined by a concentration of impurities visible in thin sections (Figure 14). Carbonate inclusions in chert nodules are very common and show the same transitional zone. In most cases they are relatively small and are completely cemented by disordered cristobalite. The outer limits of the disordered cristobalite zone toward the original carbonate sediment is rather diffuse. Foraminifera chambers are generally void or lined with disordered cristobalite spherules of the same type as those described in porcelanitic cherts. Scanning electron micrographs show that small spherules are also present within the predominantly carbonate matrix in the immediate vicinity of the transitional zone. They seem to be responsible for some cementation of the carbonate ooze around chert nodules (Figure 15).

Chert in other Types of Sediments

During Leg 17 chert has been rarely found in sediments other than clays, radiolarian-rich clay, and calcareous sediments. However, at Site 166 some rare nodules of porcelanitic chert occur in a radiolarian ooze (Core 12). They consist of disordered cristobalite. Although the radiolarian ooze appears devoid of clay minerals in smear slides, X-ray diagrams show that these are present in a noticeable amount. Therefore, these cherts seem to have the same environmental basis as those found in clays.

MECHANISM OF CHERT FORMATION

The observations reported above allow for the consideration of possible mechanisms leading to the formation of concentrations of recrystallized silica in lithologically different sediments. Again the duality observed in the mineralogy of chert shall be emphasized in order to estimate the role of the sedimentary environment in chert formation.

Origin of Silica

The origin of the silica found in chert has been often debated, and no clear answer to the problem has been provided. Generally two different origins are considered: (1) alteration of volcanic material (montmorillonite and volcanic glass) and (2) dissolution of tests of siliceous organisms (radiolarian, diatoms, and sponge spicules). A volcanic origin could be invoked to explain both the absence of radiolarian remains in some silicified claystones

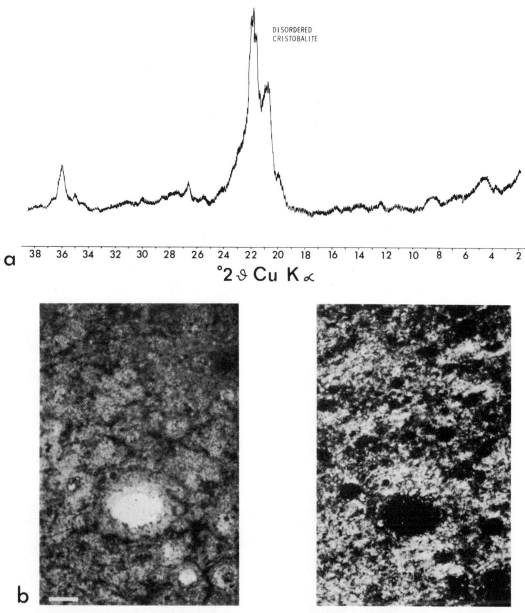

Figure 3. (a) *X-ray diffraction pattern of a porcelanitic chert devoid of quartz (164-14-1-20, 25 cm),* (b) *Aspect in thin section: most radiolarian molds have been filled with sediments and are replaced by disordered cristobalite. A large one remained unfilled and is devoid of silica. (left: plane polarized light; right: crossed nicols; white bar = 100μ).*

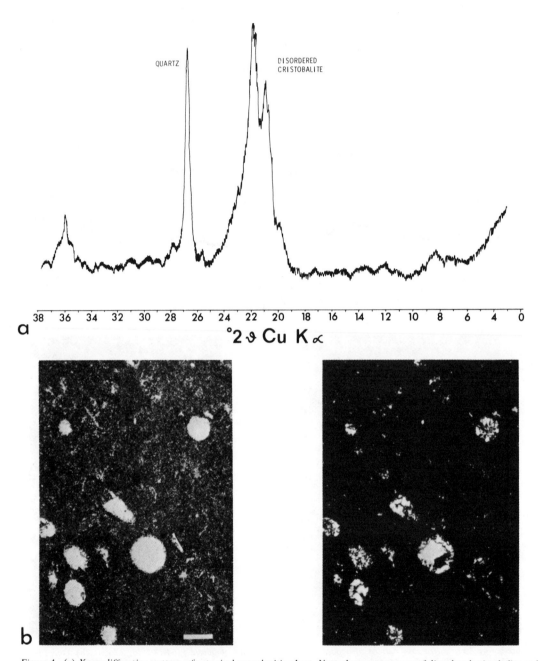

Figure 4. (a) *X-ray diffraction pattern of a typical porcelanitic chert. Note the cooccurrence of disordered cristobalite and quartz.(164-13, CC),* (b) *Aspect in thin section: chalcedony is restricted to radiolarian molds. (left: plane polarized light; right: crossed nicols; white bar = 100μ).*

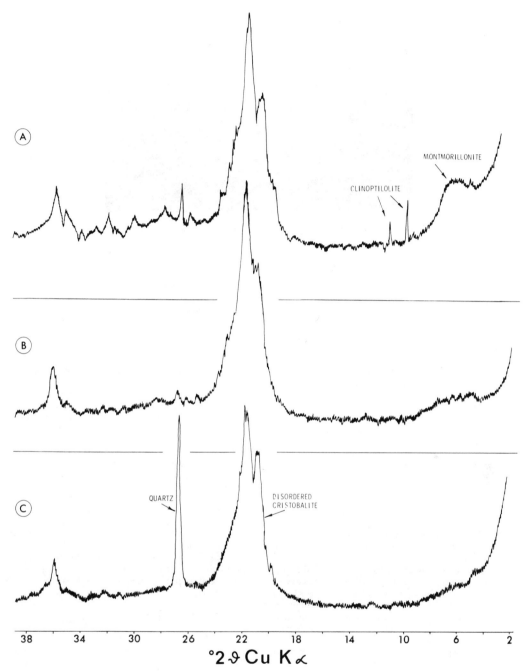

Figure 5. *X-ray diffraction analysis of the transition between a porcelanitic chert nodule and the surrounding partly silicified sediment (164-4, CC). (a) Outer zone: mainly disordered cristobalite with some remnants of sediment (montmorillonite and clinoptilolite; rare quartz is here possibly detrital), (b) Intermediate zone: disordered cristobalite with rare clay minerals, (c) Inner part of nodule: disordered cristobalite and quartz (as in Figure 4).*

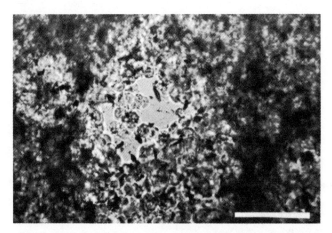

Figure 6. *Silica spherules (?disordered cristobalite) in a radiolarian mold.*
Thin section, plane polarized light; white bar = 100μ (164-14, CC).

and the occurrence of well-preserved radiolarians in radio-larian mudstones. In the first case, the absence of remains would indicate that radiolarians were never present in the sediments, while in the second, the preservation of radiolarians would rule out the possibility that they dissolve and supply the necessary silica. At any rate it is clear that the density of chert nodules is such that their formation implies considerable concentration of silica. Obviously, as already noted by Heath and Moberly (1971), some circulation of silica-rich pore waters is required in the sediment over a range of centimeters to meters. In some cases the radiolarians might have been entirely dissolved, while in others they simply have been impregnated by silica-rich solution coming from nearby layers.

Observation of the sediments recovered from the vicinity of the youngest cherts at Site 167 strongly suggests that dissolution of radiolarians tests is the most probable source of silica. An interval of about 60 meters of radiolarian-nannofossil chalk is present above the chert-bearing layers (Cores 24-32). The Radiolarians in these late Eocene sediments are thick-walled and well-preserved. In Core 32, starting at about 1 meter above the chert, radiolarians show traces of dissolution which are observed in smear slides and in scanning electron micrographs (Figure 16). The same observation applies to sediments recovered between the chert nodules.

Radiolarians, diatoms, and spicules seem to be the most abundant amorphous silica in the sediments. Studies of solubility of silica in water show that this form of silica is by far the most soluble. Of course, in the vicinity of continents and under some particular climatic conditions, it is possible that large amounts of dissolved silica could have been delivered to the oceans. Nevertheless, as discussed later in this chapter, it appears that any primary input of silica in the water column is immediately used by siliceous organisms, and it is clear that large quantities of amorphous silica of biogenic origin are available in many deep-sea sediments. The solubility of amorphous and crystalline silica in water has been studied experimentally by Wey and

Siffert (1961), and the results show (Figure 17) that at room temperature the solubility of amorphous silica reaches 120 to 140 ppm while that of opal (disordered cristobalite), cristobalite, and quartz are very close together and much lower (6 to 15 ppm). Krauskopf (1956, 1959) has shown that at lower temperatures solubility of amorphous silica decreases slightly (50 to 80 ppm at 0°) and still remains largely above that of crystalline forms. He also demonstrated that the influence of pH was negligible below a value of around 9 which can be considered as rather exceptional in deep-sea sediments.

The silica content of sea water is extremely low (around 3 ppm), but interstitial waters in the sediments show much higher values. They range from around 10 ppm in sediments devoid of siliceous organisms and silicates to 40-50 ppm in radiolarian oozes, except in the upper few meters of sediments where the silica content decreases abruptly upward in the section and reaches sea-water values close to the sea bottom (Manheim et al., 1970; Manheim and Sayles, 1971a, 1971b; Sayles and Manheim, 1971; Sayles et al., 1972). If those values are compared to the solubility curves, it is clear that interstitial waters are generally undersaturated with respect to amorphous silica and supersaturated with respect to the crystalline forms. This implies that amorphous silica can be selectively dissolved and repre-cipitated as any of the crystalline forms. As will be mentioned later, it is probable that other components of the mineral phase play an important role in the control of this transfer.

Recrystallization Processes

Environmental Conditions Versus Aging Effect

It is, of course, tempting to interpret the absence of chert in recent sediments as the result of the extreme slowness of the recrystallization of silica, but several observations lend support to the assumption that chert is a product of early diagenesis. Evidence from DSDP samples show that chert was formed in many cases prior to

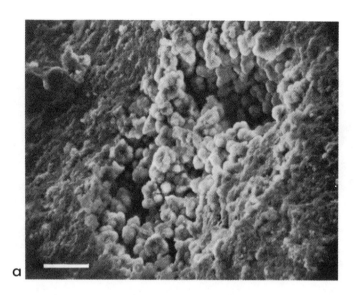

Figure 7. *Scanning electron micrographs of silica spherules (?disordered cristobalite) in radiolarian molds.* (a)*164-4, CC; white bar = 50μ.* (b) *Detail view of similar spherules (167-33-1, 104-107 cm); white bar = 1μ.*

153

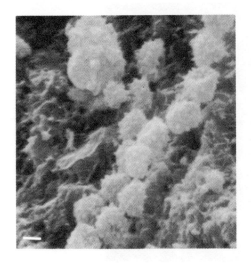

Figure 8. *Scanning electron micrographs of ?-silica spherules (164-6, CC) (cf. Fig. 7). White bar = 1μ.*

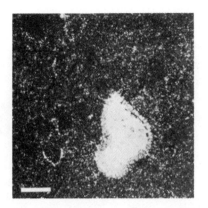
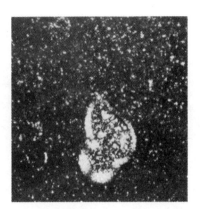

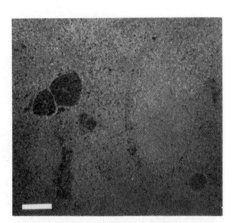
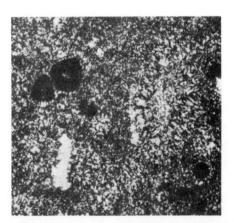

Figure 9. *Quartzose chert in calcareous sediments (thin sections).* (a) *Well-preserved foraminifera (calcitic test) filled with quartz or chalcedony. Matrix is microcrystalline quartz (167-133-1,122-126 cm). (Left: plane polarized light; right: crossed nicols; white bar = 100μ.),* (b) *Large microfossil in right part of the picture has been entirely replaced by quartz while others, probably previously filled with clayey sediment, show cristobalitic filling. Near the lower-left corner a large calcite fragment is strongly etched and partially replaced by quartz. This sample came from a turbiditic section and shows a mixture of displaced elements (165A-16, CC). (Left: plane polarized light; right: crossed nicols; white bar = 100μ).*

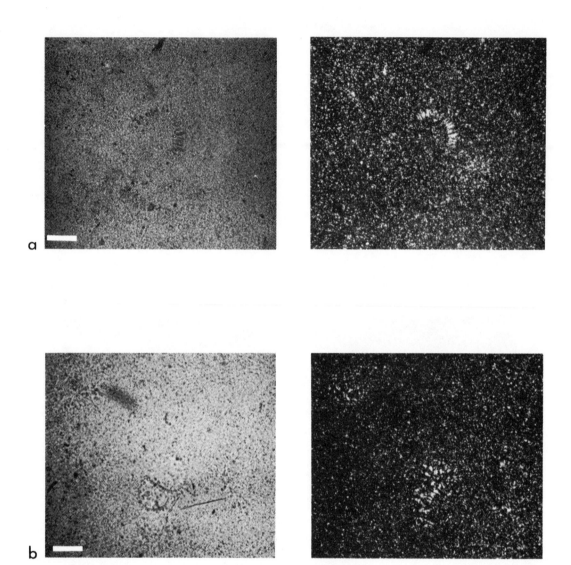

Figure 10. *Quartzose chert in calcareous sediments (thin sections).* (a) *Some remnants of a calcitic foraminifera test still visible in center of picture. Matrix is microcrystalline quartz (167-33-1, 143-145 cm: (left: plane polarized light; right: crossed nicols; white bar = 100μ). (b) Completely replaced foraminifera. Detailed structure is still barely visible in plane polarized light picture (left). Only coarser crystalline quartz indicates the presence of foraminifera ghosts in crossed nicols picture (right)(167-37, CC; white bar = 100μ).*

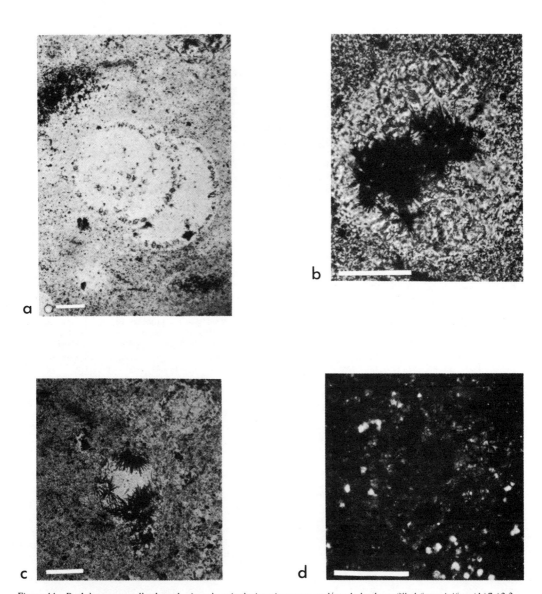

Figure 11. *Dark-brown, needle-shaped minerals as inclusions in quartz and/or chalcedony-filled foraminifera (167-62-2, 138-140 cm; [a], [b], and [c]: plane polarized light, [d] crossed nicols with condensed light, white bars = 100μ). Note that in (a) the rare small clusters seem to originate preferentially on the calcitic remnants of the foraminifera.*

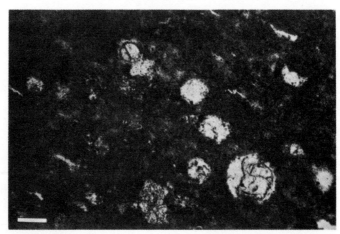

Figure 12. *Partial calcite filling (coarsely crystalline sparry calcite) in microfossils (?radiolarian molds). The larger mold is filled with both silica (on the left side) and calcite (rest of chamber). Others are filled with either silica (center of picture) or calcite (uppermost and lowermost molds). (167-61-2, 43-47 cm; thin section; crossed nicols; white bar = 100μ).*

compaction of the sediments. Figure 18 shows that small quartz veinlets in limestones from Site 167 were already hard when the compaction took place. In some cases they resisted compaction, and in others they were broken by the compaction effects and partially stacked in an "en accord-éon" manner. Furthermore, field evidence has been provided by Bernoulli (1972) that suggests very early diagenetic chert formation in upper Cretaceous limestone from Greece. Plates of broken chert occur in a slumped bed, while the overlying layers are undisturbed (Figure 19). The sharpness of the edges of the broken chert fragments suggests an early diagenetic solidification rather than a post-deformational selective silicification.

Several authors have proposed that chert formation undergoes several phases beginning with precipitation of cristobalite and slowly "maturating" into quartz (Heath and Moberly, 1971; Berger and von Rad, 1972). Although a very slow inversion from disordered cristobalite to quartz cannot be completely ruled out, some evidence found during the study of Leg 17 chert shows that it is probably only of secondary importance compared to the influence of the lithology of matrix sediments on the mineralogy of the chert. It should be pointed out that if the relationship between carbonate and quartzose chert, as well as between clayey or marly sediments and porcelanitic chert, as described here is valid, it is quite obvious that there will be statistically more quartz in the old sediments than in the more recent ones simply because most of the older sediments are rich in carbonates. This is a consequence of the sea-floor spreading mechanism and associated subsidence which brings carbonates deposited on the midoceanic ridge flanks to deeper parts of the basins where they receive a cover of noncalcareous sediments. Most of the chert-bearing carbonate layers lie presently in deep water below the carbonate compensation depth, and the majority of the

calcareous deposits presently above that level are of post-Eocene age and are, therefore, generally devoid of chert. Whenever chert is reported in these younger sediments, the patterns described earlier in this report can be observed, and the lithology of the host sediments seems to provide the main control upon its mineralogical nature. For example, at Site 76 from Leg 9 (Hays et al., 1972) chalcedonic chert occurs in calcareous sediments possibly as young as Pliocene. At Site 164 porcelanitic chert present in clays is as old as lower Cretaceous. Also the occurrence of chalcedony in the radiolarian molds of the Oligocene cherts of this same site indicates that quartz or chalcedony can primarily precipitate even in youngest chert, given the proper environmental conditions. Similar deduction can be made from Sites 62 and 63 (Winterer et al., 1971). At Site 62 Oligo-Miocene chert associated with calcareous sediments is made of chalcedony, and at Site 63 Tertiary calcareous oozes contain quartzose chert, while it is cristobalitic in the marls of the older parts of the section.

Role of Foreign Cations

The possible influence of cations able to cause disorder in the crystalline structure of silica during the growth of crystals has been summarized very well by Millot (1964). Analysis of natural opals (made of disordered cristobalite) has shown that they contain regularly an appreciable amount of alkaline and alkaline-earth elements (Buerger, 1954). Weil (1926), Flörke (1955), and Millot (1960, 1964) indicate that the disorder in the structure of cristobalite-tridymite layers found in disordered cristobalite is open enough to accommodate large cations, while the electrostatic balance is obtained by Al ions replacing Si in tetrahedrons. Millot (1964) points out that if no foreign cations are available, only quartz can precipitate, and

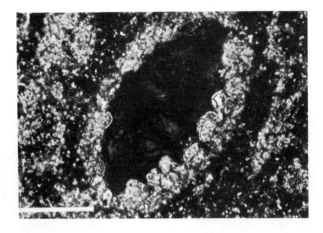

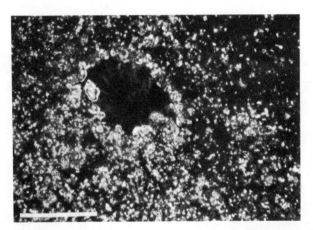

Figure 13. *Possible secondary calcite recrystallization in silica-filled microfossil molds. Calcite crystals seem to grow from the walls toward the inside and may partially replace chalcedony filling (167-75, CC). Compare to the aspect of partially replaced foraminifera from Figures 9 and 10 (thin section; crossed nicols; white bar = 100μ).*

conversely only disordered cristobalite can precipitate if these cations are present in sufficient amounts.

Of course, clay minerals are the most likely source for easily exchangeable cations. It is thus proposed here that they are responsible for the diagenetic precipitation of disordered cristobalite in clayey sediments, while only quartz can precipitate in a carbonate environment.

Role of Permeability

Disordered cristobalite is confined to sediments in which clay minerals represent a major obstacle to circulation and create a microenvironment in which pore spaces are small enough to keep the ratio of silica to metallic cations at relatively low values. Whenever the permeability increases, such as in a large void, chalcedony and quartz can precipitate because of a sharp increase of this ratio. This is well illustrated in Figure 20, where a radiolarian mold shows geopetal partial filling with sediment. The part filled with clayey sediments has been silicified by disordered cristobalite, while the void in the upper part has been filled with chalcedony.

In carbonates, both the lack of aluminum and metallic cations (except Ca and Na) and the high permeability favor

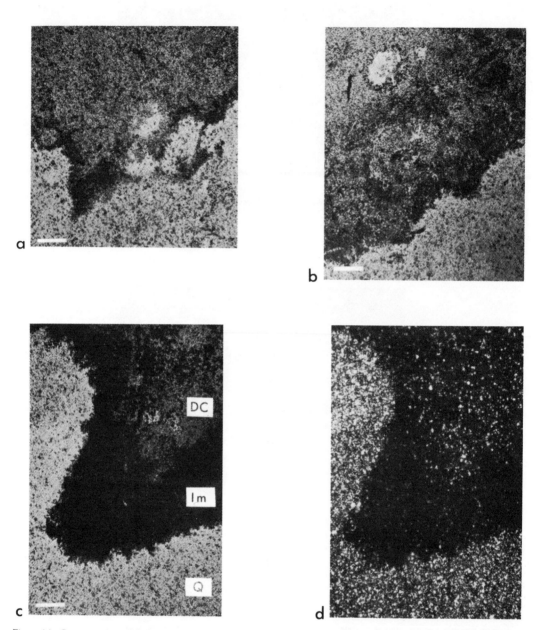

Figure 14. *Concentration of dark impurities at boundary between quartz and disordered cristobalite (thin sections).* (a) and (b): *167-33-1, 121-122 cm; quartz is in lower part, disordered cristobalite in upper part (plane polarized light; white bar = 100μ).* (c) and (d): *167-35, CC; Q = quartz, DC = disordered cristobalite, Im = impurities (*[c]: *plane polarized light;* [d]: *crossed nicols; white bar = 100μ).*

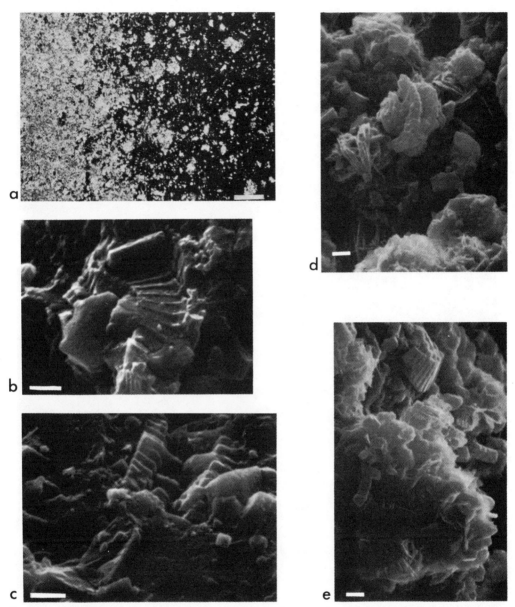

Figure 15. *Boundary between opaline rim and calcareous sediment at the periphery of quartzose chert nodules.*
(a): *165A-15, CC; thin section; crossed nicols; white bar = 100μ; calcite is at left, disordered cristobalite at right.*
(b), (c), (d), *and* (e); *167-33-1, 104-107 cm; scanning electron micrographs showing partial cementation of the*
calcareous elements by silica. Silica is either cryptocrystalline and structureless (b) *and* (c) *or in the spherulitic*
form (d) *and* (e), *white bar = 1μ.*

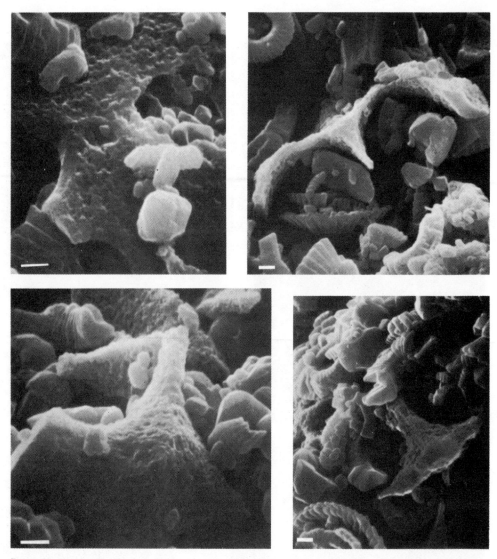

Figure 16. *Dissolution of radiolarian fragments in the vicinity of chert (167-32-3, 150 cm). Scanning electron micrographs; white bar = 1μ. Note the strongly etched surface of the fragments and their partial dissolution.*

direct precipitation of microcrystalline quartz. A good example of such a process can be observed at Site 165-16, CC where relatively coarsely crystalline quartz has precipitated in a coarse-grained limestone consisting of displaced shallow-water elements (Figure 21). The role of permeability is also emphasized by the occurrence of small silica spherules precipitating directly in the carbonate matrix (Figure 15), while in the clays they appear restricted only to the radiolarian molds. The zonation observed at the periphery of nodules in carbonates can be the result of the relatively high permeability of the sediment. It is

suggested that the development of a quartzose chert nodule in a calcareous ooze can be explained in the following way:

1) Quartz precipitates directly in the carbonate matrix where clay minerals are rare and dispersed.

2) The nodule develops outwards by accretion, and in the process all the clay minerals and dissolved cations that cannot be accommodated in the quartz structure are excluded and move along a "quartzification front." This can be achieved only because of the high permeability of the sediment, allowing relatively free circulation of the interstitial waters.

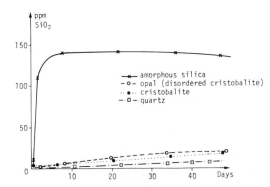

Figure 17. *Solubility of amorphous silica, opal, cristobalite, and quartz in water at room temperature (after Wey and Siffert, 1961).*

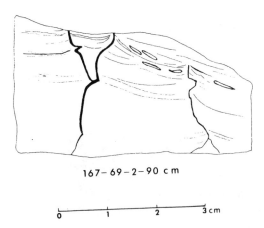

167– 69– 2– 90 c m

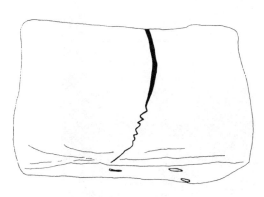

167– 69– 2– 105 cm

Figure 18. *Compaction effect on quartz veinlets. Upper: note the compaction of sediments around the veinlets which were already hard at the time of compaction. Lower: the hard quartz veinlet has been broken and folded during compaction.*

3) At the periphery of the quartz nodule the concentration of clay minerals and foreign cations increases, while that of dissolved silica decreases because of limited supply. These conditions favor precipitation of disordered cristobalite that makes the rim commonly observed around the nodule.

This concentration process can be visualized in thin sections by observing the distribution of dark impurities near the contact between quartzose chert and cristobalitic rim.

Role of Crystal Nucleation

In many samples where permeability appears high, it seems that crystal growth is favored by the presence of crystalline nuclei. This might explain the ease with which quartz seems to spread in a carbonate ooze.

Such nucleation processes that seem to characterize the most permeable environments can be observed in thin sections from brecciated cherts from Site 167 and from a "septarian nodule" from Site 166. Large quartz crystals and chalcedony seem to develop from a generation of small crystals along the edges of chert fragments (Figure 22).

Role of Time

Concepts which require that the mineralogical nature of chert is controlled by age-related "maturation" processes are not supported by the observations described here. However, it is possible that time plays a subsidiary role in the quartzification of the cherts. This process involving conversion from metastable forms of silica (disordered cristobalite) to most stable forms (quartz) could probably be favored by high temperatures found under great sediment overburden (see discussion in Heath and Moberly, 1971). The role of silica solution in that case is emphasized by the observations made on the brecciated chert from Site 167 (Core 63-1, 60-67 cm) (see color frontispiece of this volume). Brecciation is believed to have occurred after some lithification of the sediment, since no sediment filling has been observed in the fractures. Furthermore, the precipitation of large sparry calcite crystals in these fractures certainly results from late diagenetic processes since (1) it correlates with the youngest occurrence of sparry calcite inside foraminifera chambers at the same site (167, Core 62); (2) in some other samples fractures previously filled with chalcedony have been broken, and calcite has precipitated in the voids associated with this second generation of fractures (Figure 23); and (3) oxygen isotope analyses performed on the calcite crystals cementing the brecciated chert (courtesy of R. Letolle, University of Paris-VI, France) indicate a temperature of crystallization of about 30°, which can be reasonably estimated as indicative of the present in situ temperature under 870 meters of overburden, as comparable thermal gradients have been observed in sediments during Leg 19 of the Deep Sea Drilling Project (G. Bryan, personal communication). It is then remarkable to observe that some of the chalcedony has precipitated after calcite. This is suggested by the pattern of chalcedony fibers that seem to mold the large calcite crystal edges (Figure 24). It indicates that solution and reprecipitation of silica can occur very late in already lithified sediments. It is possible that, given the

Figure 19. *Broken chert layer in slumped upper Cretaceous limestones from Vigla, Western Greece (from Bernoulli, 1972). See text for explanation.*

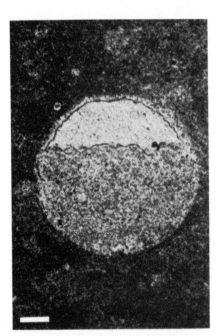

Figure 20. *Differential silica recrystallization in a partially filled radiolarian mold (thin section; 164-4, CC, left: plane polarized light; right: crossed nicols; white bar = 100μ). The partial filling gives a good geopetal indication as the upper surface is parallel to the bedding observed in the sample. The sediment-filled part has been replaced by disordered cristobalite and only the upper part, devoid of sediment, has been filled with chalcedony.*

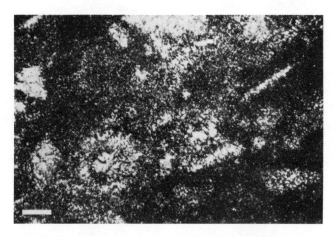

Figure 21. *Etching and replacement of large calcite fragments by quartz in a highly permeable calcareous sediment (base of a turbidite sequence, 165A-16, CC; thin section; crossed nicols; white bars = 100μ).*

necessary permeability, all the silica could be able to slowly invert to quartz by solution and reprecipitation.

The maturation ("quartzification") of chert that can be observed on chert nodules from outcrops on land has been often attributed to age effects. It is noteworthy that no cristobalitic rim can be observed on the quartzose chert nodules from the Upper Cretaceous chalks of the Paris Basin (F. Mélières, personal communication). If a rim of light-colored material is indeed present, it is exclusively made of microcrystalline quartz. As cherts from the same age recovered by deep sea drilling do not show this complete quartzification, it is believed that the age effect was not predominant and the transformation of cristobalite into quartz on land might have been caused by weathering after the sediments had been exposed subaerially.

Possible Primary Causes of Chert Formation

Although it is possible to imagine some of the processes that might have led to the recrystallization of silica in the sediments, knowing what caused this recrystallization is still an open question.

The most puzzling observation, leading to many controversies, is that apparently no chert is being formed in recent sediments. This observation has been accounted for in two different ways: (1) chert is not presently forming because environmental conditions are now different from the ones existing during early Tertiary time; and (2) formation of chert is such a slow process that it has not had enough time to form since the Oligocene. For reasons mentioned earlier, it seems that the former explanation is the most plausible.

165

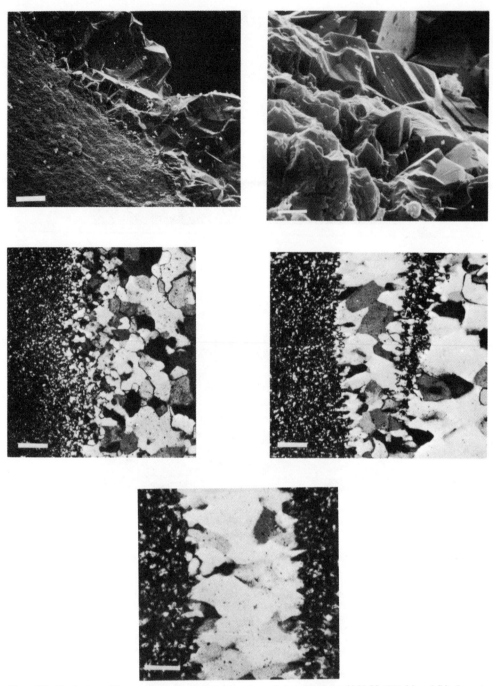

Figure 22. *Nucleation of large quartz crystals from the edges of chert fragments (166-28, CC).* (a) *and* (b): *Scanning electron micrographs; white bar = 100µ in part* (a), *= 1µ in part* (b). (c), (d), *and* (e): *Thin sections; crossed nicols; white bars = 100µ. Note the continuity between some quartz crystals in the veins and those in the chert fragments.*

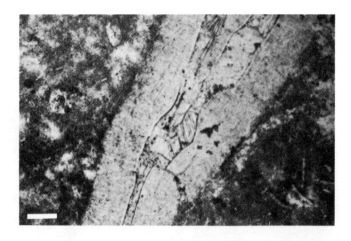

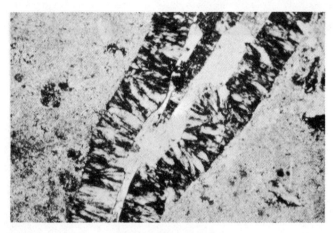

Figure 23. *Broken chalcedonic double veinlets showing post-fracturing precipitation of sparry calcite cement (especially in lower center part of the picture); 167-69-1, 58-61 cm, thin section; upper: plane polarized light; lower: crossed nicols; white bar = 100μ).*

The mechanism of silica transfer from amorphous siliceous tests to crystalline forms by solution and reprecipitation according to a differential solubility pattern is theoretically possible in recent sediments. The dissolved silica content of the interstitial waters is such that, apart from the few upper meters of sediments, where it is close to that of sea water, the transformation should indeed occur. However, it does not, at least on a scale large enough to produce appreciable concentrations of recrystallized silica. It is possible that the main control of the process lies in the nature of the silicates, and especially the alumino-silicates, present in the sediments. Such an explanation is suggested by the fact that, as far as can be seen from the rather scanty analyses of chert-bearing sediments recovered from the deep basins, the chert seems to be generally accompanied by an association of silicon-rich silicates which are not prevalent in the recent sediments. Some examples are given by the composition of sediments at Site 164 where it is clear that chert is mostly restricted to layers where clinoptilolite predominates over phillipsite, K-feldspars over plagioclases, and palygorskite over other clay minerals. These minerals are indicative of a silica-rich environment. Although not much is known of the possible diagenetic processes that could lead to transformation of phillipsite into clinoptilolite and montmorillonite into palygorskite in a silica-rich environment, it is unlikely that K-feldspar could be derived from alteration of plagioclase. It is more probable that these relatively Si-rich minerals are merely

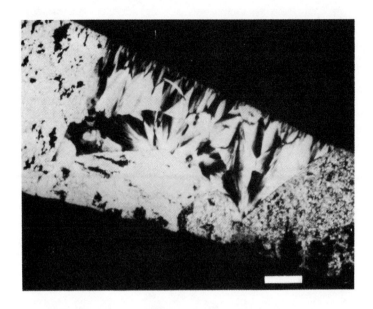

Figure 24. *Late precipitation of chalcedony (upper) and quartz (lower) as filling of voids left open by precipitation of sparry calcite in the fractures of a brecciated chert. The pattern of chalcedony fibers in the upper picture indicates that silica precipitated after calcite (167-63-1, 64 cm; thin sections; crossed nicols; white bars = 100μ). (See also color frontispiece, this volume.)*

indicative of an originally different oceanic environment that was more silicic than today. The origin of silica could be found in a different type of volcanism, as suggested, for example, for certain areas of the Indian Ocean by Venkatarathnam and Biscaye (in press), or by a large output of dissolved silica from the continents under particular climatic conditions as suggested by Millot (1964) for the peri-African realm during Eocene time.

Broecker (1971) has shown that the main control of the silica content of sea water is achieved by siliceous organisms. Any influx of dissolved silica in the water is immediately compensated by biogenic precipitation of amorphous silica in the form of radiolarians, diatoms, and sponges. A large part of that silica is rapidly dissolved and permanently recycled in the water column provided the water circulation is able to recycle enough nutrients to keep biologic productivity at a high level. The remaining silica is sedimented and buried. In the sediments, the siliceous tests dissolve slowly and release monomeric silica into the interstitial waters. The excess silica is trapped in the sediments and is available for subsequent diagenetic processes. It is possible that in post-early Tertiary sediments, Si-poor silicates could buffer this excess silica by the processes determined experimentally by MacKenzie et. al. (1967) so that the silica content of interstitial waters would reflect equilibrium with the alumino-silicates present in the sediments. If, during the early Tertiary and before, most of the alumino-silicates sedimented together with siliceous organisms were already Si-saturated, the excess of silica may have been available for direct reprecipitation into chert.

ACKNOWLEDGMENTS

This study has greatly benefitted from the cooperation of Leg 17 scientific and technical teams aboard the *Glomar Challenger* and especially from helpful discussions with J. I. Ewing, E. L. Winterer, S. O. Schlanger, and R. M. Moberly both during and after the cruise. Most of the scanning electron observations were performed at Scripps Institution of Oceanography, and I thank the direction of the Deep Sea Drilling Project for providing the necessary funds and E. Flentye for her helpful assistance. The major part of the mineralogical analyses has been done at the University of Paris-VI (Laboratoire de Geologie Dynamique and Laboratoire de Mineralogie), and I am greatly indebted to Professors L. Glangeaud and J. Wyart and to W. Nesteroff, F. Mélières, R. Letolle, and M. C. Sichère for their cooperation. Financial support during the operations at sea and during the analytical part of the work performed in France has been provided by CNEXO (Centre National pour l'Exploitation des Océans, France). This support and especially the help of J. Debyser has been greatly appreciated. Support was also provided at Lamont-Doherty Geological Observatory by the National Science Foundation under grant GA27281. J. I. Ewing, K. Venkatarathnam, and G. Carpenter made helpful comments and suggestions while critically reviewing the manuscript.

REFERENCES

Berger, W. H. and Von Rad, U., 1972. Cretaceous and cenozoic sediments from the Atlantic Ocean: Initial Reports of the Deep Sea Drilling Project, Volume XIV. Washington (U. S. Government Printing Office), p. 787-954.

Bernoulli, D., 1972. North Atlantic and Mediterranean Mesozoic facies: A comparison: Initial Reports of the Deep Sea Drilling Project, Volume XI. Washington (U. S. Government Printing Office), p. 801-871.

Bramlette, M. N., 1946. The Monterey formation of California and the origin of its siliceous rocks: U. S. Geol. Surv. Profess. Paper, 212, 57 p.

Broecker, W. S., 1971. A kinetic model for the chemical composition of sea water: Quat. Res., V. 1, p. 188-207.

Buerger, M. F., 1954. The stuffed derivatives of the silica structures: Am. Minerol., V. 39, p. 600-614.

Ernst, W. G. and Calvert, S. E., 1969. An experimental study of the recrystallization of porcelanite and its bearing on the origin of some bedded cherts: Am. J. Sci., V. 267-A, p. 114.

Flörke, O. W., 1955. Zur Frage des "Hoch"–Cristobalit in Opalen, Bentoniten und Glasern: Neues Jb. Mineral. Monatsh., V. 10, p. 217-223.

Hays, J. D., Cook, H. E., Jenkyns, D. G., Cook, F. M., Fullen, J. T., Goll, R. M., Milow, E. D., and Orr, W. N., 1972. Initial Reports of the Deep Sea Drilling Project, Volume IX. Washington (U. S. Government Printing Office), 1205 p.

Heath, G. R. and Moberly, R., 1971. Cherts from the Western Pacific, Leg 7, Deep Sea Drilling Project: Initial Reports of the Deep Sea Drilling Project, Vol. VII. Washington (U. S. Government Printing Office), p. 991-1007.

Krauskopf, K. B., 1956. Dissolution and precipitation of silica at low temperatures: Geochim. Cosmochim. Acta, V. 10, p. 1-27.

————, 1959. The geochemistry of silica in sedimentary environments: Econ. Paleontol. Mineral. Soc. Spec. Publ., V. 7, p. 4-19.

Levin, I. and Ott, E., 1932. The crystallinity of opals and the existence of high temperature cristobalite at room temperature. Am. Chem. J., V. 54, p. 828-829.

————, 1933. X-ray study of opals, silica glass and silica gel. Z. Krist., V. 81, p. 305-318.

MacKenzie, F. T., Garrels, R. M., Bricker, O. P., and Buckley, F., 1967. Silica in sea water: Control by silica minerals: Science, V. 155, p. 1404-1405.

Mallard, E., 1890. Sur la lussatite, nouvelle variété minérale cristallisée de silice. Soc. Fr. Mineral. Bull., V. 13, p. 63-66.

Manheim, F. T., Chan, K. M., and Sayles F. L., 1970. Interstitial water studies on small core samples, Deep Sea Drilling Project, Leg 5: Initial Reports of the Deep Sea Drilling Project, Volume V. Washington (U. S. Government Printing Office), p. 501-511.

Manheim, F. T. and Sayles, F. L., 1971a. Interstitial water studies on small core samples, Deep Sea Drilling Project, Leg 6: Initial Reports of the Deep Sea Drilling Project, Volume VI. Washington (U. S. Government Printing Office), p. 811-821.

————, 1971b. Interstitial water studies on small core samples, Deep Sea Drilling Project, Leg 8: Initial Reports of the Deep Sea Drilling Project, Volume VIII. Washington (U. S. Government Printing Office), p. 857-869.

McManus, D. A., Burns, R. E., Weser, O., Vallier, T., Von der Borch, C. V., Olsson, R. K., Goll, R. M., and Milow, E. D., 1970. Initial Reports of the Deep Sea Drilling

Project, Volume V. Washington (U. S. Government Printing Office) 827 p.

Millot, G., 1960. Silice, silex, silicifications et croissance des cristaux: Serv. Carte Geol. Abstr. Lorr. Bull., V. 13, p. 129-146.

⸺, 1964. Géologie des argiles, Paris (Masson et Cie), 499 p.

Rex, R. W., 1970. X-ray mineralogy studies—Leg 2: Initial Reports of the Deep Sea Drilling Project, Volume II. Washington (U. S. Government Printing Office), p. 329-346.

Sayles, F. L. and Manheim, F. T., 1971. Interstitial water studies on small core samples, Deep Sea Drilling Project, Leg 7: Initial Reports of the Deep Sea Drilling Project, Volume VII. Washington (U. S. Government Printing Office), p. 871-881.

Sayles, F. L., Manheim, F. T., and Waterman, L. S., 1972. Interstitial water studies on small core samples, Leg 9, Deep Sea Drilling Project: Initial Reports of the Deep Sea Drilling Project, Volume IX. Washington (U. S. Government Printing Office), p. 845-855.

Tracey, J. I., Jr., Sutton, G. H., Nesteroff, W. D., Galehouse, J., Von der Borch, C., Moore, T., Lipps, J., Bilal Ul Haq, U. Z., and Beckmann, J. P., 1971. Initial Reports of the Deep Sea Drilling Project, Volume VIII. Washington (U. S. Government Printing Office) 1037 p.

Venkatarathnam, K. and Biscaye, P. E. Deep Sea Zeolites: variations in space and time in the sediments of the Indian Ocean: Mar. Geol. (in press).

von Rad, U. and Rösch, H., 1972. Mineralogy and origin of clay minerals, silica and authigenic silicates in Leg 14 sediments: Initial Reports of the Deep Sea Drilling Project, Volume XIV. Washington (U. S. Government Printing Office), p. 727-751.

Weil, R., 1926. Influence des impuretés sur la température de transformation paramorphique de la cristobalite. Compt. Rend., V. 183, p. 753-755.

Wey, R. and Sifferi, B., 1961. Réactions de la silice monomoléculaire en solution avec les ions Al^{3+} et Mg^{2+}. In Genèse et Synthèse des argiles. Coll. Intern. C.N.R.S., V. 105, p. 11-23.

Winterer, E. L., Riedel, W. R., Bronnimann, P., Gealy, E. L., Heath, G. R., Kroenke, L., Martini, E., Moberly, R., Resig, J., and Worsley, T., 1971. Initial Reports of the Deep Sea Drilling Project, Volume VII. Washington (U. S. Government Printing Office) 1757 p.

Wise, S. W. and Hsu, K. F., 1971. Genesis and lithification of a Deep Sea chalk: Ecol. Geol. Helv., V. 64, p. 273-278.

Wise, S. W., Bute, B. F., and Weaver, F. M., 1972: Chemically precipitated sedimentary cristobalite and the origin of chert. Ecol. Geol. Helv., V. 65, p. 157-163.

170

8

Reprinted from *Soc. Géol. France Bull.* **16**(5):498–508 (1974)

Nature, importance relative et place dans la diagenèse des phases de silice présentes dans les silicifications de craies du Bassin océanique de Madagascar (Océan Indien) et du Bassin de Paris

par François FRŒHLICH *

Sommaire. — Les analyses minéralogiques et pétrographiques effectuées sur des échantillons de craies silicifiées de l'Océan Indien et du Bassin de Paris ont permis de préciser la structure des sphérules de cristobalite-tridymite, ainsi que la forme de trichites sous laquelle le quartz se trouve. Les proportions relatives de ces trois phases de silice et des alumino-silicates (argiles et clinoptilolite) varient en fonction de l'importance de la silicification. La cristobalite et la tridymite y jouent un rôle prépondérant, l'importance de la silicification étant liée à leur abondance. La cristallisation de ces deux minéraux est précoce, probablement contemporaine de la sédimentation. Elle est suivie par une silicification secondaire diagénétique, plus tardive, avec la formation de trichites de quartz dans les craies peu silicifiées et de calcédoine dans les cherts.

Abstract. — The mineralogical and petrographic analyses carried out on samples of silicified chalk taken from the Indian Ocean and the Basin of Paris have made it possible to determine the structure of cristobalite-tridymite spherules and quartz whiskers (« trichites »). The ratio of the three silica phases and of alumino-silicates (clay minerals and clinoptilolite) vary according to silicification intensity. Cristobalite and tridymite play there a leading part, the importance of the silicification being linked to their abundance. The cristallisation of these two minerals is early, probably as early as sedimentation. It is followed by a later secondary diagenetic silicification with the formation of quartz whiskers in slightly silicified chalk and of chalcedony in cherts.

INTRODUCTION.

Objet et limites.

A la suite des campagnes de forage du « *Glomar Challenger* » dans le cadre du Deep Sea Drilling Project, l'importance des silicifications dans les sédiments océaniques et plus particulièrement dans les craies a été mis en évidence [voir par exemple Calvert, 1971]. Des silicifications comparables existent également dans les craies déposées en milieu épicontinental [L. Leclaire, G. Alcaydé et F. Froehlich, 1973]. Plusieurs phases de silice y sont présentes : quartz, cristobalite, tridymite et silice amorphe.

Le but n'est pas ici de donner un inventaire exhaustif de toutes les formes de silicification rencontrées dans les sédiments [voir entre autres Folk et Weaver, 1952 ; Berger et Von Rad, 1972]. Ainsi

le cas des silex est exclu de cette étude car il représente un problème à part. En outre, cette étude a été volontairement limitée à quelques exemples pris dans les séries à nannofossiles. Les échantillons proviennent d'une part de l'Océan Indien (Bassin de Madagascar : Leg 25, site 245 [L. Leclaire, 1974 : R. Schlich *et al.*, 1974]) et d'autre part du Bassin de Paris (Puits G I.2 de la société « Geostock » à Gargenville, Yvelines). Echantillons prélevés dans des craies d'âge éocène *s. s.* (Océan Indien), turonien et sénonien (Gargenville). Par comparaison, quelques échantillons provenant du Turonien de Touraine, du Crétacé supérieur et

---

* Lab. de géologie du Muséum national d'histoire naturelle 61, rue Buffon, 75005 Paris. Note déposée le 27 mai 1974, présentée le 10 juin 1974.

de l'Éocène de l'Océan Atlantique (Legs 1, 3 et et 4) ont également été étudiés.

Méthodes utilisées.

Après traitement à HCl dilué, les échantillons ont été analysés en diffractométrie X, méthode indispensable pour mettre en évidence les différentes phases de silice [W. F. Berger et U. Von Rad, 1972 ; S. E. Calvert, 1971 ; O. W. Flörke, 1955 ; R. Greenwood, 1973 ; S. W. Wise, B. F. Buie et F. M. Weaver, 1972], puis comparativement en spectrophotométrie d'absorption infra-rouge [voir Calvert, 1971 ; Plyusnina, Maleyev et Yefimova, 1971] et dans certains cas en microscopie électronique à transmission et en micro-diffraction électronique. L'étude des phases siliceuses en roche totale a été faite en lames minces et au microscope électronique à balayage (MEB) avec l'apport des données de la micro-analyse X.

I. — LES CONSTITUANTS DE LA FRACTION SILICEUSE.

Les silicifications étudiées sont composées par l'association de silice et de silicates. La silice est représentée par quatre phases (fig. 1) — silice amorphe, tridymite, cristobalite et quartz — qui sont presque toujours présentes conjointement. Lui sont associés d'une part des phyllites à 10 Å (groupe de l'illite) et à 15 Å (type montmorillonite) et d'autre part un zéolite, la clinoptilolite. En raison de leur finesse, on ne peut distinguer ces minéraux en lames minces même lorsque la silicification est importante. Mais la comparaison des données obtenues par les autres méthodes permet de préciser leur forme et leur structure. L'étude a été menée sur des craies silicifiées à quartz et cristobalite-tridymite et sur des cherts proprement dits où le quartz, également associé à la cristobalite et à la tridymite se trouve essentiellement sous la forme de calcédoine.

A. — *Dans les craies silicifiées.*

— L'association **cristobalite-tridymite** est une forme de silicification classique dans les sédiments (voir aussi fig. 1). Habituellement elle se présente en fines lamelles assemblées en *sphérules* de 3 à 12 µm environ (texte-pl. II, fig. 1 à 4). A Gargenville, elle forme également des *agrégats* de fines particules de 500 Å environ qui ont parfois une structure lamellaire (texte-pl. II, fig. 5). Les sphérules sont elles-mêmes constituées par un assemblage de particules de cristobalite et de tridymite ayant cette dimension [1]. Il apparaît que ces cristallites élémentaires sont communes aux diverses formes de silicification à cristobalite-tridymite : sphérules, agrégats et même à certaines opales [O. W. Flörke, 1955]. Les

diagrammes de rayons X et les spectres infra-rouge [2] confirment cette similitude, ce qui a amené Calvert [1971] à adopter pour les sphérules le schéma cristallographique proposé par Flörke [1955] pour les opales à cristobalite : désordre unidimensionnel conduisant à l'empilement de couches de tétraèdres SiO_4 alternativement suivant l'arrangement de la cristobalite et de la tridymite. A l'image des minéraux argileux, cette forme de silice est un « interstratifié » tridymite-cristobalite. Ce désordre cristallin se concrétise sur les diagrammes X par l'absence de la plupart des raies de la cristobalite et de la tridymite (fig. 1) et sur les spectres infra-rouge par l'absence ou la faible amplitude des bandes d'absorption propres à la cristobalite. Calvert [1971] note que ces spectres, plus proches de ceux de la tridymite ou des verres de silice sont la marque du désordre important de la structure de ces cristaux.

Dans les craies peu silicifiées, les sphérules, de petite taille (environ 3 µm) sont localisées dans les loges des Foraminifères. Lorsque la silicification est plus importante, elles sont plus grosses et présentes dans tout le sédiment. L'enchevêtrement de lamelles de cristobalite-tridymite et de sphérules constitue alors le ciment des craies (texte-pl. I, fig. 4), où les sphérules peuvent englober des coccolithes (texte-pl. II, fig. 2) et remplir par coalescence les loges des Foraminifères (texte-pl. II, fig. 3).

— Le **quartz** se trouve dans les craies silicifiées en *fibres flexueuses* de quelques centaines d'Angströms d'épaisseur pour une longueur de un à plusieurs micromètres (texte-pl. I, fig. 1 et 2). La nature quartzeuse de ces fibres a été établie en premier lieu aux rayons X (fig. 2) et en infra-rouge, les préparations étant contrôlées au MEB et en micro-analyse X. Malgré leur extrême finesse elles ont pu être caractérisées également par diffraction électronique : chaque fibre est un monocristal de quartz dont la direction d'allongement est perpendiculaire aux plans 101. Ce sont donc des *trichites* (cristaux courbes) de quartz.

Ces trichites sont souvent groupées en faisceaux et peuvent tapisser entièrement l'intérieur des loges des Foraminifères (texte-pl. I, fig. 1). Dans de nombreux échantillons les sphérules sont enveloppées par de tels tissus de trichites (texte-pl. II, fig. 1).

— **Silicates associés.** Les argiles sont le plus souvent assez mal représentés. Ce sont pour la plus grande part des phyllites à 10 Å, vraisemblablement de type illite et plus rarement à 15 Å (montmorillonite). Par contre, on constate sur de nombreux échantillons l'abondance d'un zéolite, la clinoptilolite, qui est caractérisée sur les diffractogrammes X par une raie intense à 3,96 Å et deux raies importantes à 8,98 Å et 7,92 Å [B. Mason et L. B. Sand, 1960 ; F. A. Mumpton, 1960]. Elle se présente sous

172

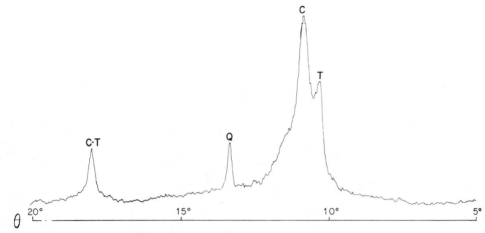

FIG. 1. — Diffractogramme X (Cu Kα) d'un chert (échantillon 4. 29 C.3.1.Top). Silicification de type III. Q = quartz ; C = cristobalite ; T = tridymite.

la forme de cristaux prismatiques d'une dizaine de micromètres environ (texte-pl. II, fig. 6) associés aux sphérules de cristobalite-tridymite.

B. — Dans les cherts proprement dits.

Dans la plupart des cas, les cherts sont constitués par un mélange de cristobalite-tridymite et de calcédoine. Dans d'autres cas, la silice massive est formée pour la quasi-totalité de cristobalite-tridymite.

— Les **cherts à cristobalite-tridymite** contiennent des coccolithes et des Foraminifères bien conservés. En lames minces la masse de cristobalite-tridymite a un relief négatif et est pratiquement éteinte entre polariseurs croisés mais avec des irisations dues aux éléments carbonatés qu'elle renferme. Ces cherts à cristobalite sont en fait des craies intensément silicifiées.

— Les **cherts à cristobalite-tridymite et calcédoine** apparaissent en lames minces formés par une matrice à relief négatif bas, sombre entre polariseurs croisés, avec de nombreux Foraminifères à test pseudomorphosé en calcédoine, l'intérieur des loges étant souvent également rempli de calcédoine fibreuse. L'ensemble n'est cependant pas homogène et peut contenir des amygdales à carbonates où le test des Foraminifères n'est pas modifié. Certains d'entre eux passent de la masse siliceuse dans les zones carbonatées : dans ce cas, seule la partie du test située dans la silice massive est pseudomorphosée en calcédoine. Ces amygdales sont également visibles au MEB et peuvent être rapportées à des craies intensément silicifiées. Quant à la masse siliceuse elle-même, le contraste entre cristobalite-tridymite et calcédoine est très faible. On peut néanmoins penser que la cristobalite-tridymite présente des surfaces comparables à celles des opales : juxtaposition de fins cristallites donnant l'aspect d'une surface granuleuse. Les surfaces calcédonieuses ont plutôt un aspect lisse et largement fissuré ou

TEXTE-PLANCHE I

FIG. 1. — Tissu de trichites de quartz dans une micro-cavité d'une craie silicifiée. Cortex d'un chert de l'Océan Indien (Éocène inférieur ; éch. 25.245.7.2.116).

FIG. 3. — Structure fibreuse d'un chert calcédonieux éocène inférieur de l'Océan Indien (éch. 25.245.5.5.14).

FIG. 2. — Silicification de type I dans une craie de la base du Sénonien du Bassin de Paris (Gargenville, — 85 m).

FIG. 5. — Assemblage sphérulitique embryonnaire des cristallites de cristobalite-tridymite dans un chert de l'Océan Indien (éch. 25.245.4.4.57).

FIG. 4. — Silicification de type III dans le cortex d'un chert de l'Océan Indien. Les lamelles de cristobalite-tridymite sont assemblées d'une manière quelconque et cimentent ici les débris d'un Foraminifère (ch. 25.245.7cc).

FIG. 6. — Croissance de lamelles de cristobalite-tridymite sur un cristal de calcite de la paroi d'un Foraminifère (intérieur de la loge) : zone non calcédonieuse d'un chert de l'Océan Indien (éch. 25.245.7cc).

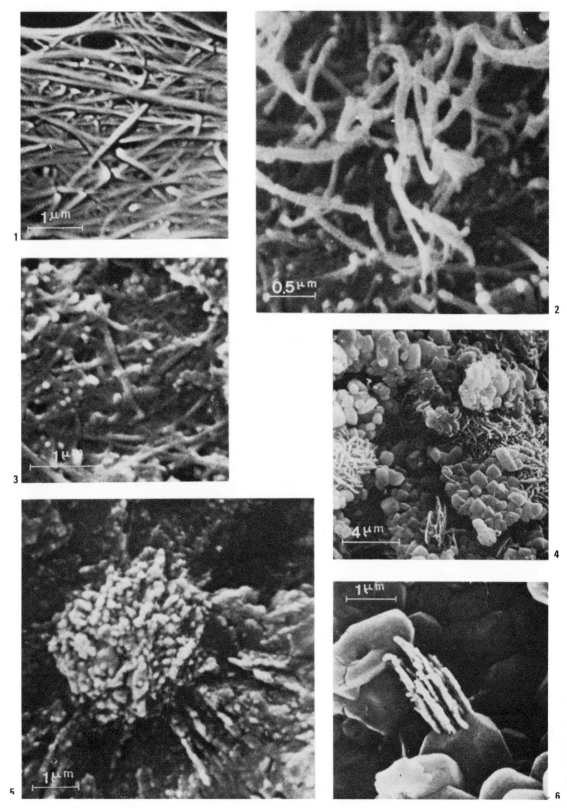

PLANCHE I.

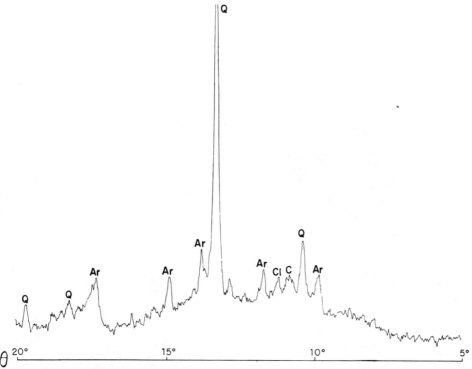

Fig. 2. — Diffractogramme X (Cu Kα) de la fraction insoluble d'une craie du Sénonien (Gargenville, — 85 m). Silicification de type I. Ar = phyllites à 10 Å (Type illite) ; Q = quartz ; C = cristobalite ; Cl = clinoptilolite.

ailleurs fibreux (texte-pl. I, fig. 3). Il semble que la matrice soit à dominante de cristobalite. On y distingue des fantômes de coccolithes dont l'aspect, lisse et fissuré, donne à penser qu'ils sont pseudomorphosés en calcédoine. Les fantômes de Foraminifères sont mieux visibles et leur paroi est souvent en calcédoine fibreuse. Les spectres infra-rouge de ces échantillons présentent en plus des bandes d'absorption propres au quartz et à la cristobalite-tridymite une bande particulière à 560 cm^{-1}, attribuée à la calcédoine [I. I. Plyusnina. M. N. Maleyev et G. A. Yefimova, 1971]. Des échantillons-témoins de calcédoine exclusivement quartzeuse, analysés dans les mêmes conditions, présentent toujours cette bande particulière.

Dans beaucoup d'échantillons, les cherts à cal-

TEXTE-PLANCHE II

Fig. 1. — Silicification de type II. Les sphérules de cristobalite-tridymite sont enveloppées par un réseau de trichites de quartz. Cortex d'un chert de l'Océan Indien (éch. 25.245.7.2.116).

Fig. 2. — Craie fortement silicifiée, de type III : un coccolithe est partiellement englobé par une sphérule. Quelques trichites de quartz sont visibles. Cortex d'un chert de l'Océan Indien (éch. 25. 245.6cc).

Fig. 3. — Remplissage d'une loge de Foraminifère par la coalescence des sphérules de cristobalite-tridymite. Craie silicifiée du Sénonien du Bassin de Paris (Gargenville, — 56 m).

Fig. 4. — Craie silicifiée de l'Éocène moyen de l'Atlantique sud (large de Rio). La silicification est de type III, mais avec une proportion importante de clinoptilolite (éch. 3.22.5.2.148).

Fig. 5. -- Agrégats de cristobalite-tridymite. Quelques lamelles sont visibles. Sénonien du Bassin de Paris (Gargenville, — 57 m).

Fig. 6. — Cristal, probablement de clinoptilolite, associé à des sphérules. Cortex d'un chert de l'Océan Indien (éch. 25.245.7cc).

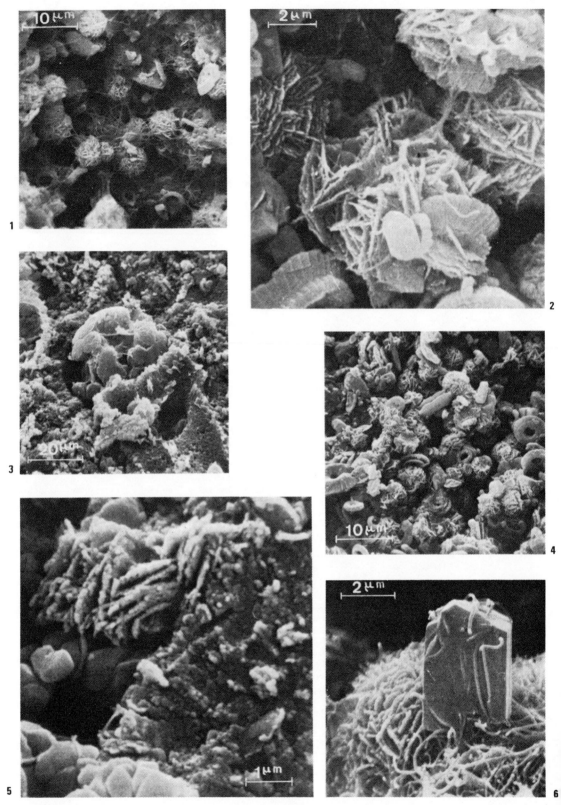

176

cédoine possèdent un cortex à cristobalite-tridymite, ou plutôt de craie intensément silicifiée. Le passage de l'un à l'autre est brutal, même aux forts grossissements, et marqué par l'apparition soudaine de la calcédoine.

II. — VARIATIONS DE L'IMPORTANCE DE LA SILICIFICATION ET DE SES CONSTITUANTS.

L'importance de la fraction siliceuse est variable dans les échantillons étudiés : de quelques pourcents à plus de 90 %. Cette variation est accompagnée d'une modification des proportions des constituants, sensible à l'observation au MEB, mais qui est surtout bien marquée sur les diffractogrammes X. Les différences sont estimées en comparant les hauteurs relatives des pics majeurs de chaque minéral : 3,34 Å pour le quartz, 4,09 Å pour la cristobalite, 8,98 Å pour la clinoptilolite (la raie à 3,96 Å étant le plus souvent oblitérée par la raie à 4,09 Å de la cristobalite). Quant aux argiles, leurs raies sont de trop faible amplitude pour que l'on puisse y déceler des variations nettes. On ne peut que constater la présence ou l'absence des raies principales. Il apparaît que la cristobalite et la tridymite sont dominantes dans les échantillons à silicification importante, alors que le quartz n'est prépondérant que dans les craies faiblement silicifiées où les alumino-silicates sont également en proportions plus importantes. On en arrive ainsi à définir pour les échantillons étudiés trois types de silicifications dans les craies, auxquels s'ajoutent les cherts à calcédoine (fig. 3).

(I) Craies peu silicifiées *à quartz dominant* sous la forme de trichites (texte-pl. I, fig. 1 et 2), avec peu de cristobalite-tridymite. Elles comportent une proportion importante d'argiles et/ou de clinoptilolite (voir fig. 2). La cristobalite s'y trouve en agrégats isolés ou en sphérules de petite taille, principalement dans les loges de Foraminifères.

(II) Craies moyennement silicifiées *à cristobalite-tridymite et quartz* en proportions comparables, avec peu d'argiles, mais souvent une forte proportion de clinoptilolite. Ce minéral est dans certains échantillons quantitativement aussi important que le quartz ou la cristobalite. Les sphérules, de 6 à 12 μm de diamètre, sont bien développées dans les Foraminifères, mais se trouvent également dans tout le sédiment en association avec des cristaux de clinoptilolite (texte-pl. II, fig. 6) et enveloppés par des faisceaux de trichites de quartz (texte-pl. II, fig. 1).

(III) Craies intensément silicifiées *à cristobalite-tridymite dominantes* (*cf.* diffractogramme de la fig. 1) et qui renferment peu de quartz. Dans la plupart des cas, on n'y trouve pas d'alumino-silicates, sinon en très faible proportion. Cette catégorie regroupe des craies à silicification encore modérée et des cherts à cristobalite massive.

(IV) Cherts à calcédoine. Ce type de chert, bien qu'étant caractérisé sur les diffractogrammes X par la présence de cristobalite-tridymite et de quartz en proportions comparables, doit être rangé dans une catégorie à part : la silice est massive, les alumino-silicates ne s'y trouvent qu'en traces et contrairement aux trois types précédents, le quartz s'y rencontre sous forme de calcédoine qui pseudomorphose les carbonates.

FIG. 3. — Variation des proportions relatives des phases de silice et des alumino-silicates en fonction de l'intensité de la silicification.

(1) Pourcentage de la fraction insoluble dans HCl dilué, par rapport à la roche totale ; (2) Pourcentage par rapport à l'ensemble des constituants de la fraction insoluble dans HCl dilué ; (3) La clinoptilolite est le plus souvent très largement dominante par rapport aux autres alumino-silicates (argiles).

On constate dans les échantillons analysés l'absence de quartz microcristallin et de structures de la silice telles que Folk et Weaver les ont décrites [1952]. L'importance que prend la cristobalite dans les craies silicifiées va à l'encontre des conclusions de Greenwood [1973] qui estimait qu'elle ne pouvait se former dans les sédiments crayeux. Par contre les variations des divers constituants observées ici correspondent assez bien à celles vues ailleurs par Berger et Von Rad [1972].

III. — Les types de silicification dans leur contexte géologique.

Jusqu'ici, la composition des silicifications et ses variations ont été envisagées globalement, sans tenir compte de l'origine des échantillons, les résultats obtenus étant communs à ceux du Bassin de Madagascar et à ceux du Bassin de Paris. Or, si l'on replace les échantillons dans leur contexte géologique, on peut poursuivre la comparaison. La répartition des types de silicification se fait en effet, aux fluctuations de détail près, dans un ordre déterminé qui correspond à un accroissement progressif de la silicification.

A. — *Bassin de Paris.*

Le puits GI. 2 à Gargenville a traversé 150 m de craie du Sénonien au Turonien supérieur. Cinquante échantillons, répartis sur toute la hauteur du puits ont été analysés. La silicification y est modérée, mais constante. Les résidus siliceux insolubles dans HCl dilué représentent 3 à 4 % de la craie en moyenne, mais peuvent aller jusqu'à plus de 10 % dans le Sénonien (— 56 m) : au-dessus et en-dessous de ce niveau existe un gradient de silicification qui est accompagné d'une variation notable de la proportion des constituants :

● jusqu'à — 45 m, craie silicifiée de type I ;

● entre — 45 m et — 55 m, silicification de type II ;

● de — 55 m à — 56 m, silicification de type III ;

● entre — 57 m et — 60 m, silicification de type II ;

● en-dessous de — 60 m, silicification de type I.

La répartition des types de silicification est symétrique par rapport au niveau — 56 m, à proximité duquel les variations deviennent plus rapides. En-dessous de — 60 m, la décroissance de la silicification se ralentit. Le passage au Turonien (— 115 m environ) n'y apporte pas de perturbation. Ces variations n'ont pas de rapport visible avec la répartition ou l'abondance des silex, mais doivent traduire une évolution des apports de silice au cours de cette période de l'histoire sédimentaire du bassin.

Des silicifications plus intenses, de type III sont connues en Touraine, sur la bordure sud du bassin [L. Leclaire, G. Alcaydé et F. Froehlich, 1973] où existent d'énormes accumulations de sphérules de cristobalite-tridymite (région de Baudres, Indre). Dans le Turonien d'Anjou, des silicifications à cristobalite et clinoptilolite ont également été décrites [J. Estéoule, J. Estéoule-Choux et J. Louail, 1971].

L'importance de ces silicifications de bordure, à proximité du continent, vient à l'appui de l'hypothèse de Leclaire [1974] qui voit dans le drainage des continents alors soumis à des climats hydrolisants l'origine d'une grande partie des apports de silice dans les océans.

B. — *Bassin de Madagascar.*

Les échantillons étudiés viennent tous du site 245 (Océan Indien, au large de Madagascar). Les 300 m de série concernés sont constitués par des boues à nannofossiles et Foraminifères, de plus en plus lithifiées vers la base où l'on trouve des craies proprement dites [R. Schlich et al., 1974]. Cette série est d'âge éocène inférieur. On y compte en tout une douzaine de niveaux de cherts ou de craies silicifiées de quelques centimètres d'épaisseur. L'étude a porté sur onze de ces niveaux [3]. La silicification est dans ces niveaux plus importante que dans le centre du Bassin de Paris, mais compte tenu du petit nombre d'échantillons étudiés et de leur faible volume, on ne peut avoir une vue générale de l'évolution de la silicification dans la série. D'après Schlich, Simpson et al. [1974], il y a sur plusieurs dizaines de mètres une augmentation progressive vers le bas de la teneur en silice. L'observation au MEB d'échantillons de boues à nannofossiles (échantillons 5. 4. 83 et 5. 4. 97) montre qu'ils contiennent de la silice en faible proportion : trichites de quartz associés à des particules argileuses et à des cristaux, probablement de clinoptilolite (silicifications de type I).

Des variations rapides de la silicification sont observables dans les échantillons de cherts où l'on peut distinguer quatre zones successives. En partant du chert proprement dit, de type IV, se succèdent des zones de craies silicifiées formant le cortex du chert : de type III, puis II et enfin I. Il existe une discontinuité très nette entre le cortex où les variations de la silicification sont progressives et le chert où apparaît brusquement la calcédoine.

Quelques échantillons de cherts provenant d'autres campagnes du « Glomar Challenger » ont également été étudiés par comparaison avec ceux du site 245 [4]. Des silicifications comparables y ont été observées, dont la plupart se sont révélées être de type III : cherts massifs à cristobalite-tridymite dominantes avec parfois des sphérules bien formées dans les micro-cavités. L'échantillon 3. 22. 5. 2. 148, une craie à cristobalite et clinoptilolite, a une structure comparable à celles de l'Éocène de l'Océan Indien (site 245 : texte-pl. II, fig. 4). Enfin, un échantillon (4. 29 C. 2. 1. Base) de l'Éocène de la mer des Caraïbes s'apparente aux silicifications de type IV, avec une matrice composée d'un mélange de cristobalite et de calcédoine renfermant de nombreux radiolaires à test calcédonieux.

IV. — Interprétation.

Il apparaît en premier lieu que les silicifications étudiées sont indépendantes du degré de lithification des craies, ainsi que de leur âge et de la profondeur de dépôt.

La proportion de cristobalite augmente schématiquement en même temps que l'importance de la silicification. Parallèlement, la proportion de carbonates — donc de coccolithes et de Foraminifères — décroît, jusqu'à ne représenter dans les cas extrêmes que des éléments isolés dans une masse de cristobalite-tridymite. Il est à remarquer que les tests calcaires sont conservés en parfait état, sans modification chimique. L'invasion progressive de la silice dans le sédiment est bien visible au MEB dans les échantillons de l'Océan Indien. La cristobalite ne s'étant pas développée au détriment des carbonates, mais étant présente à côté d'eux, on est amené à en faire un élément constitutif originel du sédiment, au même titre que le calcaire. Si l'on s'en tient à cette hypothèse et en admettant que la vitesse de sédimentation des carbonates est à peu près constante, on peut admettre que les silicifications les plus intenses se sont mises en place pendant une période de temps relativement courte et que les variations de l'importance de la silicification reflètent les fluctuations des apports de silice dans l'eau de mer.

Lorsque la teneur en silice dissoute est assez faible, la cristobalite ne pourrait se développer que dans les Foraminifères, c'est-à-dire dans des micromilieux enrichis en silice. A cet égard, les cristaux de calcite de leur paroi paraissent jouer un rôle important en tant que support : le développement des lamelles de cristobalite-tridymite semble être orienté par rapport aux faces des cristaux de calcite (texte-pl. I, fig. 6). Faut-il interpréter ce phénomène comme une épitaxie ? Lorsque la teneur en silice dissoute augmente, la cristobalite peut cristalliser partout et la taille des sphérules devient plus importante pour atteindre une valeur limite (une douzaine de micromètres). Dans le cas des silicifications massives, la cristobalite et la tridymite n'ont pu s'organiser en lamelles ou en sphérules et les cristallites s'amassent sans ordre.

Le même raisonnement ne peut être tenu en ce qui concerne le quartz qui ne forme pas de silicifications massives dans les craies. Alors que des échantillons peuvent contenir plus de 90 % de cristobalite et de tridymite, ils ne renferment jamais plus de 10 % de quartz. Les relations géométriques existant entre les trichites de quartz et les sphérules de cristobalite-tridymite montrent que les trichites ne se sont pas formées en même temps que les sphérules, mais postérieurement. On est ainsi amené à voir la formation des trichites de quartz dans

un deuxième temps de silicification, peut-être dû à la circulation dans le sédiment d'eaux interstitielles moins riches en silice. En effet, si la solubilité de la silice décroît depuis la silice amorphe (environ 140 ppm) jusqu'au quartz (7 ppm), inversement, pour alimenter la croissance de cristaux de quartz, il suffirait d'une solution moins riche en silice que pour la cristobalite et la tridymite [K. B. Krauskopf, 1956, 1959]. En outre, la cristallisation séparée de phases métastables et d'une phase stable est chimiquement plus logique. La formation des trichites de quartz a pu se faire soit immédiatement après la mise en place de la cristobalite, soit plus tard au cours de la diagenèse. Mais il ne semble pas y avoir eu remobilisation de la silice des sphérules dont les structures fines ne sont pas altérées.

Si le quartz est postérieur à la cristobalite, il l'est vraisemblablement aussi à la clinoptilolite dont les cristaux sont souvent recouverts par des trichites et qui est d'une manière générale associée à la cristobalite. Par ailleurs, le fait que sphérules et cristaux de clinoptilolite soient toujours intimement mêlés, notamment à l'intérieur des Foraminifères, rend plausible l'hypothèse d'une cristallisation conjointe de cristobalite-tridymite et de clinoptilolite, ce dernier minéral fixant une grande partie de la silice dans le cas de silicifications modérées. Quant aux argiles, aucun critère ne permet de les situer par rapport au quartz. On peut néanmoins penser que leur formation est contemporaine de celle de la cristobalite-tridymite et de la clinoptilolite. En comparant les proportions relatives des trois composants de ce premier temps de silicification, on constate que lorsque la silicification augmente, argiles, clinoptilolite et cristobalite-tridymite apparaissent successivement comme les constituants principaux de la fraction siliceuse. On peut relier cette variation à une évolution du rapport Si/Al dans l'eau de mer. Lorsque la teneur en silice augmente par rapport à celle de l'alumine, il y aurait préférentiellement formation d'argile, puis de clinoptilolite et enfin de cristobalite et de tridymite. Il est à noter que la clinoptilolite, avec un rapport silice/alumine voisin de 9 [B. Mason et L. B. Sand, 1960 ; F. A. Mumpton, 1960] est l'un des alumino-silicates les plus riches en silice.

Quant à la calcédoine, elle est strictement localisée dans les niveaux siliceux massifs. Des observations recueillies sur ces échantillons, il ressort qu'elle résulte d'une évolution diagénétique postérieure au dépôt de la cristobalite-tridymite, avec pseudomorphose des carbonates. Les silicifications de type IV proviendraient ainsi d'une resilicification de craies silicifiées de type III. Ce phénomène s'effectue en « tache d'huile » en épargnant des îlots qui sont des témoins du sédiment originel. On rejoint ici Millot [1960] qui estimait que « les carbonates sont particulièrement disposés à se transformer en calcédoine ». Autant qu'on puisse en juger, il ne semble pas ici non plus que la calcédoine résulte d'une évolution de la cristobalite et de la tridymite. Il existe en outre une ressemblance troublante entre les fibres de la calcédoine (texte-pl. I, fig. 3) et les trichites de quartz. Millot écrivait en 1960, en s'appuyant sur une interprétation cristallographique de la calcédoine fibreuse proposée par Pelto [1956] : « incapable de se développer, le quartz s'organise en longues fibres spiralées séparées par des limites désordonnées où l'eau et les cations s'emprisonnent ». Si les fibres de la calcédoine et les trichites de quartz sont identiques — ce qui reste à démontrer — on peut voir dans la cristallisation massive de trichites de quartz la formation de la calcédoine fibreuse. Dans le cas de trichites isolées et dans celui de la calcédoine, la cristallisation serait postérieure au dépôt de la cristobalite.

Remerciements.

L'auteur adresse ses plus vifs remerciements aux responsables du « Deep Sea Drilling Project » ainsi qu'à la société « Geostock » qui ont bien voulu lui

fournir les échantillons qui sont à la base de ce travail. Il remercie tout particulièrement M. le Professeur R. Laffitte, MM. L. Leclaire et G. Alcaydé pour leurs précieux conseils au cours de la rédaction de cette note. M. le Professeur J. Nicolas, directeur du Laboratoire de géologie appliquée (Paris VI) et M. A. Rimsky, du Laboratoire de minéralogie (Paris VI) qui lui ont offert les facilités de leurs laboratoires ont toute la reconnaissance de l'auteur. Il adresse enfin ses plus vifs remerciements à M. C. Willaime et à Mlle M. Gandais, du Laboratoire de minéralogie de Paris VI.

1. Résultats obtenus par diffraction électronique au Labo-de minéralogie de Paris VI.
2. Analyses effectuées au Laboratoire de géologie appliquée de Paris VI.
3. Échantillons 4.4.57 ; 5.2.135 ; 5.2.139 ; 5.4.1 ; 5.4.8, 5.5.14 ; 6.2.105 ; 6cc ; 7.2.116 ; 7cc ; 9.2.132.
4. Échantillons obtenus auprès de la Scripps Institution of Oceanography et du Lamont Doherty Geological Observatory : Leg 1, site 4 (chert dans une craie du Crétacé inférieur au large de la Floride) ; Leg 1, site 5 A (chert dans une argilite du Crétacé supérieur au large de la Floride) ; Leg 3 ; site 13 A (chert dans des boues à nannofossiles de l'Éocène moyen au large de Rio) ; Leg 4, site 29 C (série massive de cherts de l'Éocène moyen de la mer des Caraïbes).

Bibliographie

Berger W. F. et Rad U. von (1972). — Cretaceous and cenozoic sediments from the Atlantic Ocean. Initial Report of the Deep Sea Drilling Project, vol. XIV. Washington, US Government Printing Office.

Calvert S. E. (1971). — Nature of Silica phases in deep sea cherts of the North Atlantic. *Science*, vol. 234, 13, p. 133-134.

Estéoule J., Estéoule-Choux J. et Louail J. (1971). — Sur la présence de clinoptilolite dans les dépôts marno-calcaires du Crétacé supérieur de l'Anjou. *C. R. Ac. Sc.*, Paris, t. 272, D, p. 1569-1572.

Flörke O. W. (1955). — Zur Frage des « hoch » Cristobalit in Opalen, Bentoninen und Gläsern. *Neues Jb. Miner. Monat.*, 10, p. 217-223.

Folk R. L. et Weaver C. E. (1952). — A study of the texture and composition of chert. *Amer. J. Sci.*, vol. 250, p. 498-510.

Gigout M., Estéoule J., Estéoule-Choux J. et Rasplus L. (1969). — La majeure partie des argiles à silex de Touraine doit être considérée comme un faciès du Sénonien. *C. R. Ac. Sc.* Paris, t. 268, p. 471-474.

Greenwood R. (1973). — Cristobalite : its relationship to chert formation in selected samples from the Deep Sea Drilling Project. *Jour. Sed. Petr.*, V. 43, 3, p. 700-708.

Jones J. B., Sanders J. V. et Segnit E. R. (1964). — Structure of Opal. *Nature*, 204, p. 990-991.

Krauskopf K. B. (1956). — Dissolution and precipitation of silica at low temperatures. *Geoch. Cosmo. Acta*, V. 10, n° 1/2, p. 1-27.

Krauskopf K. B. (1959). — Geochimistry of Silica. *Soc. Econ. Pal. Miner.*, Spec. Publ. n° 7, p. 4-19.

Leclaire L., Alcaydé G. et Froehlich F. (1973). — La silicification des craies : rôle des sphérules de cristobalite-tridymite observées dans les craies des bassins océaniques et dans celles du Bassin de Paris. *C. R. Ac. Sc.*, Paris, t. 277, D, p. 2121-2123.

Leclaire L. (1974). — Hypothèses sur l'origine des silicifications dans les grands bassins océaniques. Le rôle des climats hydrolisants. *Bull. Soc. géol. Fr.*, (7), XVI, p. 214-224.

Leclaire L. (1974). — Late cretaceous and cenozoic pelagic deposits. Paleoenvironment and paleooceanography of the Central Western Indian Ocean. In Schlich et al., Initial Report of the Deep Sea Drilling Project, vol. XXV Washington, US Government Printing Office.

Mason B. et Sand L. B. (1960). — Clinoptilolite from Patagonia. The relationship between clinoptilolite and heulandite. *Amer. Miner.*, vol. 45, p. 340-350.

Millot G., Radier M., Muller-Feuga R., Deffossez M. et Wey R. (1959). — Sur la géochimie de la silice et les silicifications sahariennes. *Bull. Serv. Carte Géol. Als. Lorr.*, t. 12, 2, p. 3-14.

Millot G. (1960). — Silice, silex, silicifications et croissance des cristaux. *Bull. Serv. Carte Géol. Als.-Lorr.*, t. 13, 4, p. 129-146.

Millot G. (1962). — Silicifications et néoformations argileuses : problèmes de genèse. *In* : Coll. Intern. C.N.R. S., n° 105 (« Synthèse et genèse des argiles »). Paris, 1961.

Mumpton F. A. (1960). — Clinoptilolite redefined. *Amer. J. Sci.*, vol. 45, p. 351-369.

Oehler J. M. (1973). — Tridymite-like cristals in cristobalitic « cherts ». *Nat. Phys. Sci.*, 241, n° 17, p. 64-65.

Pelto C. R. (1956). — A study of chalcedony. *Amer. J. Sci.*, vol. 254, p. 32-50.

Plyusnina I. I., Maleyev M. N. et Yefimova G. A. (1971). — Infrared spectroscopic investigation of cryptocrystalline varieties of silica. *Intern. Geol. Rev.*, 13, n° 11, p. 1750-1754.

Reynolds R. C. et Anderson D. M. (1967). — Cristobalite and clinoptilolite in bentonite group, norhtern Alaska. *Jour. Sed. Petr.*, 37, 3, p. 966-969.

Schlich R. *et al.* (1974). — Initial Report of the Deep Sea Drilling Project, vol. XXV, part one. Washington, US Government Printing Office.

Wise S. W. et Hsu K. J. (1971). — Genesis and lithification od deep sea chalk. *Eclog. Geol. Helv.*, vol. 64/2, p. 273-278.

Wise S. W., Buie B. F. et Weaver F. M. (1972). — Chemically precipitated sedimentary cristobalite and the origin of chert. *Eclog. Geol. Helv.*, vol. 65/1, p. 157-163.

Wise S. W. et Weaver F. M. (1972). — Ultramorphology of deep sea cristobalitic chert. *Nat. Phys. Sci.*, vol. 237, n° 73, p. 56-57.

9

Reprinted from *Neues Jahrb. Mineralogie Monatsh.* **8**:369–377 (Aug. 1975)

Opal-CT crystals

By O. W. Flörke, Bochum, J. B. Jones, Adelaide, and
E. R. Segnit, Port Melbourne

With 11 figures in the text*

Flörke, O. W., Jones, J. B. & Segnit, E. R.: Opal-CT crystals. – N. Jb. Miner. Mh., **1975**, H. 8, 369–377, Stuttgart 1975.

Abstract: Opal-CT in the form of lepispheres is shown to be characteristic of many opal deposits in addition to its recorded occurrence in "deep sea cherts". A similar morphology has been found in laboratory-prepared opal-CT. Widely different conditions of temperature, pressure and pH can result in precipitation of silica in this form and inferences as to the nature of the environment of formation from the characteristic morphology are premature at this stage.

K e y   w o r d s : Opal (CT), cristobalite, deep sea, chert, synthesis, scanning electron microscopy, genesis.

Introduction

Most opaline silicas, both in hand specimen and under the optical microscope, have an amorphous or structureless appearance. With the higher magnifications permitted by the transmission and scanning electron microscopes, structures can be observed in many specimens but these, for the most part, appear to be related to some previous structure or their mode of deposition rather than to the atomic structure of the opal itself (Segnit, Anderson & Jones, 1970). Some varieties of opaline silica, such as the original lussatite of Mallard (opal-CT), do show, under the optical microscope, a finely crystalline structure and distinct birefringence which is a direct result of the atomic structure of the material. In fact, with the exception of most precious opals, hyalite, geyserite and unaltered silica of biogenic origin, opaline silicas regardless of morphological appearance, have been shown to be crystalline by X-ray diffraction (Flörke, 1962; Jones & Segnit, 1971). Although some deductions as to crystallite morphology can be made from the plate-like nature and single crystal behaviour of crushed grains in transmission electron microscopy, and from theoretical considerations (Jones & Segnit, 1972), direct observation of this morphology would be of considerable interest.

---

*All figures are scanning electron micrographs.

"Deep sea chert"

Recently it has been recognized that much of the so-called chert found in cores obtained in the deep sea drilling programme is cristobalitic in nature (REX, 1969, 1970) and is in fact opal-CT (CALVERT, 1971). In some samples of such sediments, aggregates of small rosettes have been found

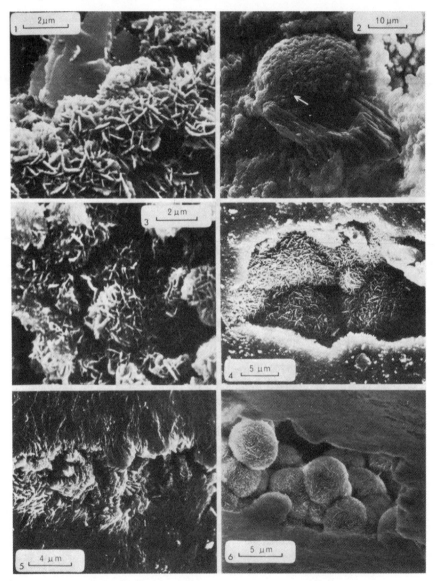

which on examination by scanning electron microscopy prove to be composed of small thin platelets apparently of opal-CT (WISE & HSÜ, 1971; WISE & KILTS, 1971; BUIE & WEAVER, 1972; WEAVER & WISE, 1972; VON RAD & RÖSCH, 1972; BERGER & VON RAD, 1972). ERNST & CALVERT (1969) in their study of porcellanites of the Monterey Formation, California and HEATH & MOBERLY (1971) following their examination of samples recovered from the Pacific drilling programme have suggested that the originally non-crystalline biogenic silica in these sediments had been subsequently recrystallized to what we would term opal-CT with further recrystallization leading to quartz.

Non oceanic occurrences

However, it should not be thought that opal-CT in the form of spherules or rosettes is confined to deep sea oozes. WISE, BUIE & WEAVER (1972) report spherules of identical appearance and size in the "flint clays" (Cretaceous-Tertiary) from the Coastal Plain of Georgia (U.S.A.). Similar rosettes are an important component of the siliceous material of Palaeogene age which is widespread over much of Southern Russia and is termed opoka by Russian workers. Fig. 1 and 2 illustrate the nature of the spherules. X-ray diffraction shows the bulk material to be a mixture of opal-CT and quartz.

Prior to this proliferation of records of silica in these rosette-like bodies, we had also observed rosettes composed of very thin plates in our early work on opal. The most interesting of these occur in cavities in an opal which has formed and possibly still is forming in crevices in the lower part of a Pliocene diatomite deposit near Lillicur, Victoria, Australia. The unaltered diatomite is in the form of opal-A whereas the silica formed as a result of solution and redeposition in joints is massive opal-CT with the characteristic rosettes in cavities (Fig. 3).

Fig. 1. Typical aggregate of lepispheres in "opoka" U.S.S.R.

Fig. 2. Cauliflower-like aggregate of lepispheres. The large well cleaved crystal on which the aggregate rests is thought to be clinoptilolite by analogy with the findings and figures of WISE & KELTS (1972).

Fig. 3. Lepispheres of opal-CT associated with diatomite, Lillicur, Victoria, Australia.

Fig. 4. Typical opal-CT crystallites coating interior of walls of a cell in opalized wood.

Fig. 5. Thicker plates of opal-CT growing from walls of cavity in opalized wood.

Fig. 6. Spherical aggregates of thin platelets of opal-CT in opalized asbestos from Tumby Bay, South Australia.

Formation of such crystallites also appears to be involved in the formation of many other massive opals whose origin is by no means as clear as the above example. Thus SEGNIT, JONES & ANDERSON (1973, Plate 22, Fig. 3) have illustrated a sample of massive opal in which small vugs

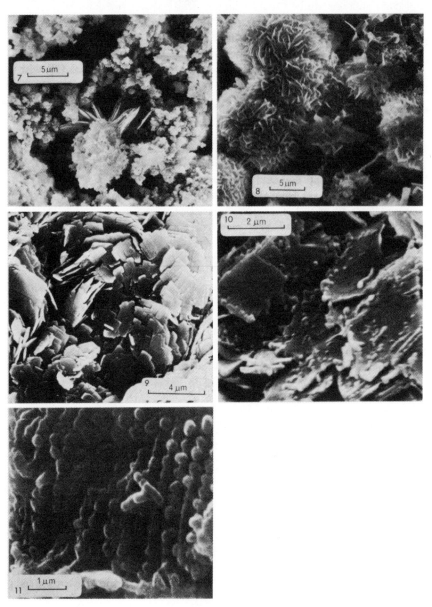

(0.1–1 mm) are coated with plate-like crystallites which can be traced into featureless opal away from the cavities. In the same paper an example of an opal which is composed of a massive accumulation of such small rosettes is shown in Plate 21, fig. 4. G. SCURFIELD (private communication, 1973) has found rosettes composed of similar platelets filling wood cells which have been replaced by opal (Figs. 4 and 5). Spherical structures composed of thin plates of opal-CT have also been found in opalized asbestos (Fig. 6).

A further example of such structures is probaly present in the opal from Nova Ves, CSSR, shown in fig. 9 of SEGNIT, ANDERSON & JONES (1970). This may be compared with Plate 32, fig. 6 of BERGER & VON RAD (1972) showing a sample of Eocene "chert" from the Atlantic. The morphology, apart from a difference of scale, is strikingly similar. Laminar crystallites were observed also in opal-C from an Icelandic lava (SEGNIT, ANDERSON & JONES, 1970). Such crystallites are probably widespread in both massive and friable opal doposits but have hitherto not been recognized either due to their submicroscopic size or perhaps more commonly because of concealment by infilling with later silica. The characteristic X-ray pattern of opal-CT lends support to this contention (JONES, SEGNIT & FLÖRKE, to be published).

Synthetic products

Recently OEHLER (1973) has described the synthesis of similar structures under hydrothermal conditions although it is not possible to be certain from his description that the material is opal-CT. Whereas OEHLER prepared his material at 150 °C and 2000 bars pressure we have prepared undoubted opal-CT with a similar morphology (Fig. 7) at only 50 bars pressure and in the temperature range 150–200 °C. The starting material was pure silica obtained by flame decomposition of $SiCl_4$ (Aerosil, Degussa) and the reaction solutions were 2 mole per cent sodium, potassium or cesium hydroxide in water. The initial opal-A type material was converted to kenyaite in the NaOH runs or to silica-X in KOH and CsOH runs

Fig. 7. Small lepispheres of synthetic opal-CT. The larger plates are thought to be kenyaite.

Fig. 8. Rosettes of silica-X formed as an intermediate product in the hydrothermal synthesis of opal-CT.

Fig. 9. Magadiite from Trinity Co., California, U.S.A.

Fig. 10. Silhydrite from Trinity Co., California, U.S.A.

Fig. 11. Detail from Fig. 2 (arrowed). Compare with the spherical nodes on the plates of silhydrite in Fig. 10.

after three days. Opal-CT was formed as intermediate product in all runs after 6–9 days, and was replaced by the stable phase (quartz) after 12 days.

Transformation sequences can be illustrated as follows:

$$\text{opal-A} \rightarrow \begin{cases} \text{SiO}_2\text{-X (rosettes)} \\ \text{Kenyaite} \end{cases} \longrightarrow \text{opal-CT (rosettes)} \rightarrow \text{chalcedony}$$
$$\longrightarrow \text{opal-CT (rosettes)} \rightarrow \text{chalcedony}$$
$$\longrightarrow \text{chalcedony}$$

These results are similar to those obtained by HEYDEMANN (1964) using similar conditions with KOH. She also obtained SiO_2-X as an intermediate product but describes the next step as α-cristobalite. In the absence of diffraction data it is impossible to be certain, but it seems likely that this may have, in fact, been opal-CT. From these experiments it is clear that this rosette habit of opal-CT can be developed by two different mechanisms. First, they may form as paramorphs after SiO_2-X, which originally formed platelets in rosette-like configuration (Fig. 8) (FLÖRKE, JONES & Köster, unpubl. 1973). Second they may form directly as platelets of opal-CT, the mechanism of which has been discussed in terms of growth activation by the action of stacking faults (FLÖRKE, 1972).

Furthermore, the morphology (as seen in the optical microscope) of much chalcedony is very similar to that of much opal-CT, and may possibly be considered as paramorphic.

It should also be pointed out that the crystal habits are similar in both natural opal-CT which has unquestionably formed by aqueous deposition at low temperature, and in synthetic opal-CT formed from vapour transport of silica.

These experiments will be described and discussed in detail in another paper.

X-ray pattern

The well-known anomalous X-ray pattern of opal-CT may now be understood as being due to the combined action of stacking disorder and the effect of anisotropic crystallite shape. The latter growth habit is the reason for the non-appearance of certain diffraction lines in the X-ray pattern of opal-CT. These lines, however, can be observed in X-ray patterns of cristobalite-tridymite having a comparable degree of stacking disorder, and in which the crystallites are nearly equidimensional. This material was prepared under non-hydrous conditions. Details will be published in another paper.

Genetic implications

Some recently discovered high silica alkali silicates (namely kenyaite, $NaSi_{11}O_{20.5}$ $(OH)_4 \cdot 3H_2O$ – mentioned above – and magadiite, $NaSi_7O_{13}$ $(OH)_3 \cdot 3H_2O$ (EUGSTER, 1967) and the closely related but non-alkaline silhydrite, $3SiO_2 \cdot H_2O$ (GUDE & SHEPPARD, 1972)) have a similar morphology (Fig. 9 and 10) and it is tempting to suggest as EUGSTER has already done (1967, 1969) that these minerals are the precursors of much "chert". Some support for this idea may be given by Fig. 11 which is a part of Fig. 2 at higher magnification – the similarity with the silhydrite of Fig. 10 is striking although nothing is known at present concerning the composition of the material.

However, it is clear from the foregoing that silica in the form of rosettes of thin plates may form by a variety of processes and under different sets of conditions. EUGSTER (1969) reasons convincingly that the cherts of the highly alkaline Lake Magadi have formed from magadiite either directly or via kenyaite and seeks to extend this mechanism to account for some of the banded iron formations of the Precambrian. However, conditions on the ocean floor appear to be quite unsuitable for the rosettes to form by this process and in any case these appear to be composed of opal-CT whereas the silicates in Lake Magadi convert directly to quartz under natural conditions (EUGSTER, 1969). Although the hydrothermal experiments produced opal-CT of the right morphology, the conditions of relatively high temperature and pH are not such as would be expected on the ocean floor. Moreover, the occurrence at Lillicur, which appears to be a straightforward solution and crystallization from presumably fairly pure rainwater at the earth's surface, makes it clear that such silica is readily formed under normal surface conditions. Similar conditions may have prevailed during the precipitation of much such silica.

Nomenclature

With this uncertainty concerning the origin of these "cherts" and indeed chert itself, it is essential that nomenclature be as precise as possible. Chert in its time-honoured usage is a very fine grained q u a r t z - rich rock and is thus inapplicable to the materials discussed in this paper with the exception of the Magadi cherts. As noted above, it is likely that the genesis of these latter is different from that of the Palaeogene and Tertiary "deep sea cherts".

Porcellanite has been suggested by several workers as a suitable term but it is ill defined and unsuitable on that score.

Cristobalite chert has also been used (WISE, BUIE & WEAVER, 1972) but is even more undesirable. The material is not cristobalite but a disordered interstratification of cristobalite and tridymite layers on an atomic scale. As shown above, the material has all the characteristics of friable to lithified opal-CT and is indistinguishable by objective mineralogical criteria from many silica deposits of presumed lacustrine origin (Lillicur, Chowilla) or which are now on land (opoka) and probably never were in deep sea conditions. The material is opal-CT and should be termed such. The term lepispheres proposed by WISE & KELTS (1972) seems very appropriate to describe the rosettes where the opal-CT occurs in this morphology.

Acknowledgements

We wish to thank Dr. A. J. GUDE and Dr. R. A. SHEPPARD for the specimens of silhydrite and magadiite. Dr. C. A. ANDERSON took the photographs of Figs. 3, 4 and 5 on a Cambridge stereoscan and Dr. G. J. BIJVANK those of Figs. 7 and 8 on a JEOL Scanning microscope, J. B. J. produced the remainder of the figures with an Etec Autoscan.

References

BERGER, W. H. & VON RAD, V. (1972): Cretaceous and Cenozoic Sediments from the Atlantic Ocean. – In: HAYES, D. E. et al., Initial Reports of the Deep Sea Drilling Project, vol. XIV. Washington (U.S. Government Printing Office).

CALVERT, S. E. (1971): Nature of silica phases in deep sea charts of the North Atlantic. – Nature Physical Science 234, 133–134.

–   (1971): Composition and origin of North Atlantic deep sea cherts. – Contr. Mineral. Petrol. 33, 273–280.

ERNST, W. G. & CALVERT, S. E. (1969): An experimental study of the recrystallisation of porcellanite and its bearing on the origin of some bedded cherts. – Amer. Sci. 267-A, 114–133.

EUGSTER, H. P. (1967): Hydrous sodium silicates from Lake Magadi, Kenya. Precursors of bedded chert. – Science 157, 1177–1180.

–   (1969): Inorganic bedded cherts from the Magadi Area, Kenya. – Contr. Mineral. Petrol. 22, 1–31.

FLÖRKE, O. W. (1962): Untersuchungen an amorphem und mikrokristallinem SiO_2. – Chem. Erde 22, 91–110.

–   (1972): Transport and deposition of SiO_2 with H_2O under supercritical conditions. – Kristall & Technik 7, 159–166.

GUDE, A. J. & SHEPPERD, R. A. (1972): Silhydrite, $3SiO_2 \cdot H_2O$, a new mineral from Trinity County, California. – Amer. Miner. 57, 1053–1065.

HEATH, G. R. & MOBERLY, R. (1971): Cherts from the Western Pacific, Log 7 deep sea Drilling Project. – In: WINTERER et al., Initial Reports of the Deep Sea Drilling Project, vol. VII. Washington (U.S. Government Printing Office).

HEYDEMANN, A. (1964): Untersuchungen über die Bildungsbedingungen von Quarz im Temperaturbereich zwischen 100 °C und 250 °C. – Beitr. Miner. Petr. 10, 242–259.

JONES, J. B. & SEGNIT, E. R. (1971): The nature of opal I. Nomenclature and constituent phases. – J. Geol. Soc. Aust. 18, 57–68.

JONES, J. B. & SEGNIT, E. R. (1972): Genesis of cristobalite and tridymite at low temperatures. – J. Geol. Soc. Aust. **18**, 419–422.

OEHLER, J. H. (1973): Tridymite-like crystals in cristobalitic "cherts". – Nature Physical Science **241**, 64–65.

REX, R. W. (1969): X-ray mineralogy studies. – Log. 1. – In: EWING, M. et al., Initial Reports of the Deep Sea Drilling Project, vol. 1. Washington (U.S. Government Printing Office), 354.

– (1970): X-ray mineralogy studies – Log. 2. – In: PETERSEN, M. M. A. et al., Initial Reports of the Deep Drilling Project, vol. 2. Washington (U.S. Government Printing Office), 329.

SEGNIT, E. R., ANDERSON, C. A. & JONES, J. B. (1970): A scanning microscope study of the morphology of opal. – Search **1**, 349–351.

SEGNIT, E. R., JONES, J. B. & ANDERSON, C. A. (1973): Opaline silica from the Murray River region west of Wentworth, N.S.W., Australia. – Mem. Nat. Mus. Vic. **34**, 187–194.

VON RAD, V. & RÖSCH, H. (1972): Mineralogy and origin of clay minerals, silica and authigenic silicates, in Log 14 sediments. – In: HAYES, D. E. et al., Initial Reports of the Deep Sea Drilling Project, vol. XIV. Washington (U.S. Government Printing Office).

WEAVER, F. M. & WISE, S. W. (1972): Ultramorphology of deep sea cristobalitic chert. – Nature Physical Science **237**, 56–57.

WISE, S. W., BUIE, B. F. & WEAVER, F. M. (1972): Chemically precipitated sedimentary cristobalite and the origin of chert. – Eclogae Geol. Helv. **65**, 157–163.

WISE, S. W. & HSÜ, K. J. (1971): Genesis and lithification of a deep sea chalk. – Eclogae Geol. Helv. **64**, 273.

WISE, S. W. & KELTS, K. M. (1971): Submarine lithification of Middle Tertiary chalks in the South Atlantic Ocean Basin. – In: Program with Abstracts (VIII Internat. Sediment Congr. 1971, Heidelberg).

– – (1972): Inferred diagenetic history of a weakly silicified deep sea chalk. – Trans. Gulf Coast Ass. Geol. Soc. **22**, 177.

191

Part IV

ORIGIN OF DIAGENETIC SILICA

Editor's Comments
on Papers 10, 11, and 12

One of the fundamental questions and controversies in silica diagenesis concerns the identity of the direct source of diagenetic silica. Two sources have been proposed, volcanic and biogenic. This controversy is as old as the chert problem itself (Shepherd 1972). The results from deep-sea drilling have revived some of the old arguments and also added new ones. Papers 10 and 11 and the following discussion (Papers 12A, B) serve as a good illustration of the fundamental controversy. The arguments center around the Eocene siliceous sediments in the western Atlantic Ocean, the Caribbean Sea, and the stratigraphic correlatives of the Atlantic and Gulf Coastal Plain. The siliceous horizons in the Atlantic (Horizon A) and the Caribbean (Horizon A') were discovered by the DSDP.

Gibson and Towe (Paper 10) base their arguments in favour of a volcanic origin of the silica on the statement that "the presence of an authigenic mineral suite containing clinoptilolite-heulandite, opal-cristobalite, and montmorillonite in varying proportions has usually been considered indicative of the alteration of fine-grained volcanic glass" (p. 197). They realize that this mineralogic suite is commonly associated with sediments rich in diatoms and radiolarians, but they reason that increased productivity of siliceous organisms is a direct result of the availability of excess silica from altered pyroclastic volcanic material. The authors stress that the siliceous sediments being discussed were deposited in different environments, from deep-oceanic to restricted near-shore and that the sediments represent a restricted time interval. They are of the opinion that all these aspects could best be explained by assuming an (ultimate) volcanic source for the silica.

In an earlier paper Reynolds (1970) had described the Lower Tertiary siliceous sediments of the Atlantic Coastal Plain in Alabama. He noted that some strata contain the mineral assemblage montmorillonite, zeolites (heulandite and clinoptilolite), and cristobalite, while other strata are monomineralic, composed of only cristobalite. The first assemblage he explained as alteration products of rhyolitic volcanic ash. He considers the monomineralic cristobalite sediments as "being a primary product precipitated from silica saturated shallow estuarine waters having a high concentration of silica due to nearby volcanism" (p. 837). The author based his conclusions almost exclusively on analytical data rather than on microscopic or electron-microscopic observations.

A few years later Weaver and Wise (Paper 11) published a paper in which they postulate a biogenic origin for the siliceous sediments being discussed. They refer to the papers by Gibson and Towe (Paper 10) and by Reynolds (1970). They argue that the coastal plain "opaline claystones" are altered diatomites, deposited in an open-marine environment rather than in a restricted coastal environment. They base their arguments on SEM studies that showed molds of abundant diatoms and some sponge spicules.

The latter point is an important one because such molds can be observed only with an SEM on fracture surfaces of the sedimentary rocks. The siliceous microfossils themselves generally have disappeared, and their biogenic opal has been replaced by opal-CT lepispheres.

The authors point out that most molds are of marine diatoms, interspersed sponge spicules, and radiolarians. They suggest that most of the montmorillonite, if present, may be detrital. They also note that zeolites are often rare or absent, and therefore, the mineral assemblage, thought to be characteristic of altered rhyolitic ash, is far from ubiquitous in the sediments discussed.

As to the ultimate origin of the silica, Weaver and Wise suggest that during the Eocene profuse blooms of siliceous plankton occurred at various times, the nutrients being provided by favorable ocean-current systems. Finally, they remark that if volcanic ashes were deposited at all, they were volumetrically insignificant and were incidental rather than causative to biogenic silica-fixation.

A year later Gibson and Towe (Paper 12A) took up the argument again. They define their point of view as a "dual and partially sequential cause; direct volcanic contributions to help explain the presence of smectite and zeolite but with accompanying increased nutrients (phosphorus, iron, and silica from dissolved fine pyroclastics) which would increase the productivity of siliceous organisms above normal background levels, providing increased contributions to the sediments'' (p. 207). The authors are of the opinion that oceanic circulation alone cannot be sufficient for increased productivity. Volcanic ash would have to be supplied to the currents. They accuse Weaver and Wise of using misleading and highly selective data.

Weaver and Wise's reply (Paper 12B), rejects the accusation of using misleading and highly selective data by pointing out that their arguments were based on a far broader range of evidence than Gibson and Towe (Paper 10) originally used. They reiterate most of their earlier arguments, the most important one being that many of the opaline sediments contain hardly any of the so-called alteration products of volcanic ash, viz. zeolites and montmorillonite. The authors consider the question of whether volcanic activity was the cause of increased production of siliceous plankton "a highly speculative matter" (p. 210). They point out that many authors have questioned this relationship and that some have suggested specific climatic conditions (lateritization) for increased silica supply to the oceans during Eocene time. (The latter theory, incidentally, will be taken up again in the commentaries for Papers 17 and 18 in this volume.)

Supported by many references, Weaver and Wise argue that the alteration of volcanic ash on the sea floor cannot have supplied sufficient silica and nutrients to the surface waters to affect the productivity of siliceous plankton.

No doubt the controversy about the volcanic versus biogenic silica source will be with us for quite some time. The subject comes up in most papers dealing with silica diagenesis. Table 2 in the Introduction will assist the reader in finding other papers in this volume that comment on this controversy. Increased detailed observations and geochemical work and experiments no doubt will lead to the eventual resolution of this controversy.

REFERENCES

Reynolds, W. R. (1970) Mineralogy and stratigraphy of Lower Tertiary clays and claystones of Alabama. *Jour. Sed. Petrology* **40:** 829–838.

Shepherd, W. (1972) Flint—*Its origins, properties, and uses.* Faber and Faber, London, 255 pp.

10

Reprinted from *Science* **172**:152–154 (Apr. 9, 1971)

Eocene Volcanism and the Origin of Horizon A

Thomas G. Gibson and Kenneth M. Towe

Abstract. *A series of closely time-equivalent deposits that correlate with seismic reflector horizon A exists along the coast of eastern North America. These sediments of Late-Early to Early-Middle Eocene age contain an authigenic mineral suite indicative of the alteration of volcanic glass. A volcanic origin for these siliceous deposits onshore is consistent with a volcanic origin for the cherts of horizon A offshore.*

Widespread sedimentary deposits throughout the western North Atlantic area contain mineralogical suites indicative of altered pyroclastic material and can be interpreted as the result of a series of Late-Early to Early-Middle Eocene volcanic events (*1*). Subsequent drilling during legs of the JOIDES (Joint Oceanographic Institutions for Deep Earth Sampling) program has shown the presence of hard, siliceous beds of common age in many of the deep oceanic areas sampled. These radiolarian-diatom and chert deposits, which often inhibit drilling operations, have been dated as Early to Middle Eocene (*2*). We suggest that these essentially time-equivalent deposits have resulted from the marine diagenetic alteration of volcanic material coupled with an increase in the productivity and preservation of siliceous microplankton in response to an increase in both silica and the nutrient phosphorus that result from the ash alteration process. Similar chert deposits have been found to be seismic reflecting horizons in the ocean basins, and the evidence is compelling that horizon A in the western North Atlantic is a result of this Eocene volcanism.

We originally found siliceous material in Eocene sediments while examining dredge hauls off Long Island in the Hudson Canyon, in other nearby canyons, and on the continental slope to the east along the Atlantic margin of the United States. The rocks are primarily chalks and limestones containing as much as 54 percent SiO_2 (by weight); most is in the form of opal-cristobalite with little of terrigenous origin. The samples of latest Early Eocene age contain the largest amounts of silica, and those of Paleocene, earliest and latest Eocene, and Oligocene age in the same area contain considerably less silica (*3*). Additional mineralogical examination showed the presence of the zeolite clinoptilolite and clay minerals of the montmorillonite group. In addition to other textural considerations the authigenic nature of the zeolites is demonstrated in Fig. 1, where crystal growth of clinoptilolite has incorporated a coccolith. Although alternative explanations have been proposed (*4*), the presence of an authigenic mineral suite containing clinoptilolite-heulandite, opal-cristobalite, and montmorillonite in varying proportions has usually been considered indicative of the alteration of fine-

197

Fig. 1. Electron micrograph of siliceous marine rock dredged from the "70-30" Canyon in the Atlantic continental slope off Long Island. The rocks are the same age as seismic reflector horizon A in the western North Atlantic. The authigenic nature of the zeolite is shown by the clinoptilolite crystal that has incorporated a coccolith element during crystal growth ($\times$ 20,000).

grained volcanic glass (5).

Examination of more samples from the western North Atlantic region and of published reports reveals the widespread nature of this Eocene mineralogical suite. Evidence is found in New Jersey in the Manasquan formation (Ash Marl) from mineralogical and paleontological data (6). In a study of Eocene clay minerals in the coastal plain of South Carolina, Heron (7) reported opal-cristobalite and montmorillonite but discounted a volcanic source in the absence of zeolites. New analyses (8) on coarser fractions in which zeolites are more common showed the presence of clinoptilolite in these sediments. Weaver (9) has reported this suite from the

JOIDES cores taken on the Blake Plateau off the coast of Florida. Reynolds (*10*) has reported this suite from the Eocene of the Alabama Coastal Plain, and it has been observed, together with bentonites, in Mississippi by Grim (*11*) and by Wermund and Moiola (*12*) in beds as much as 30 m thick. The siliceous Toledo member of the Eocene Universidad formation in Cuba averages 10 to 15 m in thickness (*13*). Samples of these Cuban rocks obtained from the collection of Brönnimann and Rigassi in the Museum of Natural History at Basel, Switzerland (*14*), were x-rayed and found to contain major quantities of clinoptilolite, opal-cristobalite, and montmorillonite. Although somewhat difficult to evaluate because of the dilution effect of terrigenous material in some areas, there appears to be a general decrease in the amounts of siliceous materials from south to north.

The age relations inferred from the stratigraphy and the associated microfossils attest to the relatively restricted and similar time interval involved over the entire geographic range. The planktonic foraminiferal assemblages can be referred to the *Globorotalia aragonensis–Globorotalia palmerae* zone, and the nannoplankton assemblages belong to the *Discoaster lodoensis* and *Marthasterites tribrachiatus* zones. Chert beds of horizon A encountered in JOIDES drilling operations have been assigned to similar Early to Middle Eocene ages (*2*).

The Hudson Canyon and nearby offshore samples are dated by both planktonic foraminifera (*15*) and nannoplankton (*16*) and fall within these zones. The New Jersey strata contain planktonic foraminifera belonging to this interval (*6*), as do the cores from the JOIDES holes off Jacksonville, Florida (*17*). Both foraminifera and nannoplankton assign the Cuban material to these zones (*13*), whereas the Tallahatta formation in Mississippi and Alabama contains sparse amounts of nannoplankton assigned to this interval (*18*). The Black Mingo, Warley Hill, and Congaree formations in South Carolina are Late Paleocene through Middle Eocene in age (*7, 19*), but, because of deposition under shallow, nearshore conditions, planktonic zonations are not available.

The geographically widespread and closely time-equivalent nature of the siliceous sediments from varied depositional environments is thus established. Mineralogical evidence indicates that much of the siliceous composition is derived from the alteration of volcanic material with probable additional contributions in the deep sea from abundant siliceous microfossils. The association of siliceous sediments with radiolarian-diatom deposits is a common one (*20*), and the relationship here deserves discussion. The explanation that the silica for these cherts and related rocks is derived primarily from the dissolution of biogenic opals lacks, not only an extensive radiolarian-diatom source of silica, but also a reason for their increased productivity.

The explanation that the presence of abundant silica in the water can account for the increased productivity of siliceous organisms is inadequate in itself, although increased silica abundance should act to slow down the dissolution rate of dead tests, thus increasing their apparent concentration in sediments (*21*). This process could also be influenced by the presence of heavier tests in Eocene radiolaria relative to those found in the Quaternary (*22*). Pyroclastic volcanic material, however, would not only provide additional silica but also the nutrient phosphorus. The average phosphorus content of volcanic rock types ranges from 0.10 to 0.45 percent P_2O_5 (*23*). Although meaningful quantitative estimates would be difficult to make, the relatively high solubility of the volcanic glass coupled with its fine particle size would undoubtedly have made available additional phosphorus for biogenic consumption over the geologic interval under discussion. The increased productivity would account for high concentrations of radiolarian-diatom oozes containing more robust tests, both resulting from the concomitant increase in available silica.

It has been suggested (*24*) that the initiation of cold, deep water circulation in the North Atlantic as a result of seafloor spreading in the early Cenozoic might augment the distribution of nutrients and thus the productivity of siliceous organisms. Although such changes in oceanic circulation might help to explain the oceanic cherts, they cannot also explain the extensive and time-equivalent nearshore sediments of volcanic origin on the continents or the relatively restricted time interval involved.

Sources for the volcanic materials are unknown, but, in view of the apparent general decrease in the thickness of siliceous deposits in the northerly direction together with the absence of significant Eocene volcanic activity in this same direction, the most likely source appears to be the then active Middle American and Caribbean region (*25*). In addition to atmospheric dispersal, the northward movement of the Florida and Greater Antillean currents and the Gulf Stream would distribute volcanic material in this general direction along the east coast of the United States and into the oceanic areas with relative uniformity when considered over the geologic time interval involved.

Clear evidence of volcanism in the Late-Early and Early-Middle Eocene exists in the sediments all along the Atlantic and eastern Gulf coast of the United States. The presence of widespread siliceous deposits of similar age in the oceanic areas of the western North Atlantic cannot be considered fortuitous and of a different origin. Ewing *et al.* (*26*) have commented on the distribution and synchroneity of horizon A and speculated about the presence of correlative deposits on continental areas. The evidence presented here provides information on such deposits, information that bears on the origin of the cherts in question.

A volcanic origin provides a consistent explanation for the presence of these siliceous deposits in a wide variety of environments ranging from brackish and nearshore to deep marine over a wide geographic area. It also provides an explanation for the relatively restricted time interval involved.

References and Notes

1. K. M. Towe and T. G. Gibson, *Geol. Soc. Amer. Annu. Meeting Program*, 1968, p. 299.
2. M. Ewing *et al.*, *Initial Reports of the Deep-Sea Drilling Project* (Government Printing Office, Washington, D.C., 1969), vol. 1; M. N; A. Peterson *et al.*, *Initial Reports of the Deep-Sea Drilling Project* (Government Printing Office, Washington, D.C., 1970), vol. 2; S. Gartner, Jr., *Science* **169**, 1077 (1970).
3. T. G. Gibson, *Geol. Soc. Amer. Bull.* **81**, 1813 (1970).
4. G. Brown, J. A. Catt, A. H. Weir, *Mineral. Mag.* **37**, 480 (1969).
5. M. N. Bramlette and E. Posnjak, *Amer. Mineral.* **18**, 167 (1933); K. S. Deffeyes, *J. Sediment. Petrol.* **29**, 602 (1959); J. C. Hathaway and P. L. Sachs, *Amer. Mineral.* **50**, 852 (1965); R. L. Hay, *Geol. Soc. Amer. Spec. Pap. 85* (1966).
6. R. L. C. Enright, Jr., personal communication (1968); thesis, Rutgers University (1969).
7. S. D. Heron, Jr., *Geol. Soc. Amer. Annu. Meeting Program* (1962), p. 71A; S. D. Heron, Jr., G. C. Robinson, H. S. Johnson, Jr., *S.C. State Develop. Board Div. Geol. Bull. 31* (1965).
8. We thank S. D. Heron, Jr., for restudying these sediments at our request and providing new x-ray data on their mineralogy.
9. C. E. Weaver, *Southeast. Geol.* **9**, 57 (1968).
10. W. R. Reynolds, *J. Sediment. Petrol.* **40**, 829 (1970)
11. R. E. Grim, *Miss. State Geol. Surv. Bull. 30* (1936).
12. E. G. Wermund and R. J. Moiola, *J. Sediment. Petrol.* **36**, 248 (1966).
13. P. Brönnimann and D. Rigassi, *Eclogae Geol. Helv.* **56**, 193 (1963).
14. We thank Dr. E. Gasche for providing these samples for study.

15. T. G. Gibson, J. E. Hazel, J. F. Mello, *U.S. Geol. Surv. Prof. Pap. 600-D* (1968), p. 222.
16. M. N. Bramlette, personal communication (1969).
17. JOIDES, *Science* **150**, 709 (1965).
18. M. N. Bramlette and F. R. Sullivan, *Micropaleontology* **7**, 129 (1961).
19. W. K. Pooser, *Univ. Kans. Paleontol. Contrib. Arthropoda*, Art. 8 (1965).
20. N. L. Taliaferro, *Calif. Univ. Dep. Geol. Sci. Bull.* **23**, 1 (1933); M. N. Bramlette, *U.S. Geol. Surv. Prof. Pap. 212* (1946).
21. W. R. Riedel, *Soc. Econ. Paleontol. Mineral. Spec. Publ.* **7**, 80 (1959).
22. T. C. Moore, Jr., *Geol. Soc. Amer. Bull.* **80**, 2103 (1969).
23. R. A. Daly, *Igneous Rocks and the Depths of the Earth* (McGraw-Hill, New York, 1933).
24. R. S. Dietz and J. C. Holden, *J. Geophys. Res.* **75**, 4939 (1970); W. A. Berggren and J. D. Phillips, *Symp. Geol. Libya*, in press.
25. H. J. MacGillavry, *Proc. Koninklijke Ned. Akad. Wetensch. Ser. B* **73**, 64 (1970).
26. J. Ewing, C. Windisch, M. Ewing, *J. Geophys. Res.* **75**, 5645 (1970).
27. We thank R. Cifelli, J. E. Hazel, and W. G. Melson for reviewing the manuscript. M. N. Bramlette provided valuable assistance in nannoplankton identification and biostratigraphy.

11

Reprinted from *Science* **184**:899–901 (May 24, 1974)

Opaline Sediments of the Southeastern Coastal Plain and Horizon A: Biogenic Origin

Fred M. Weaver and Sherwood W. Wise, Jr.

Abstract. *Scanning electron microscope techniques show that Eocene opaline claystones (fuller's earth and buhrstone) of the Atlantic and Gulf Coastal Plain, deposits long considered volcanic in origin, are actually highly altered diatomites formed as transgressive facies in normal marine continental shelf environments. These findings are in agreement with a biogenic origin for time-equivalent horizon A and A″ deep-sea cherts of the North Atlantic and Caribbean.*

Opaline (cristobalite-rich) Eocene claystone deposits of the Atlantic and Gulf Coastal Plain have recently been cited in *Science* (*1*) and elsewhere (*2*) as examples of altered rhyolitic ashes which accumulated in nearshore or brackish coastal environments. Such ashes are also thought to have been distributed by atmospheric and water currents into the North Atlantic Ocean basin where they were presumably responsible for the formation of the cristobalite-rich, horizon A Eocene cherts (*1, 3*). We present evidence here to show that opaline claystones of the coastal plain are altered diatomites, not ashes, and that they formed in normal marine rather than in restricted coastal environments. Our evidence is compatible with a biogenic rather than a volcanic origin for the horizon A cherts and their Caribbean equivalents (horizon A″).

Opaline claystones are unusually porous, lightweight siliceous rocks which possess oil clarification properties (*4*). Accordingly, they have been referred to locally as fuller's earth (*5*)

or buhrstone (*6*). Scanning electron microscopy of fracture surfaces of opaline claystones from 14 Southeastern Coastal Plain localities (Mississippi to South Carolina; see Table 1) reveals siliceous microfossils which occur as molds in 90 percent of the samples examined. The fossils are most abundant in samples which contain 60 to 90 percent SiO_2. The opaline material is unidimensionally disordered alpha-cristobalite (*7*) in the form of bladed microspherulites (*8*). Most of the microfossil molds are of marine diatoms including large and small centrics (Fig. 1A), pennates and forms which resemble *Triceratium* (Fig. 1B), and *Actinoptychus* (Fig. 1C). Sponge spicule (Fig. 1D) and radiolarian molds (*9*) are interspersed in the South Carolina and Alabama material. Clearly, the opaline claystones represent highly altered diatomite deposits rather than ash beds. Most microfossils in the deposits, however, have been completely destroyed by dissolution.

Siliceous microfossils have not been reported previously in South Carolina

Table 1. Opaline claystone samples which contain siliceous microfossil molds. The Black Mingo and McBean units were collected by S. D. Heron. All other samples were collected by the authors.

Formation	Samples and localities	Age
Nanafalia (Grampian Hills member)	GH-1 (Wilcox County, Ala.)	Late Paleocene
Black Mingo (opaline facies)	9-6-1 (Sandy Run Creek, S.C.); 9-9-1 (Big Beaver Creek, S.C.); 6-10-4, 9-11-4 (Little Beaver Creek, S.C.); 9-18-1, 9-18-2 (Bates Mill Creek, S.C.); 9-67-1, 9-67-2, 9-67-3 (Thelma Hill property, Calhoun County, S.C.); 9-68-2 (Dicks Swamp, S.C.); A-183-1 (Williamsburg Bridge, S.C.); 43-6-1, (Tavern Creek, S.C.); 43-7-5 (Holy Cross Church, Sumter County, S.C.)	Early-Middle Eocene
McBean	A-3-1, A-3-2 (Early Branch, S.C.)	Middle Eocene
Tallahatta	T-3 [Choctaw County, Ala.; locality 135 of Toulmin and LaMoreaux (17)]; I-10-12, I-10-13, 33-1 (U.S. Highway I-10, Meridian, Miss.)	
Barnwell (Twiggs Clay member)	KL-1 (Georgia-Tennessee Clay Corporation pit, Wrens, Ga.)	Late Eocene

and Georgia fuller's earth deposits to our knowledge, although they are mentioned in an 1894 report on the Tallahatta buhrstones of Alabama (6). The significance of that early observation of siliceous microfossils, however, was overlooked during later mineralogical studies (10), particularly x-ray studies (2) which established the presence of zeolite (clinoptilolite), montmorillonite, and alpha-cristobalite in the Tallahatta formation. Mineralogists concluded (1, 2, 4) that the clinoptilolite, cristobalite, and at least a portion of the montmorillonite in Eocene coastal plain sediments represent a suite of alteration products from devitrified volcanic glass. Most of the montmorillonite, however, may be detrital [(4); see also (11)], and zeolites are generally rare or absent in the cristobalite-rich beds of the Tallahatta formation (2) and of the Black Mingo formation of South Carolina (4). Instead, zeolites are concentrated in the

soft clay units of the Alabama material (2), and no zeolites are present in the Twiggs Clay (Barnwell formation) of Georgia (12).

Early diagenesis of the Eocene diatomites of the coastal plain probably followed a pattern recently postulated for deep-sea siliceous ooze diagenesis, that is, in situ dissolution of biogenous opal with silica reprecipitated inorganically as authigenic disordered alpha-cristobalite (8, 13). In places where the opal contents of deep-sea oozes are extremely high, molds of microfossils may not be preserved. Significantly, once diagenesis begins, practically all available biogenous opal in the affected material may be converted to cristobalite with little trace remaining of the original substance (14). Radiolarians and certain diatom taxa (for example, *Actinoptychus*) which occur in our coastal plain material do not tolerate fresh or brackish water; therefore, these deposits must have formed in

203

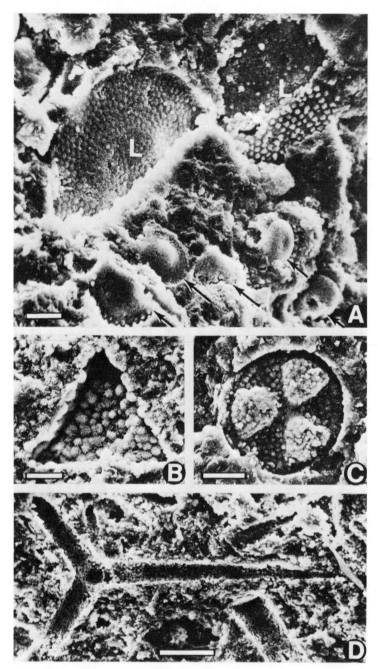

Fig. 1. Microfossil molds in Eocene opaline claystones of the Southeastern Coastal Plain. (A) Centric diatoms (arrows indicate small centrics; *L*, large centrics); sample KL-1, Twiggs Clay member (scale, 20 μm). (B) *Triceratium* sp. (diatom). Note bladed microspherulites of cristobalite which line mold interior; sample 43-7-5, Black Mingo formation (scale, 5 μm). (C) *Actinoptychus* sp. (diatom); sample KL-1, Twiggs Clay member (scale, 5 μm). (D) Sponge spicule; sample 9-67-3, Black Mingo formation (scale, 50 μm).

open marine, nonrestricted shelf environments. Abundant pteropods in the buhrstones of the Tallahatta formation (*15*) and the lithostratigraphies of the opaline claystone formations in question also suggest normal marine environments. At all outcrop localities studied, opaline claystones overlie quartz sands or montmorillonite-rich clays, or both, which can be interpreted as beach, tidal flat, or offshore bar facies of transgressive sequences [for example, see (*16*)]. The opaline sediments represent near- to offshore, normal marine facies of these transgressive sequences. In Alabama (*17*) and Georgia (*12, 18*), opaline claystones are overlain by regressive sand units which complete the record of what may be considered classic transgressive-regressive depositional cycles.

Our evidence makes it necessary to revise presently accepted paleoenvironmental interpretations and models of Eocene coastal lithofacies. For instance, Reynolds' (*2*) proposal of restricted back-bay coastal lagoons as sites of cristobalite deposition is incompatible with the open marine environment we demonstrate. His model of direct chemical precipitation of cristobalite from circulating bottom waters, therefore, is invalid and unnecessary in view of the fact that the immediate silica source was diatomite rather than volcanic ash. Similarly, the Twiggs Clay of Georgia should be considered not a regressive unit composed of detrital clastics (*18*) of a deltaic complex (*19*) but rather a diatomaceous member of a transgressive sequence which also includes the time-equivalent outer shelf marls and limestones illustrated by Carver (*12*) in his figure 3. We believe such reinterpretations will aid (i) location of additional deposits of economically important fuller's earth in the coastal plain and (ii) more faithful reconstruction of Eocene paleoenvironments of deposition in the western Atlantic-Caribbean area.

With respect to the source of the diatomite deposits, profuse blooms of siliceous plankton along the continental shelves at various times during the Eocene could have been stimulated by nutrients supplied by favorable ocean current systems. Ramsay (*20*) presents a paleocurrent model which explains high Eocene siliceous plankton productivity not only in the Gulf of Mexico and Caribbean but also in the North Atlantic where the horizon A cherts formed. Our data, which include a failure to detect textural or structural features that would suggest ash deposition, are compatible with Ramsay's model. Some zeolites do occur in Eocene coastal plain deposits, and these may owe their origin to the deposition of various types of volcanic ash; nevertheless, these ashes were certainly not as volumetrically important as mineralogists (*1, 2, 10*) have suggested. Any ash deposition was apparently incidental to rather than causative of a general pattern of biogenic silica deposition (*21*), as indicated by the fact that the times of formation of the Paleocene Grampian Hills member of the Nanafalia formation and of the Upper Eocene Twiggs Clay (15 m thick) are not coincident with the schedule of rhyolitic volcanic activity in the Caribbean defined by Gibson and Towe (*1*) and Mattson and Pessagno (*3*). We conclude, therefore, that the opaline claystones of the coastal plain, which are all essentially identical in hand specimen, mineral content, and microstructure, owe their unusual character to a biogenic mode of deposition.

References and Notes

1. T. G. Gibson and K. M. Towe, *Science* **172**, 152 (1971).
2. W. R. Reynolds, *J. Sediment. Petrol.* **40**, 820 (1970).

3. P. H. Mattson and E. A. Pessagno, Jr., *Science* **174**, 138 (1971). Strictly speaking, the horizon A material is not true chert because it is opaline-rich rather than quartz-rich.
4. S. D. Heron, Jr., *S.C. State Dev. Board Div. Geol. Geol. Notes* **13**, 27 (1969).
5. E. Sloan, *S.C. Geol. Surv. Bull. (Ser. 4) 2* (1908); C. W. Cooke, *U.S. Geol. Surv. Bull. 867* (1936); H. X. Bay, *U.S. Geol. Surv. Bull. 901* (1940), p. 83.
6. C. Lyell, *Quart. J. Geol. Soc. Lond.* **1**, 429 (1845); E. A. Smith, L. C. Johnson, D. W. Langdon, *Ala. Geol. Surv. Spec. Rep. 6* (1894).
7. O. W. Floerke, *Ber. Deut. Keram. Ges.* **10**, 217 (1955).
8. S. W. Wise, B. F. Buie, F. M. Weaver, *Eclogae Geol. Helv.* **65**, 157 (1972).
9. Illustrations of radiolarian molds and of other diatoms will appear in an extended paper (S. W. Wise and F. M. Weaver, *Trans. Gulf Coast Assoc. Geol. Soc.*, in press).
10. R. E. Grim, *Miss. State Geol. Surv. Bull. 30* (1936).
11. P. W. Biscaye, *Geol. Soc. Am. Bull.* **76**, 803 (1965).
12. R. E. Carver, *Fla. Dep. Nat. Resour. Bur. Geol. Spec. Pub. 17* (1972), p. 91. S. M. Pickering [*Ga. Geol. Surv. Bull. 81* (1970)] found no textural or structural evidence of ash deposition in the Twiggs Clay but did discover abundant benthonic foraminifers. He assumed that "the Twiggs Clay was precipitated as colloidal particles formed by the action of salt water on normal terrestrial clay minerals."
13. G. R. Heath and R. Moberly, in *Initial Reports of the Deep Sea Drilling Project* (Government Printing Office, Washington, D.C., 1971), vol. 7, p. 991; F. M. Weaver and S. W. Wise, *Nature (Lond.)* **237**, 56 (1972).
14. F. M. Weaver and S. W. Wise, *Antarct. J. U.S.*, in press.
15. J. Gardner, *Geol. Soc. Am. Mem. 67* (1957), vol. 2, p. 573.
16. L. D. Toulmin, *Trans. Gulf Coast Assoc. Geol. Soc.* **19**, 465 (1969), figure 3.
17. ——— and P. E. LaMoreaux, *Ala. Geol. Surv. Map 8* (1953).
18. R. E. Carver, *Southeast. Geol.* **7**, 83 (1966).
19. J. F. L. Connell, *ibid.* **1**, 59 (1959).
20. A. T. S. Ramsay, *Nature (Lond.)* **233**, 115 (1971).
21. Gibson and Towe's interesting suggestion (*1*) that rhyolitic volcanic activity could possibly stimulate high productivity of siliceous plankton in deep-sea areas has yet to be substantiated by independent evidence. Nevertheless, any such activity on a regional scale in the Caribbean, even if it did somehow serve as a stimulus, would be entirely inadequate to account for the extensive Eocene chert deposits which have now been encountered by the Deep Sea Drilling Project in the Pacific, Indian, and Southern ocean basins as well as in the Atlantic and Caribbean.
22. We thank S. D. Heron (Duke University) for numerous opaline claystone samples, L. D. Toulmin (Florida State University) for helpful discussion and guidance in the field, D. S. Cassidy (Florida State University) for photographic enlargements, and W. I. Miller and P. F. Ciesielski (Florida State University) for participation in the electron microscope studies. Support for this research was provided by a Sigma Xi research grant to F.M.W., a Florida State University faculty research grant to S.W.W., and a grant from the Donors of the Petroleum Research Fund administered by the American Chemical Society.

12A

Reprinted from *Science* **188**:1221 (June 20, 1975)

ORIGIN OF HORIZON A: CLARIFICATION OF A VIEWPOINT

Thomas G. Gibson

U.S. Geological Survey

Kenneth M. Towe

Smithsonian Institution

As shown in earlier studies (*1*), Weaver and Wise (*2*) report that siliceous microfossils (diatoms, sponge spicules, radiolarians) occur in some of the high-purity Tertiary opaline deposits of the Atlantic and Gulf Coastal Plain. They conclude therefore that these deposits as well as horizon A must be biogenic in origin. We wish to comment on this as it regards their interpretation of our earlier report (*3*) on this subject.

As a result of the JOIDES (Joint Oceanographic Institutions for Deep Earth Sampling) drilling program, a prominent and widespread oceanic seismic reflector known as horizon A has been shown to consist of hard, siliceous beds containing diatoms and radiolarians and to have a narrowly defined age from late Early to early Middle Eocene (*4*). A principal problem has been to explain the origin of this geographically widespread and non-linear siliceous horizon and not the other scattered and discontinuous siliceous deposits situated stratigraphically higher or lower.

Impressed by the presence of siliceous microfossils, Dietz and Holden (*5*), Berggren and Phillips (*6*), and Ramsay (*7*) have offered explanations based on several oceanic circulation models to explain horizon A as an entirely biogenic deposit. A major biogenic role in the formation of horizon A is undeniable. But no oceanic circulation model *alone* can explain the occurrences of smectites and zeolites that are found associated with almost all of the sediments correlating with this unusual horizon and found not only in the Atlantic and Gulf Coastal Plain but also on the shelf and in the Caribbean. Accordingly, Gibson and Towe (*3*), supported by Mattson and Pessagno (*8*), considered that a combined volcanic and biogenic explanation for the time-equivalent deposits was more consistent with *all* of the facts than a strictly biogenic explanation. The widespread distribution of horizon A in the western North Atlantic and the composition of the deposits themselves led us to a dual and partially sequential cause: direct volcanic contributions to help explain the presence of smectite and zeolite but with accompanying increased nutrients (phosphorus, iron, and silica from dissolved fine pyroclastics) which would increase the productivity of siliceous organisms above normal background levels, providing increased contributions to the sediments. We wish to clarify that we did not state that the entire source of the relevant deposits was altered volcanic ash. We did not extend our conclusions to other siliceous deposits of different ages in the Atlantic and Gulf Coastal Plain nor did we extend them to siliceous deposits in other oceans, as Weaver and Wise have implied (*2*).

Silica is constantly being mobilized and deposited by diatoms in the world oceans, and few will argue about this biogenic contribution. But for horizon A, an oceanwide "chert" deposit, some mechanism is needed to raise the siliceous productivity and the deposition and preservation above the normal background level over a wide area in the Atlantic region. Changes in sediment dilution or in oceanic circulation patterns can be invoked to explain only part of the deposit, since such changes do not normally also provide a mechanism for zeolitic and smectite clays. However, wind and ocean currents can distribute soluble, *fine* pyroclastics and thus add potential planktonic nutrients that would contribute to the formation of the varied deposits observed. The relationship between siliceous organisms and volcanism has been noted from the time of Lyell (*9*) up to the present (*10*). In support of this concept, Lisitsyn (*11*) has presented consistent evidence for the *indirect* influences of volcanism in the active Bering Sea region on such nutrients as iron and phosphorus and the importance of these elements to plankton, notably diatoms (*12*). More recently, Huang *et al.* (*13*) have provided still further support for this viewpoint.

We noted for horizon A the consistent occurrence of correlative deposits indicative of both volcanic and biogenic activity (*3*), although the degree of influence of one aspect or the other varies from place to place as might be expected. We believe this to be a noncoincidental cause-and-effect relationship, and we may be wrong; but, be that as it may, in order that any alternative explanation be correct, it must be based on all the relevant data rather than the misleading and highly selected data chosen by Weaver and Wise (*2*).

References and Notes

1. R. E. Grim, *Miss. State Geol. Surv. Bull. 30* (1936); S. D. Heron, Jr., *Geol. Soc. Am. Spec. Pap. 73* (1962), p. 171; B. F. Buie and C. H. Oman, *ibid.*, p. 2; E. A. Smith, L. C. Johnson, D. W. Langdon, *Ala. Geol. Surv. Spec. Rep. 6* (1894); E. N. Lowe, *Miss. State Geol. Surv. Bull. 41* (1919); B. F. Buie and L. R. Gremillion, *Ga. Mineral Newsl.* **16**, 20 (1963).
2. F. M. Weaver and S. W. Wise, Jr., *Science* **184**, 899 (1974).
3. T. G. Gibson and K. M. Towe, *ibid.* **172**, 152 (1971).
4. M. Ewing *et al.*, *Initial Reports of the Deep Sea Drilling Project* (Government Printing Office, Washington, D.C., 1969), vol. 1; M. N. A. Peterson *et al.*, *ibid.*, vol. 2; S. Gartner, Jr., *Science* **169**, 1077 (1970); J. Ewing, C. Windisch, M. Ewing, *J. Geophys. Res.* **75**, 5645 (1970).
5. R. S. Dietz and J. C. Holden, *J. Geophys. Res.* **75**, 4939 (1970).
6. W. A. Berggren and J. D. Phillips, in *Symposium on the Geology of Libya* (University of Libya, Tarabulus, 1971).
7. A. T. S. Ramsay, *Nature (Lond.)* **233**, 115 (1971).
8. P. H. Mattson and E. A. Pessagno, Jr., *Science* **174**, 138 (1971).
9. C. Lyell, *Principles of Geology* (Appleton, New York, ed. 11, 1887), vol. 1, p. 645.
10. H. Blatt, G. Middleton, R. Murray, *Origin of Sedimentary Rocks* (Prentice-Hall, Englewood Cliffs, N.J., 1972), p. 540.
11. A. P. Lisitsyn, *Recent Sedimentation in the Bering Sea* (English translation, Department of Commerce, Washington, D.C., 1969).
12. Lisitsyn (*11*) is, however, careful to deny any *direct* chemogenic relationship between major siliceous deposits and volcanism, a viewpoint popular in some earlier geologic literature.
13. T. C. Huang, R. H. Fillon, N. D. Watkins, D. M. Shaw, *Deep-Sea Res.* **21**, 377 (1974), and references therein.
14. We acknowledge advice and assistance from J. E. Hazel, R. F. Fudali, T. E. Simkin, and W. Poag.

12B

Reprinted from *Science* **188**:1221–1222 (June 20, 1975)

ORIGIN OF HORIZON A: CLARIFICATION OF A VIEWPOINT—A REPLY

Fred M. Weaver
Sherwood W. Wise, Jr.

Florida State University, Tallahassee

We are pleased that Gibson and Towe (1) accept our documentation (2) of biogenic opaline silica deposits within the Southeastern Coastal Plain but are mystified that they consider our data "misleading and highly selected" since these data cover a far broader range of evidence than they themselves are willing to consider. We have demonstrated (2, 3) a historic pattern of intermittent biogenous silica deposition in coastal plain sediments ranging from Paleocene to Eocene in age [it should be noted that, in South Carolina and elsewhere, these opaline lithologies are found well into the Miocene (for example, Coosawhatchie clay of the Hawthorne Formation)]. Until recently, practically all of these deposits have been variously classified as bentonites or altered volcanic ash deposits, a notion that we hope we have laid to rest through the presentation of fossil evidence.

We see no reason to assume that our evidence for a biogenic origin for the opaline facies of the early Middle Eocene Black Mingo, McBean, and Tallahatta formations is not compatible with the often-postulated (4) biogenic origin for the time-equivalent horizon A "cherts." Although we consider Gibson and Towe's postulation (1, 5) of a volcanic origin for this deposition interesting and encourage further research into the matter, we do not find present evidence for their speculation compelling.

Whereas various quantities of zeolite, montmorillonites, or unaltered ash indicative of volcanic activity are present in various portions of the Paleogene sequences in the areas in question, we see little extraordinary about the quantity of such materials associated with the late Early to early Middle Eocene portions of those sections. Indeed, Mattson *et al.* (6) have observed that montmorillonite occurs in the JOIDES (Joint Oceanographic Institutions for Deep Earth Sampling) cores at levels above the reflecting horizons and is less abundant in the cherty horizons than in the clays. We are impressed by the amount of biogenic silica in these sequences. For some deposits such as the Black Mingo Formation of South Carolina, zeolites indicative of ash deposition are extremely rare; however, we have demonstrated (2, 3) for the first time the presence of siliceous microfossils in that material. Opaline sediments of similar lithology and slightly younger age in Georgia (Twiggs Clay of the Barnwell Formation) contain no evidence of volcanic material but have yielded abundant siliceous microfossil remains (2). We conclude that the immediate source of silica for these high-purity opaline deposits is biogenic silica, and that contributions of silica to these deposits arising from the decomposition of volcanic ash are relatively minor or, in some cases, completely lacking. Thus the volcanic ashes are not as

209

volumetrically important as previously assumed, and, as stated earlier (2, p. 901), "any ash deposition was apparently incidental to rather than causative of a general pattern of biogenic silica deposition."

Similar conclusions can be drawn in regard to the origin of the high-purity layer A "cherts." We have commented elsewhere (7) on the apparent low yield of chert-forming silica from deep-sea bentonites. The question of whether volcanic activity was the cause of enhanced plankton production and silica deposition during the late Early to early Middle Eocene is, of course, a highly speculative matter. Calvert's (8) tabulations suggest that, for the present, the amounts of silica supplied annually to the oceans by submarine volcanism are insignificant by comparison with that delivered in solution by streams. After assessing data from modern ocean basins, Garrison (9) and Lisitsyn (10) strongly argue against a direct connection between volcanism and the formation of pelagic sediments, such as volcanically induced chemical precipitation or plankton blooms. For the volcanically active Bering Sea region, Lisitsyn (10, p. 117) finds that "The hydrochemical characteristics established during the last twenty years do not provide any indications of any appreciable influence of volcanism on the water masses, although in many cases investigations were performed during subaerial and submarine volcanic eruptions."

This is not to say that the supply of silica to the general reservoir of the Atlantic, Gulf, and Caribbean could not have been increased during the Early Tertiary. Frakes and Kemp (11) have summarized evidence for the Paleocene-Eocene poleward expansion of warm and humid climates. This resulted in intense weathering and laterization of soils as far north as 55°N and as far south as 45°S during the Early and Middle Eocene, and probably increased the quantitatively significant input of silica to the oceans from streams. Early Tertiary volcanism did contribute ash to the deep-sea floor, but, as noted by

Riedel (12), liberation of silica into the bottom water via dissolution of any of this ash prior to burial cannot affect the production of siliceous organisms at the surface unless a circulation mechanism [for examples, see (4)] is available. In addition, a supply of nutrient is necessary. Gibson and Towe (1, 5) discount the various circulation models, reasoning (5, p. 153) that a circulation model would not explain the presence of "extensive and time-equivalent nearshore sediments of volcanic origin on the continents." We (2, 3) have shown these deposits to be biogenic rather than volcanic. Gibson and Towe instead assume dissolution of vast quantities of ash in surface waters, a fact not yet demonstrated. They further assume (5) the release of nutrient phosphorus from this ash. The efficiency of this process, however, is not clear. Recent studies by Berner (13) suggest that the reaction of iron oxides and phosphates released by submarine volcanism along the East Pacific Rise causes a net removal of phosphorus from seawater. Thus the release of iron and phosphorus from volcanogenic materials by marine waters may possibly produce a phosphorus sink for the area in question, thereby depriving plankton of nutrient necessary for proliferation.

Gibson and Towe (1) cite studies by Huang et al. (14) as support for their viewpoint. Huang et al. (14, 15) show a correlation between maxima in the species diversity of radiolarians and intense volcanic episodes recorded in two deep-sea cores. One should realize that maxima in the species *diversity* are not synonymous with maxima of *abundance*. Huang et al. make clear that radiolarian abundances in their cores correlate with climatic events rather than with volcanic episodes or maxima in the species diversity. They also note judiciously (16) that changes in species diversity, which they suspect may be related to the release of volcanogenic silica and metals, could perhaps result from the selective dissolution of the radiolarians studied.

In view of the speculative and uncertain

nature of present knowledge about the influences of volcanic activity on plankton productivity and chert formation, it may be premature to label anyone's contribution of data on the subject "misleading and highly selected." Perhaps, however, this uncertainty will stimulate others in the vigorous pursuit of additional data on this intriguing subject.

References and Notes

1. T. G. Gibson and K. M. Towe, *Science* **188**, 1221 (1975).
2. F. M. Weaver and S. W. Wise, Jr., *ibid.* **184**, 899 (1974).
3. S. W. Wise, Jr., and F. M. Weaver, *Trans. Gulf Coast Assoc. Geol. Soc.* **23**, 305 (1973).
4. R. S. Dietz and J. C. Holden, *J. Geophys. Res.* **75**, 4939 (1970); W. A. Berggren and J. D. Phillips, in *Symposium on the Geology of Libya* (University of Libya, Tarabulus, 1971); A. T. S. Ramsay, *Nature (Lond.)* **233**, 115 (1971).
5. T. G. Gibson and K. M. Towe, *Science* **172**, 152 (1971).
6. P. H. Mattson, E. A. Pessagno, C. E. Helsley, *Geol. Soc. Am. Mem.* *132* (1972), p. 57.
7. S. W. Wise, Jr., and F. M. Weaver, *Int. Assoc. Sediment. Spec. Publ.* **1**, 301 (1974).
8. S. E. Calvert, *Nature (Lond.)* **219**, 919 (1968).
9. R. E. Garrison, *Int. Assoc. Sediment. Spec. Publ.* **1**, 367 (1974).
10. A. P. Lisitsyn, *Recent Sedimentation in the Bering Sea* (English translation, Department of Commerce, Washington, D.C., 1969); see also reference 12 in (*1*).
11. L. A. Frakes and E. M. Kemp, in *Implications of Continental Drift to the Earth Sciences*, D. H. Tarling and S. K. Runcorn, Eds. (Academic Press, London, 1973), vol. 1, pp. 539–559.
12. W. R. Riedel, *Soc. Econ. Paleontol. Mineral. Spec. Publ.* *7* (1959), p. 80.
13. R. A. Berner, *Earth Planet. Sci. Lett.* **18**, 77 (1973).
14. T. C. Huang, R. H. Fillon, N. D. Watkins, D. M. Shaw, *Deep-Sea Res.* **21**, 377 (1974).
15. T. C. Huang, N. D. Watkins, D. M. Shaw, J. P. Kennett, *Earth Planet. Sci. Lett.* **20**, 119 (1973).
16. T. C. Huang, N. D. Watkins, D. M. Shaw, *Antarctic J. U.S.* **9**, 257 (1974).
17. Support provided by a grant from the Donors of the Petroleum Research Fund administered by the American Chemical Society. Drs. J. W. Morse and G. W. Brass of Florida State University provided helpful discussion.

Part V

GEOCHEMICAL ASPECTS OF DIAGENESIS

Editor's Comments
on Papers 13 Through 16

13   MATTER, DOUGLAS, and PERCH-NIELSEN
*Fossil Preservation, Geochemistry, and Diagenesis of Pelagic
Carbonates from Shatsky Rise, Northwest Pacific*

14   NEUGEBAUER
Some Aspects of Cementation in Chalk

15   KNAUTH and EPSTEIN
*Hydrogen and Oxygen Isotope Ratios in Silica from the JOIDES
Deep Sea Drilling Project*

16   KASTNER, KEENE, and GIESKES
*Diagenesis of Siliceous Oozes. I. Chemical Controls on the Rate
of Opal-A to Opal-CT Transformation—An Experimental Study*

Papers 13–16 were selected because they represent the recent
trend to shift the focus of diagenesis research from SEM observations and
mineralogical analyses (principally X-ray diffraction) to geochemical
and thermodynamic considerations. Studies of stable isotope compositions and the chemistry of interstitial liquids have opened especially
promising avenues in understanding diagenetic processes. The first two
papers concern themselves mainly with carbonate diagenesis while the
last two deal with silica diagenesis.

Matter et al. (Paper 13) studied biogenic sediments from Shatsky
Rise in the northwest Pacific. This rise is an oceanic plateau, having a
geologic setting comparable to the Manihiki Plateau and the Ontong-
Java Plateau (see the volume editor's comments on Paper 5). At each of
these three Pacific oceanic plateaus, the sediments consist of biogenic
carbonates, with chert horizons in the lower parts of the sediment column. At each locality the sediments directly overlying oceanic crust are
of Aptian or older age. As stated earlier, these three localities lend
themselves well to comparative studies (Papers 2, 5, 7, and 13).

Matter et al. first discuss fossil preservation. They give special attention to the relationships between nannofossil morphology and selective
dissolution and overgrowth. Their observations in broad outline confirm
the findings by other workers (e.g., Papers 3, 5, and 14).

Discussing the geochemistry of the sediments, the authors consider the elements calcium, magnesium, and strontium, both in the sediments and in the pore liquids, as well as the oxygen and carbon isotopes. There seems to be a change in the geochemistry with depth, starting at about 50 m subbottom depth. This level coincides with the start of diagenetic alteration. The changes in geochemistry seem to reflect the progressive and selective dissolution and reprecipitation of calcite. The authors draw the important conclusion that this trend is inconsistent with cementation near the sediment-water interface but is consistent with normal continuous sedimentation and progressive lithification under a normal geothermal gradient. The dissolution and reprecipitation process involves "a considerable ion exchange as well as an isotopic exchange between solids and interstitial waters" (p. 231).

It is still very difficult to quantify diagenetic processes. Matter et al., in discussing this problem, suggest possible lines of attack.

The observation that chalk remains soft and porous, even under considerable overburden, has puzzled many scientists (Neugebauer 1973). A widely accepted explanation was that chalk, because of its unique composition (almost exclusively nannofossils and foraminifera) lacked sufficient metastable aragonite that could act as a source for cement (see also Bathurst 1975). Neugebauer (1973) reviews this and other theories but rejects them as being untenable in the light of modern data. He goes on to point out that, theoretically, a chalk should lithify through pressure solution, given sufficiently large overburden pressures. However, observations show some minor initial pressure solution, welding sharp grain contacts, thus establishing a fairly rigid framework that can resist crushing through overburden. After that the chalk remains soft and porous for depths as great as 1500 m. To explain this, Neugebauer develops a new theory which is based on the fact that chalk, both epicontinental and oceanic, consists of low-magnesium calcite. In that respect it differs from most other (shallow-marine) carbonates. Sea water, trapped as pore fluid, has a high magnesium content. It is supersaturated with respect to low-magnesium calcite, and "low-magnesium calcite can only dissolve when the solubility surpasses the concentration of the pore fluid through pressure at the contact" (p. 239). As a consequence, pressure solution of low-magnesium calcite can take place only once a threshold value of 250 to 1000 atmospheres has been surpassed. For very sharp point contacts, this threshold can be reached at about 100 m depth (initial establishment of a grain framework), but for lithification to become significant, an overburden of 1000 m or more may be required.

In his 1973 paper Neugebauer refers only briefly to some early DSDP results. Only a year later, he was able to publish another paper in which he enlarged on his theory by drawing extensively on newer and more abundant data from deep-sea drilling. This paper is reproduced here (Paper 14). Neugebauer's two papers clearly illustrate the rapid increase in our knowledge thanks to the DSDP. The major advance is that many in situ data on the chemical composition of pore liquids are now available.

According to Neugebauer, the DSDP data show a slight decrease in magnesium content in pore liquids, associated with a slight increase in calcium content, up to a depth of 500 to 700 m. He explains this phenomenon by the formation of some cement with a higher magnesium content at the expense of low-magnesium calcite. At greater depths, the role of pressure solution becomes more important. A more rapid formation of cement more quickly depletes the magnesium in the pore liquids. However, to change a chalk into a limestone at depths smaller than a few thousand meters, very low values of magnesium concentration in the pore liquids would be required. Various lines of evidence suggest that the critical value might be 0.01 M magnesium.

In the last part of his paper, Neugebauer discusses the deposition of calcite cement in chalk. This deposition takes place selectively, depending on crystal size (solubility product) and crystal shape (specific surface energy).

The importance of Neugebauer's paper is that it tries to derive a theory on the basis of empirical data. As such it will provide more orientation to future research. Incidentally, it is interesting to compare his theory on the role of magnesium in the diagenesis of deep-sea carbonate sediments with the experimental findings of Kastner et al. (Paper 16). It seems that magnesium also plays an important role in silica diagenesis.

It is important to note here that the occurrence of limestone at sub-bottom depths shallower than 1000 m (e.g. 600–800 m at DSDP site 167, Paper 2; 800 m at DSDP site 288, Paper 5; and 600 m at DSDP site 305, Paper 13) seems to contradict Neugebauer's theory. However, at these sites there appears to be a correlation between the lithification of the carbonate sediment and the presence of porcelanite and chert. This supports the theory by the volume editor that silica diagenesis influences carbonate diagenesis (see comments on Paper 5). An important consequence of this theory would be that, if some chert is early-diagenetic (see Papers 7 and 15), some advanced carbonate lithification may also be early-diagenetic.

Knauth and Epstein (Paper 15) discuss the hydrogen and oxygen isotope ratios of diagenetic silica in deep-sea sediments. They first review the existing ideas about the various diagenetic silica phases and their transformations (maturation process). Of interest is their remark that although several authors have accepted the opal-CT to quartz transition to be a solid-solid process, the actual mechanism is not known. The authors call the final (stable) silica phase "granular microcrystalline quartz." They do not consider chalcedony to be part of the typical deep-sea silica paragenetic sequence but rather a result of direct growth from solution after burial.

The authors draw some important conclusions from their stable isotope analyses. The main conclusion is that opal-CT was formed before or during shallow burial. The $\delta^{18}O$ values obtained at these shallow depths are preserved to burial depths of at least 1000 m. But granular microcrystalline quartz, formed progressively during increasing burial depths, has $\delta^{18}O$ values indicative of the higher temperatures at its depth of formation. The two silica phases, having different oxygen-isotope values, can exist together.

Knauth and Epstein's initial results offer great hopes for the future. Stable isotope studies could well provide the answers to some basic diagenetic problems such as the maturation versus direct-precipitation origin of quartz, the early- versus late-diagenetic origin of chert, and maybe the volcanic versus biogenic origin of diagenetic silica.

At the 9th International Sedimentological Congress in Nice, Kastner and Keene presented a paper describing their experiments on silica diagenesis. Their experiments were designed specifically to test some of the controversial theories, such as the maturation theory and Lancelot's theory (Paper 7). Their paper received much attention (Friedman 1976) and stimulated animated discussions. Joined by Gieskes, they recently wrote an expanded version of their paper to the Sedimentological Congress. We asked permission to reproduce the original paper as it was presented at the Congress, but the authors preferred to have their latest paper reproduced in this volume, notwithstanding the fact that it was still in manuscript form. We are very grateful for their generous attitude. Their paper rounds off the research on silica diagenesis in deep-sea sediments based on the data collected during the seven years of the DSDP. As is the case with the other three papers in this section, their work suggests fruitful new lines of research for IPOD, the present phase of the DSDP.

Kastner et al. (Paper 16) start their paper by reviewing earlier papers on silica diagenesis, covering a period of about 60 years. From this review they carefully define major problem areas around which their experiments were designed. These problem areas are the role of the host-sediment chemistry and the testing of the maturation theory.

Their experiments suggest that both magnesium content and alkalinity strongly affect the rate of opal-A diagenesis, much more so than any other chemical parameter of seawater. It seems immaterial whether

217

the alkalinity is caused by calcium carbonate or other chemicals in solution. In experiments with the correct magnesium concentration and alkalinity, siliceous tests corrode, and embyronic opal-CT lepispheres form. During this process, both magnesium concentration and alkalinity decrease (are being "consumed"). When the alkalinity has been consumed, no new embryonic lepispheres can form, but the existing ones continue to grow into well-developed lepispheres. The authors explain this phenomenon by the theory that the formation of lepispheres start with the formation of magnesium-hydroxyde nuclei, which can attract silanol groups. The depletion of magnesium and the decrease in alkalinity stop the process of magnesium-hydroxyde nucleation. Another important conclusion from the experiments is that the growth of lepispheres seems to be retarded by the presence of reactive clay minerals such as montmorillonite.

Kastner, Keene, and Gieskes's results explain why the opal-A to opal-CT transformation is much faster in carbonate sediments than in clay-rich sediments. The authors suggest that similar geochemical conditions could also influence the transition from opal-CT to quartz. They accept the theory that most quartz in deep-sea sediments had an opal-CT precursor, but they cannot exclude the possibility that some quartz could have crystallized directly. Though the authors accept the maturation process in principle, their experiments indicate that time and temperature (burial depth) are not the only controlling factors. The chemistry of the host sediments and the pore fluids also strongly affects silica diagenesis. This can even result in reversals in the overall maturation sequence.

Kastner, Keene, and Gieskes's paper seems to establish a bridge between the maturation theory and Lancelot's "direct precipitation" theory. However, it should be realized that the experiments deal with only the first step in silica diagenesis, the transition from opal-A to opal-CT. No answers exist as yet for the formation of diagenetic quartz. However, the authors have shown that the chemistry of the host sediments and the pore fluids have a fundamental influence on the rate of silica diagenesis, and their assumption that the final step in silica diagenesis might be controlled by the same parameters seems a reasonable one. There is little doubt that these experiments have advanced our knowledge substantially and fundamentally. They clearly point the way to further, more integrated research.

REFERENCES

Bathurst, R. C. C. (1975) Carbonate sediments and their diagenesis. *Developments in Sedimentology*, No. 12 (2d enlarged ed.). Amsterdam: Elsevier. 658 pp.

Friedman, G. M. (1976) In Nice—International Congress of Sedimentology. *Geotimes* (February 1976): 18–19.

Neugebauer, J. (1973) The diagenetic problem of chalk—The role of pressure solution and pore fluid. *Neues Jahrb. Geologie u. Paläontologie Abh.* **143:** 223–245.

13

Reprinted from pp. 891–921 of *Initial Reports of the Deep Sea Drilling Project,* Volume
32, R. L. Larson *et al.,* Washington, D.C.: U.S. Government Printing Office,
1975, 980 pp.

FOSSIL PRESERVATION, GEOCHEMISTRY, AND DIAGENESIS OF PELAGIC
CARBONATES FROM SHATSKY RISE, NORTHWEST PACIFIC

Albert Matter, University of Berne, Berne, Switzerland,
Robert G. Douglas, University of Southern California, Los Angeles, California
and
Katharina Perch-Nielsen, Federal Polytechnic Institute, Zürich, Switzerland, and
Institut for historic geologi og paleontologi, Kobenhavn, Denmark

INTRODUCTION

The vast scientific effort devoted to the study of the
sea floors during the past 20 years has resulted in a rapid
increase in our understanding of many aspects of marine
sedimentation, including the origin, facies distribution,
and diagenesis of marine sediments.

Much of this effort has focused on shallow-water carbonates. Therefore, most ideas about cementation and
lithification of carbonate rocks are based on these
studies. This explains why, until the mid-1960's, it was
generally believed that stabilization (conversion of
metastable aragonitic and magnesian-calcitic particles
to low magnesian calcite) accompanied by thorough
lithification of a carbonate sediment takes place mainly
in the subaerial, fresh water, and intertidal environments (Bathurst, 1971, p. 323).

This somewhat biased view had to be revised when,
during the past 10 years, more and more examples of
both shallow- and deep-water submarine cementation
and lithification of carbonate sediments were reported.
All examples of lithified carbonates recovered from the
deep sea, prior to deep-sea drilling techniques, are lithic
layers or crusts which are generally cemented by magnesian calcite (for a detailed review, see Bathurst, 1971;
Milliman, 1974).

Studies of pelagic carbonates based on sections drilled
by the Deep Sea Drilling Project have documented in
situ cementation and lithification of pelagic carbonates
which increases with depth of burial and age. The most
important sedimentological results gained through
DSDP up to 1973 are summarized by Davies and Supko
(1973).

In all pelagic carbonate sections drilled by *Glomar
Challenger*, a change from soft ooze to friable chalk to
limestone is observed with increasing depth. It is accompanied by an increase in density and seismic velocity
and a simultaneous decrease in porosity. These lithologic changes are caused by progressive selective dissolution of the more soluble planktonic foraminifera,
delicate coccoliths, and supersoluble micritic carbonate
grains. The latter are formed by the breakdown of
calcareous skeletons and the dissolved $CaCO_3$ is reprecipitated as overgrowth cement on the more robust discoasters and coccoliths (Matter, 1974; Schlanger and
Douglas, 1974). Hence, the large amounts of carbonate
necessary to cement a carbonate ooze are not introduced
from an outside source, but rather are derived from the
surrounding material.

It follows from the overgrowth cementation and the
large reduction of porosity with depth observed in the
pelagic carbonate sequences that compaction must be
important in pelagic carbonates, a point suspected by
Tracey et al. (1971), and Cook and Cook (1972), and
others. However, because compaction features in
micrites have been only rarely observed, gravitational
compaction was held negligible in carbonate muds
(Pray, 1960; Bathurst, 1971) until recently.

Because foraminiferal tests possess a large intrabiotic
void space, their dissolution would result in a large reduction of the porosity of the bulk sediment. Schlanger
et al. (1973) and Schlanger and Douglas (1974) have
shown that dissolution of foraminifera might indeed account for the entire porosity decrease, from 80% to 40%,
which is observed in the transition of ooze to limestone.
They observed that the percentage of broken benthonic
foraminifera rapidly increases with depth of burial.
Dissolution and subsequent breakdown also affects the
tests of planktonic foraminifera. However, because of
their greater solubility, fragments of planktonic
foraminifera are dissolved, so little evidence of fractured tests is preserved. The paucity of benthonic foraminifers in deep-sea sediments and therefore the relative
scarcity of crushed benthonics is another reason why
compaction was considered unimportant. On a
macroscopic and microscopic scale, compaction also
affected shelf coccolith oozes of the Irish and English
Chalk (Scholle, 1974; Kennedy and Garrison, in press).

The gross aspects of pelagic carbonate diagenesis have
been described. However, many details, such as the influence of siliceous biogenic and terrigenous components, amount of organic matter, water depth, etc, remain to be solved. The operative processes and the interactions of the solids with the interstitial fluids remain
almost totally unknown.

In this paper we shall first provide an estimate of the
abundance and preservation of foraminifera and nannofossils at Sites 305 and 306. Secondly, we shall discuss
the Ca^{2+}, Mg^{2+}, and Sr^{2+} distribution in the solids and
interstitial waters. Combining all this information and
linking it with stable isotope, porosity, and accumulation rate data, we will discuss the downhole evolution of
diagenetic textures and provide a quantitative measure
for the diagenetic potential.

REGIONAL SETTING AND NATURE
OF SEDIMENTS

The present study is based on the cores recovered during DSDP Leg 32 on Shatsky Rise at Sites 305 and 306.
Detailed description and discussion of the structure,
stratigraphic relationships, and lithology are given in the

site chapters in this volume. Here only data pertinent to the diagenesis of the cored sediments are summarized.

Sites 305 and 306 are located on the southern edge of Shatsky Rise, an irregularly shaped, large, elongated plateau in the northwestern Pacific (Figure 1). It trends north-northeast to south-southwest for about 1400 km and has a maximum width measured northwest-southeast perpendicular to the strike of the feature of about 500 km. It rises more than 2000 meters above the surrounding deep-sea floor, with large parts of the plateau lying in water depths between 3500 and 2000 meters.

Like other elevated plateaus in the Pacific, such as the Ontong-Java and Manihiki plateaus, Magellan and Hess Rise, Shatsky Rise is capped by a thick (close to 1000 m) sedimentary cover. Shatsky Rise has been a site of biogenic, mainly carbonate, deposition at least since Late Jurassic time.

The water depth at Site 305 is 2903 meters. The drilling was abandoned at 640.5 meters below the sea floor. Site 306 was drilled on a ledge on the upper slope of the rise, 37 km to the west-southwest of Site 305, in an area where upper Albian strata crop out. At Site 306 upper Albian sediments are covered only by a thin unconsolidated veneer which contains mixed Albian to Recent faunas. The fauna and presence of a few graded beds indicate that these thin sediments were derived by erosion of older strata which crop out further upslope. Site 306 was abandoned at 475 meters subbottom depth before reaching the crystalline basement.

The lithologic sequences of both sites and the age relationships are shown in Figure 2. The two sections can be correlated both on a litho- and a biostratigraphic basis and are therefore combined into one single composite section (Figure 3). The following five different lithologic units are recognized from top to bottom.

Unit 1

Siliceous foram-rich nanno oozes of early Miocene to Quaternary age occur in the uppermost 52 meters of the section. They consist on the average of 60% to 80% nannofossils and foraminiferal tests, 5% to 20% siliceous fossils, and a minor terrigenous admixture (Plate 1, Figure 1). The oozes are very soft; the GRAPE porosity decreases from 68% at the top of the unit to about 60% at 50 meters depth. Because of large pore spaces, these sediments still appear texturally as a rather loose accumulation of calcareous and siliceous microfossils which have been only moderately affected by dissolution (Plate 1, Figure 1). A hiatus representing part of the early Miocene and late Oligocene was found at the base of this unit.

Unit 2

An almost pure foraminiferal nanno ooze, ranging in age from late Maestrichtian to late Oligocene, is present from 52 to 148.5 meters. The carbonate content exceeds 95%. These oozes have a pale orange color which is caused by the presence of a small amount of ferroman-

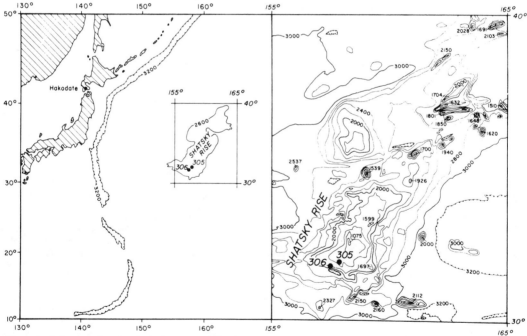

Figure 1. *Index map showing the bathymetry of Shatsky Rise and location of drill sites.   Simplified contours are in fathoms from map by Chase et al. (1971).*

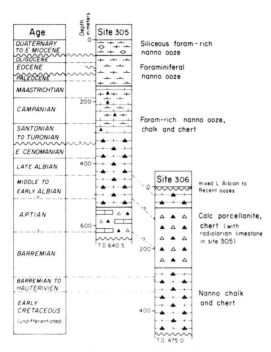

Site 305

Figure 2. *Stratigraphic sequences at Sites 305 and 306. Age correlations are based on the data presented in biostratigraphy chapters (this volume).*

ganese oxides. According to Broecker (1974) manganese nodules occur in sediments which have an average accumulation rate of 3 m/10⁶ yr. The sediments in the upper half and lower third of Unit 2 accumulated at a rate of about 2 m/10⁶ yr and 4 m/10⁶ yr, respectively (Figure 3); low for deep-sea carbonate oozes. The slow rate of accumulation is the major reason for the relatively high concentration of ferromanganese oxides in this unit.

Only minor compositional changes were observed within Unit 2. This unit consists predominantly of calcareous nannofossils with variable minor admixtures of foraminiferal tests and traces of phillipsite and clay minerals. The GRAPE porosity decreases to 50% at about 125 meters, however, in the lowest part of the unit, it increases slightly to a more or less constant value of 50% to 55% (Figure 3).

The compositional and textural characteristics of Unit 2 are shown on the SEM micrographs of two samples taken at 86.3 meters and 119.9 meters subbottom depth (Plate 1, Figures 2, 3). Both samples are nannofossil oozes. However, compare the tighter fabric of the deeper sample with the open framework fabric of the shallower one. This difference is expected from the porosities mentioned above. Also, note the abundant isolated distal shields of coccoliths, whose delicate central structures have been dissolved, and the heavy calcite overgrowths, mainly on discoasters.

Unit 3

Foram-rich nanno ooze, chalk, and chert (Aptian to middle Maestrichtian) comprise this unit. The upper part of this unit consists of alternating soft to stiff ooze and semilithified chalk layers made up of nannofossils and foraminifera in varying proportions (Plate 1, Figure 4). Thin-walled echinoid spines are a typical constituent, in trace amounts, of these sediments. The carbonate content is close to 100%. Chert occurs as irregular stringers and thin layers in the upper part of this section. Notice that the packing of the nanno chalk (shown in Plate 1, Figure 4) is less dense than in the shallower sample shown in Plate 1, Figure 3.

Below 242 meters, only rock fragments were recovered for each 9-meter cored interval. Besides abundant chert pieces, only a few chalk fragments other than chalky void fillings in chert or chalk crusts on chert were retrieved. This explains the lack of control points in Figure 3. Thus, reconstruction of the lithologic sequence remains ambiguous. Most likely the sequence is dominated by chalk with lesser chert nodules or layers.

Below about 365 meters Radiolaria were again observed and silicification becomes more important. The original opaline radiolarian tests are altered and filled mainly with fibrous chalcedony, as are the chambers of foraminifera.

The chalks become harder and more porcellanite-like towards the base of the unit (Plate 1, Figure 5), although the porosity does not change. The rather advanced induration apparently results from the combination of two effects: progressive deposition of carbonate cement and silicification.

Unit 4

Unit 4 consists of radiolarian limestone, calcareous porcellanite, and chert of Barremian to Aptian/Albian age. This unit was recovered in the lowermost part of Site 305 and at Site 306 (Figure 2). The foram nanno chalks of Unit 3 grade into the radiolarian nanno limestones in Unit 4 by a gradual increase of the abundance of Radiolaria and progressive lithification. The radiolarian opal is replaced by chalcedony, disordered cristobalite, and occasionally by calcite. In the lower part of this unit, walls of the foraminifera are recrystallized and their chambers are filled with one or several large sparry calcite crystals and occasionally some euhedral barite crystals (Plate 2, Figure 6). The fabric of the radiolarian limestones is fairly dense, as shown in Plate 1, Figure 5.

The radiolarian limestones grade into harder porcellanites which is indicated by a decrease in the carbonate content to below 50%, an increase in the silicification, and a minor increase of the clay mineral content. The porcellanites have duller lusters and higher densities than the radiolarian limestones. Plate 2, Figures 1 and 2 are SEM photomicrographs of an acid-leached and an untreated fragment of the same sample. About 60% by weight of the sample is noncarbonate (mainly silica) which was remobilized during diagenesis and reprecipitated partly in the pore space and partly as a replacement of small carbonate particles and fragments of calcareous nannofossils. Only the sturdiest coccoliths escaped silicification.

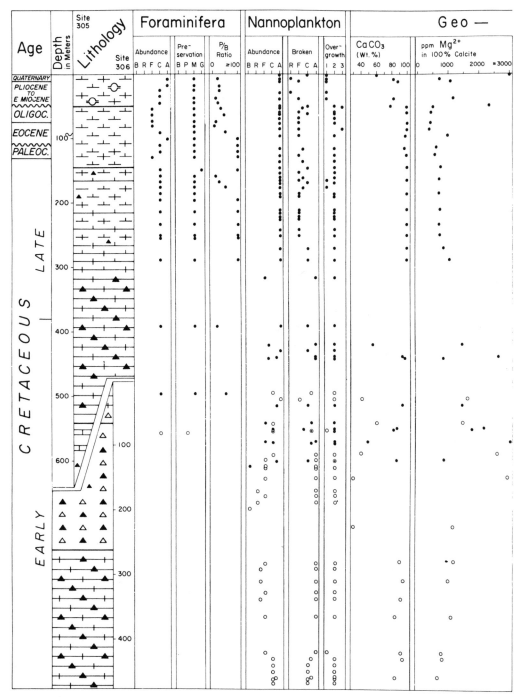

Figure 3. *Composite stratigraphic section of Sites 305 and 306 with foraminifera-nannofossil data, geochemical data, porosities, and sedimentation rates. Site 305 is indicated by closed circles, Site 306 by open circles.*

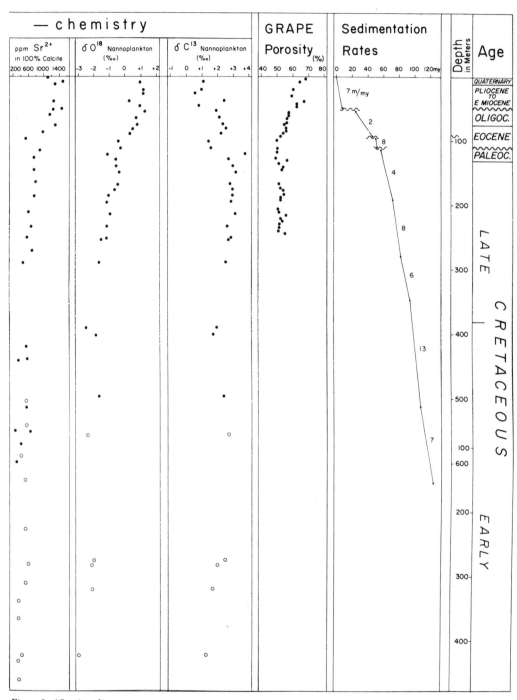

Figure 3. *(Continued).*

224

A more detailed description of the porcellanites and cherts appears in the lithology section of the site chapters and in Keene (this volume).

Unit 5

Nanno chalk and chert of Early Cretaceous to Barremian age comprise Unit 5. Because only core-catcher samples containing a few rock fragments were retrieved, it is impossible to construct a reliable lithologic sequence for this unit. Most likely, however, it consists mainly of grayish, faintly laminated, nannofossil chalk with chert nodules and irregular chert layers. Bioturbation, which destroyed the laminations, was frequently noticed.

The chalk is composed of nannofossils, foraminifera, Radiolaria, small amounts of dolomite rhombs (Plate 2, Figure 5), pyrite, and clay minerals. The dolomite crystals, which are frequently twinned, average about 50 microns in size and may be as large as 150 microns. According to X-ray analysis and using the curve of Goldsmith et al. (1961), the average composition of the dolomite is $Ca_{54} Mg_{46} (CO_3)_2$.

Plate 2, Figures 3 and 4, show the textural aspect of the chalks from Unit 5. The packing of the coccoliths, which have accumulated abundant secondary calcite, is fairly dense.

Large calcite crystals tend to overgrow fragments of coccoliths, thereby occluding part of the pore space. Notice that the chalks at the bottom of the hole are less indurated than those in Unit 4 above it.

METHODS OF STUDY

The samples were studied in smear slides and the results of the visual estimates are shown in Site Report chapters, Sites 305, 306 (this volume). The mineralogy was determined by X-ray diffraction technique and some additional samples were analyzed by Matter. The SEM work on sediment textures was carried out with a Cambridge stereoscan mark IIA instrument at the Geology Department of Berne University, using osmic acid prior to the normal preparatory technique as described previously by Matter, 1974. Abundance of nannofossils, the amount of broken specimens, and the presence of overgrowths were estimated from smear slides, at a magnification of 1000×, and sediment surfaces prepared for the study of sedimentary textures with the SEM. Additional samples were suspended in distilled water and a drop of the suspension was then sedimented on a glass-covered SEM stub, dried, and carbon and gold coated. Ultrasonic and centrifuge treatment, both routine techniques in calcareous nannofossil investigations, were deliberately not used in order to avoid any possible changes of preservation and of the ratio of whole to broken nannofossils.

A total of 40 sediment samples from Sites 305 and 306 was selected for geochemical analysis. The samples were finely ground, stirred for 5 min in distilled water, and then filtered and washed three times to avoid contamination with pore fluids. One-hundred milligrams from the oven-dried samples were dissolved in $2N$ hydrochloric acid and immediately filtered after effervescence had stopped. The sample was then diluted to 100 ml with distilled water.

Calcium was analyzed by EDTA titration using HHSNN as an indicator. Both strontium and magnesium were determined on a Perkin-Elmer Model 303 atomic absorption spectrophotometer using the "standard procedure" recommended for the instrument. The standard solutions were prepared from commercially obtained stock solutions (Merck).

Quadruplicate analyses of a few samples showed that the analytical errors are ±0.2% for Ca^{2+}, ±2% for Mg^{2+}, and ±5% for Sr^{2+} by weight. A strontium solution prepared from a NBS $SrCO_3$ standard was analyzed and had a far better precision than the samples.

The techniques used for isotopic analysis of oxygen and carbon are described in Douglas and Savin (this volume). The results are given in per mil deviations (δ) from the PDB-1 standard.

PRESERVATION OF FOSSILS

Foraminifera

Planktonic foraminifera decrease from abundant in the Quaternary and late Pliocene to few in the Oligocene to middle Eocene (50 to 80 m) (Figure 3). In the lower Tertiary the abundance of planktonic foraminifera varies between few to abundant, whereas in the Cretaceous they are generally common.

A qualitative system for fossil preservation, such as the one shown in Figure 3, assesses only the average preservation state of a planktonic foraminiferal assemblage. Hence, the preservation state of the planktonic foraminifera is indicated as moderate throughout the section. However, a detailed inspection reveals differential preservation of different species as well as a range of preservation states within each species. This is in agreement with studies of Recent planktonic foraminiferal assemblages which demonstrate that the tests are selectively dissolved and fragmented.

The ratio of the number of planktonic to benthonic foraminifera is another measure of preservation. Studies of Recent foraminiferal assemblages have shown that this ratio is mainly controlled by selective dissolution of planktonic foraminifera. After burial the planktonic to benthonic foraminiferal ratio is subject to further change as the result of diagenesis. The planktonic to benthonic foraminiferal ratio is low (15 to 25) in Quaternary to middle Miocene sediments. In early Eocene to late Maestrichtian and early Campanian and Santonian sediments the values are around 200, whereas in middle Maestrichtian sediments they are below 50. The high values found in early Campanian and Santonian sediments may not be reliable because of the very few benthonic foraminifera present. The planktonic to benthonic foraminiferal ratios suggest excellent preservation of the Campanian and lower Maestrichtian, but probably better than actually exists. With increasing lithification, larger numbers of crushed and partly dissolved tests were observed in thin sections and prepared samples (Plate 1, Figure 4). In the Early Cretaceous some benthonic foraminifera are overgrown with chamber infillings of calcite spar and some barite crystals (Plate 2, Figure 6).

Calcareous Nannofossils

Optical studies of the nannofossil samples from Sites 305 and 306 are shown in Figure 3 together with the lithological, physical, and geochemical data. At Site 305 calcareous nannofossils are abundant, from the Pleistocene to Santonian, independent of the degree of dissolution of foraminifera. Nannofossils vary from abundant to few in the chert-bearing sequence below the Santonian.

The number of broken specimens is taken as a measure of the degree of dissolution. "Broken coccoliths" in Figure 3 includes coccoliths whose proximal and distal shields (see Figure 4) are detached as well as specimens with broken shields and walls. The number of broken nannofossils is related to lithology and depth of burial at Site 305. In the first 40 meters broken nannofossils are generally rare. However, in the foraminiferal nanno ooze and the top part of the foram-rich nanno ooze and chalk, they become few to common. Below the chert-bearing sediments of Site 305, broken nannofossils are common to abundant (Figure 3).

The degree of overgrowth on the nannofossils increases with depth of burial. The first 40 meters of sediment show rare overgrowths on nannofossils, whereas deeper in the section secondary calcite is usually important. It must be noted, however, that it is difficult to judge the degree of overgrowth in Quaternary and pre-upper Paleocene sediments lacking discoasters. Overgrowths are only recognizable in such samples by the occurrence of heavy-appearing dissolution-resistant coccoliths such as *Watznaueria barnesae*, *Micula staurophora*, and others. These samples have to be studied with the aid of the SEM to fully evaluate the degree of secondary calcite deposition on coccoliths.

The abundance of coccoliths at Site 306 decreases from abundant in the Cenomanian to Albian chalks to rare or barren in the Barremian to Aptian porcellanites. Coccolith abundance decreases with the decrease in total carbonate content (Figure 3). In the nanno chalks below the porcellanites, coccoliths increase in abundance. Often these Cretaceous assemblages contain coccoliths without central structures, and the wall or the shield are overgrown. Rhabdoliths rarely have their shields attached.

The following brief description of selected samples summarizes the observations made with the aid of the SEM.

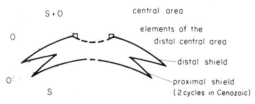

Figure 4. *Cross-section of a placolith similar to* Watznaueria *(Cretaceous),* Ericsonia, *and* Coccolithus *(Tertiary) showing major structural features and sites of preferential deposition of overgrowth cement (O) or of dissolution (S).*

305-1-2, 20 cm

The soft ooze contains disaggregated specimens of *Emiliania huxleyi* and many single-shielded *Coccolithus pelagicus* and *Cyclococcolithus leptoporus*. The details of the structure are generally poorly visible. Dissolution is important whereas deposition of overgrowth cement is negligible and diffuse (Plate 3, Figures 1, 2). Poorly preserved diatoms and silicoflagellates are also found.

305-4-3, 140 cm

The discoasters have accreted some overgrowth cement, but their arms are still slender. *Discoaster pentaradiatus* shows less overgrowth than *Discoaster surculus*. Minor amounts of cement were also deposited on a few single elements of the distal shields of coccoliths, whereas proximal shields are etched (Plate 3, Figures 3, 4).

305-6-4, 92 cm

Discoasters display dissolution and overgrowth (Plate 3, Figures 5, 6). Sphenoliths generally have lost their base due to dissolution, but they have acquired calcite overgrowths on the apical segments or the spine (Plate 3, Figures 7, 8). Distal and proximal shields of the Prinsiaceae are still entirely individualized (Plate 3, Figure 9).

305-6-5, 135 cm

In the 2-meter interval separating this and the previous sample, a hiatus was found (Figure 2). Optical investigation revealed a marked difference in the preservation between the two samples (Figure 3), which is obvious in the SEM micrographs (compare Plate 3, Figures 5, 6, with Figure 10). Discoasters in Sample 6-5, 135 cm show thick overgrowths with well-developed crystal faces. Occasionally overgrowth and dissolution etching are coexistent on the same specimen. Sphenoliths have accreted thick calcite overgrowths on proximal as well as on apical elements (Plate 3, Figure 11), but both kinds of elements are found disconnected. Well-preserved proximal shields of *Coccolithus pelagicus* and *Ericsonia ovalis* together with partly dissolved ones were observed (Plate 3, Figures 12, 13). Generally, the coccoliths from this sample appear to be less affected by dissolution than in the previous sample.

305-9-3, 100 cm

Surprisingly, many empty specimens of *Reticulofenestra umbilica*, whose central areas had been dissolved away, were observed in this sample. When well preserved, *R. umbilica* has a very delicate grid that spans the central area. However, the grid can also be overgrown (Perch-Nielsen, 1971, 1972). Proximal and distal shields of Prinsiaceae are either individualized or partly merged. Here, a varying number of elements have grown by calcite accretion and have bridged the intershield space, welding elements of the proximal and distal shield. The proximal shield of *Ericsonia ovalis* often is incomplete due to dissolution. The elements of the central area of *Dictyococcites bisectus* show moderate overgrowth. In general, however, discoasters and *Isthmolithus recurvus* are almost indeterminable because of thick calcite overgrowths masking the original structure (Plate 4, Figures 1, 2).

305-11-4, 90 cm

Almost all discoasters show heavy overgrowths (Plate 4, Figures 3, 7). Considerable amounts of secondary calcite were also deposited on the elements of the central area and on the distal shield of *Ericsonia ovalis* (Plate 4, Figure 8). Specimens of *Campylosphaera dela* usually have lost their central cross by dissolution (Plate 4, Figure 6). *Chiasmolithus eograndis* and other *Chiasmolithus*, which, when well preserved, display a fine net between the arms of the central cross, have lost this structure in this sample.

305-13-4, 90 cm

A dissolution-overgrowth pattern similar to that of the previous sample was observed here. *Discoaster multiradiatus* and some fasciculiths show heavy overgrowth (Plate 4, Figures 9, 10) whereas the proximal shield of *Chiasmolithus* is mainly affected by dissolution. *Ericsonia ovalis* clearly demonstrates varying degrees of preservation in the same sample: some specimens are moderately preserved with both shields still connected (Plate 4, Figure 11); others have partly welded proximal and distal shields with segments bridging the intershield space (Plate 4, Figure 9); and other specimens show both dissolution and overgrowth features (Plate 4, Figure 10). A large number of tiny carbonate crystals, ranging from about 0.1μm to a few μm in size, are scattered between broken and whole coccoliths. Schlanger et al. (1973), Matter (1974), and others have shown that these anhedral crystals originate from the disaggregation of coccoliths and foraminifera caused by dissolution. Nannofossil diagenesis in Tertiary sediments at Site 305 consists of: (1) dissolution of delicate coccoliths, the proximal shields and central areas of Prinsiaceae, *Chiasmolithus*, and other dissolution-resistant forms; (2) deposition of calcite overgrowth cement on discoasters, sphenoliths, and the central areas and distal shields of coccoliths. The pattern of diagenesis changes somewhat in Cretaceous sediments because different types of calcareous nannofossils are present.

305-16-5, 23 cm

This sample comes from the uppermost chalk layer recovered at Site 305. Many coccoliths are broken and often lack their central structures. Overgrowths occur, in the absence of discoasters and sphenoliths in this Maestrichtian assemblage, mainly on zygoliths' walls and on indeterminable fragments. Micritic crystals are abundant, but no interstitial cement has yet been deposited (Plate 5, Figure 1).

305-19-2, 130 cm

Specimens lacking a central area are abundant in this sample. The delicate laths of the central process of *Stradneria limbicrassa* are still well preserved, but the terminal structure on the same process shows heavy overgrowth (Plate 5, Figure 2). A considerable amount of secondary calcite has been deposited on the nannofossils, such as on the central area of *Arkhangelskiella*

cymbiformis, on the distal and proximal shields of *Prediscosphaera cretacea* which are welded, and on *Micula staurophora* (Plate 5, Figures 3-5).

305-27, CC

Preservation in this and the previous sample are similar. For example, *Broinsonia parca* (Plate 5, Figure 6) shows calcite overgrowths on laths within the central structure. However, coccospheres of *Watznaueria barnesae* (Plate 7, Figures 1, 2) are common. It has been shown by Okada and Honjo (1970) that coccospheres of most species are common in the water column down to about 100 meters and then decrease in abundance until almost none are preserved below 200 meters.

305-31, CC

With the exception of a slightly more advanced state of overgrowth, the degree of nannofossil preservation in this sample is much the same in Cores 19, 27, and 31. *Micula staurophora*, a very dissolution-resistant species is relatively more abundant here than in the other samples. This is an indication that most other more delicate species have been dissolved to a greater extent than in the previous samples perhaps furnishing the carbonate required for the very heavy calcite overgrowths observed on the remaining coccoliths which renders them taxonomically indeterminable (Plate 5, Figure 7).

305-34, CC

Watznaueria barnesae is almost the only abundant nannofossil species preserved in this chalk sample taken from a coating on a chert nodule (Plate 5, Figure 9). Overgrowths are well developed on the few other resistant species present, such as on the distal part of the central process of *Prediscosphaera cretacea* (Plate 5, Figure 8), whereas the type of central processes which occurs on *Eiffellithus turriseiffeli* are still void and the laths are thin. Micritic crystals which were anhedral in the younger samples show crystal faces (Plate 5, Figure 9).

305-46, CC

In contrast to the low diversity encountered in Core 34, the assemblage in this sample is more diverse due to the decreased effects of dissolution and overgrowth.

305-59-1, 140 cm

Overgrowth features are prominent in this sample. Almost all shields of *Watznaueria barnesae* are welded and many euhedral crystals have reached a considerable size. Some of these envelope coccoliths to various degrees, as shown by sutures of coccolith elements exposed on crystal surfaces, yet other overgrowths have grown by deposition of calcite cement on micritic seed crystals (Plate 5, Figures 10-12).

305-64-1, 100 cm

In this sample the central processes of the *Eiffellithus turriseiffeli*-type show signs of dissolution and diffuse overgrowth. Disaggregated coccoliths and micrite crystals are abundant, but there is less overgrowth cement than in the previous sample (Plate 5, Figure 13).

Nannofossil preservation in chalk layers from the porcellanite, chalk, and chert units at Site 306 (Figure 3) is similar to that described at Site 305. An advanced stage of calcite overgrowth development is observed on dissolution-resistant *Watznaueria barnesae*. Proximal and distal shields are entirely welded (Plate 6, Figure 10), and the central areas are filled with secondary calcite (Plate 6, Figure 6). Pressure dissolution features are rare at Sites 305 and 306. An exception is found in the Aptian to Hauterivian chalk of Site 306 (Plate 6, Figure 9).

Molds of Coccoliths

Molds and replicas of coccoliths are observed in the calcareous porcellanites from Cores 306-3, 6, 8, 9, and 10 (Plate 6, Figures 1 to 5, 7, and 8). Deposition of silica in the pore space and partial silicification of nannofossils has formed a mosaic of opal-CT and quartz which tightly fits the contours of surviving *Watznaueria barnesae* specimens. Two dissolution episodes are recognized: dissolution seems to predate silicification because many broken coccoliths are observed to have continued after silicification as indicated by molds of entire specimens and of internal parts of partly silicified coccoliths (Plate 6, Figures 2, 4, 7, 8).

In Barremian to Hauterivian sediments from Site 306 dolomite crystals are present which have incorporated relatively well-preserved coccoliths and micritic grains (Plate 6, Figure 11). Coccoliths partly engulfed by dolomite crystal are illustrated on Plate 6, Figure 11 and suggests that the coccoliths were not dissolved during crystallization. We did not determine whether these coccoliths are still unaltered or whether they have been dolomitized.

Preservation of *Coccolithus pelagicus-Ericsonia ovalis* and *Watznaueria barnesae*

We have studied the detailed preservation of *Coccolithus pelagicus* and *Ericsonia ovalis* from the Tertiary and *Watznaueria barnesae* from the Cretaceous of Sites 305 and 306. All three species are dissolution-resistant forms likely to be abundant in deep-sea carbonate sediments. In the Tertiary forms, selective dissolution first attacks the proximal shield, which disintegrates into micrite-sized elements. These tiny particles are preferentially dissolved, because of their small size, during progressive diagenesis until nothing remains of the proximal shield (Plate 1, Figure 2). In these species overgrowth cement is deposited first on single elements and, with progressive diagenesis, on all elements of the distal shield (Plate 3, Figure 13) and central area (Figure 4, Plate 4, Figure 8; Perch-Nielsen, 1972). Calcite cement is preferentially deposited on the proximal side of the distal shield until the two shields are welded. This is the general pattern for placoliths, as shown by Matter (1974). Partly dissolved proximal shields which have been subsequently overgrown, can be found (Plate 4, Figure 10).

In *Watznaueria barnesae* dissolution removes the elements surrounding the central area (Plate 7, Figures 2, 6, 7). Overgrowth cement may be deposited on the same specimen lacking central areas. The overgrowths appear first on single elements (Plate 6, Figure 7) and then progressively cover the entire shield (Plate 7, Figures 3-5) until the two shields become welded. The same pattern was also recognized on some other species, such as *Prediscosphaera cretacea*, which have two shields.

Geochemistry

Calcium

Variations of the calcium carbonate content of the sediments are closely related to changes in the ratio of calcareous microfossils to siliceous fossils plus diagenetic silica. Other constituents such as terrigenous and authigenic silicates, volcanic glass, and iron oxides occur in very minor amounts. Figure 4 shows the relationship between HCl-leachable calcium, plotted as $CaCO_3$ and lithology. At Site 305 the lowest carbonate content in the uppermost 52 meters corresponds with siliceous foram nanno oozes. The foram nanno ooze and the chalks below it are almost pure carbonate. In the siliceous chalks and calcareous porcellanites, carbonate values range from 50% to 98%. A similar variation occurs in the siliceous sediments at Site 306 (Table 1).

Magnesium

In Quaternary to Miocene siliceous foram nanno oozes the values show a range from about 1000 to 5150 ppm magnesium. Recent planktonic foraminifera contain between 500 to 2400 ppm magnesium (Savin and Douglas, 1973), and coccolith oozes show values ranging from 1500 to 1700 ppm Mg^{2+} (Milliman, 1974). The anomalously high values exceeding 2000 ppm Mg^{2+} are caused by magnesium ions which were leached from clay minerals during sample preparation. At 50 meters the magnesium content drops off sharply to 400 ppm, the lowest magnesium content found. This change coincides with a lithologic boundary and an unconformity (Figure 2). No further change is observed down to 90 meters. This interval of low values correlates well with the low abundance of planktonic foraminifera.

The chalks and oozes between 100 and 250 meters have a slightly higher magnesium content (600 to 700 ppm) which coincides with more abundant planktonic foraminifera. In the lower part of Site 305 and in the upper part of Site 306 the magnesium content of the sediments varies and may reach values exceeding 2000 ppm magnesium. Dolomite crystals were noted in small amounts and may be the source of magnesium in these sediments.

Strontium

The strontium content of the Neogene sediments (0 to 50 m) averages about 1300 ppm. A sharp decrease from 1300 ppm to about 700 ppm is observed in the mid-Tertiary between 50 and about 100 meters. The decrease correlates with deposition of overgrowth cement on nannofossils which is accompanied by a rapid change in the ratio of carbon and oxygen isotope. Note the coincidence with the lowest accumulation rate found at Site 305. Below 100 meters the strontium content decreases in a regular fashion to about 350 ppm at the bottom of Hole 306. The progressive lithification is accompanied

A. MATTER, R. G. DOUGLAS, K. PERCH-NIELSEN

TABLE 1
Total Carbonate Content and Concentration of Strontium and Magnesium
in Sediment Samples from Sites 305 and 306

Sample (Interval in cm)	Depth (m)	CaCO₃ (wt %)	Sr²⁺ (ppm)	Sr²⁺(ppm) per 100% calcite	Mg²⁺ (ppm)	Mg²⁺(ppm per 100% calcite
Site 305						
1-1, 20	1.70	60.94	720	1180	3136	5146
1-6, 140	7.90	82.87	1250	1508	575	694
2-3, 30	11.30	88.4	1150	1300	922	1043
5-3, 50	39.00	83.47	1050	1258	935	1120
6-4, 92	50.42	78.37	1200	1536	1815	2316
6-5, 135	52.35	99.29	1230	1238	488	491
7-4, 100	60.00	100.09	1150	1150	392	392
9-3, 100	77.00	100.19	1300	1300	374	374
10-3, 130	86.3	99.59	980	984	368	368
11-4, 90	96.9	98.19	530	540	923	940
13-4, 90	115.9	97.99	900	918	565	576.
14-5, 120	127.2	98.96	750	760	530	535
16-5, 23	145.23	99.90	750	750	738	738
18-5, 148	165.48	99.96	798	798	687	687
20, CC	186.0	99.99	750	750	669	669
23-5, 120	212.20	100.59	600	600	705	705
25, CC	233.00	100.32	670	670	684	684
27, CC	251.50	100.9	586	586	699	699
29, CC	270.0	99.68	700	702	826	826
31, CC	289.50	91.9	400	435	923	1004
44, CC	419.0	56.55	300	530	814	1439
46, CC	438.0	94.08	550	585	2472	2628
47-1, 128	439.28	98.38	340	345	796	809
54, CC	513.0	92.38	500	541	1329	1439
58, CC	550.5	88.28	220	250	1887	2138
59-1, 140	551.9	82.67	550	6ʳ5	1430	1730
61-1, 135	570.8	49.34	200	₊ᵤ5	2653	5376
64-1, 100	599.0	87.60	280	320	726	829
Site 306						
3, CC	28.0	43.09	240	557	694	1611
5, CC	66.0	61.36	350	570	892	1454
8-1, 38	113.38	41.04	180	439	1061	2585
10-1, 69	151.69	28.83	150	520	832	2885
14, CC	226.0	30.6	160	523	344	1124
21-1, 62	281.62	91.08	550	604	1049	1152
24-1, 140	310.40	97.29	530	545	935	961
26, CC	337.0	94.08	340	361	2834	3012
29, CC	365.0	85.48	300	350	886	1037
36, CC	421.0	92.08	400	434	684	743
37, CC	430.5	93.88	330	352	735	783
40, CC	459.0	84.58	320	378	524	620

by a loss of strontium, as shown on Figure 3. The strontium depletion takes place through dissolution of Sr²⁺-richer skeletal calcite and reprecipitation primarily on nannofossils of an Sr²⁺-poorer calcite cement. The sediments at the bottom of Hole 306 have a strontium content of about 350 ppm which is in the range of ancient limestones on land (Bathurst, 1971).

Calcium, Magnesium, and Strontium in the Pore Fluids

The interstitial water samples cover that part of the sequence where compaction and lithification are most pronounced, yet the pore water data (Table 2) show little change. The only compositional differences occur at about 50 meters depth. The significant changes are: (1) Lower Ca²⁺ and partly lower Sr²⁺ concentrations in the siliceous foram-rich nanno oozes. (2) An increase in

alkalinity and salinity with depth to about 50 meters and then a decrease in alkalinity from 50 to 100 meters. (3) Slight decrease of Sr²⁺ from 52 to 239 meters. A calculation shows that the differences in alkalinity between samples are caused by different amounts of total dissolved calcium and magnesium. For example, the alkalinities in Cores 3 and 6 differ by 0.73 meq/kg and the sum of Ca²⁺ and Mg²⁺ by 0.4 mmoles/l which is equivalent to 0.8 meq/kg. Compared with average seawater, Ca²⁺ is enriched in the pore fluids by about 12%, whereas magnesium is depleted on the average by 5.5%. Strontium, however, whose concentration in average seawater is 8 ppm, is enriched by 25% to 43% in the interstitial water. Neugebauer (1974) shows that at some drill sites small changes of the concentrations of magnesium and calcium in the pore waters of oozes and

229

TABLE 2
Concentrations of Calcium, Magnesium, and Strontium Ions in Pore Fluids
From Site 305 and Their Alkalinity and Salinity

Sample (Interval in cm)	Depth (m)	Ca^{2+} (ppm)	Mg^{2+} (ppm)	Sr^{2+} (ppm)	$\dfrac{mMg^{2+}}{mCa^{2+}}$	$\dfrac{mSr^{2+}}{mCa^{2+}} \cdot 10^{-2}$	Alkalinity (meq/kg)	Salinity ($^{o}/_{oo}$)
1-5, 144-150	7.5	405	1274	7.8	5.18	0.88	2.34	34.9
3-2, 144-150	20.0	405	1289	10.4	5.25	1.17	2.30	35.2
6-5, 144-150	52.5	453	1270	11.5	4.62	1.16	3.03	35.5
11-5, 144-150	99.0	449	1270	10.8	4.66	1.10	2.83	35.2
16-5, 144-150	146.5	453	1274	10.8	4.64	1.09	2.48	35.8
21-4, 144-150	192.0	453	1282	10.3	4.66	1.04	2.48	35.5
26-4, 144-150	239.0	453	1274	10.0	4.64	1.01	2.49	35.5

Note: Calcium and magnesium data through courtesy of J. B. Keene.

chalks are noticed but large changes occur at other sites. For example, Manheim and Sayles (1971) found pore waters enriched in calcium to over 1200 ppm, and in strontium to 119 ppm. They also recorded, at the same drill sites, decreases of magnesium to less than 400 ppm with depth. As pointed out by Manheim and Sayles (1971) and Sayles et al. (1974), the departures of calcium, magnesium, and strontium concentrations from those of normal seawater suggest that calcite is dissolved and a calcite cement containing less strontium and more magnesium is precipitated.

Stable Oxygen and Carbon Isotopes

The isotopic compositions of nannofossil calcite are plotted versus age in Figure 3. The data from Sites 305 and 306 confirm the general decrease of the O^{18} content in pelagic carbonates with age, which has been noted on land, and at other DSDP drill sites (Anderson and Schneidermann, 1973; Douglas and Savin, 1971; Coplen and Schlanger, 1973; and Lawrence, 1973). It is critical to the interpretation of oxygen isotope data from biogenic calcite to find out whether the values have subsequently been altered. Oxygen isotope composition of pelagic carbonates may change during diagenesis, but there is no agreement regarding the time and environment of the isotope adjustment. It may occur at the sea floor or during early or late diagenesis.

Lithified pelagic carbonate rocks have been found to be in isotopic equilibrium with the ambient bottom seawater (Milliman, 1966, 1971). These lithified layers occur in nondepositional environments which indicates that lithification and isotopic reequilibration is a slow process at the open oceanic sea floor (Milliman, 1974). Wise and Hsü (1971) described a monospecific semilithified *Braarudosphaera* chalk of Oligocene age from the South Atlantic. This chalk occurs as thin bands within unconsolidated oozes. The unusually high δO^{18} values (up to $+3.3^{o}/_{oo}$) of the chalk are interpreted to reflect isotopic adjustment and precipitation of overgrowth cement in equilibrium with cold bottom waters (Wise and Hsü, 1971; Lloyd and Hsü, 1972).

Studies of the isotope composition of a series of samples from several DSDP boreholes revealed a downhole decrease of δO^{18} from generally about $+1^{o}/_{oo}$ close to the water-sediment interface to about $-4^{o}/_{oo}$ at the bottom of the holes. Anderson and Schneidermann

(1973) noted an unusually steep δO^{18} gradient ($-4^{o}/_{oo}$ to $+7^{o}/_{oo}$) in Upper Cretaceous limestones from near the basement of Sites 146 and 153. However, no decrease of the O^{18} content with depth was observed in the overlying unconsolidated oozes.

It is obvious that the limestones with δO^{18} values of up to $-7^{o}/_{oo}$ are inconsistent with lithification at the sea floor, because the lowest observed values correspond to an equilibrium water temperature of about 50°C. This high reading rules out bottom water as the equilibrating fluid. According to Anderson and Schneidermann (1973), the strongly negative isotope shift is a result of cementation and recrystallization of limestones during periods of abnormally high heat flow related to volcanic activity, an interpretation which has been questioned by Coplen and Schlanger (1973). From the absence of compaction features and dissolution welding, Anderson and Schneidermann (1973) conclude that cementation occurred during early diagenesis and in the shallow burial realm.

Isotope Data from Sites 305 and 306

Calcareous nannoplankton and planktonic foraminifera initially have similar O^{18}/O^{16} ratios because they both live in the near-surface waters. However, during diagenesis nannofossils act as receptors of calcite cement whereas planktonic foraminifera are donors of carbonate. Assuming that the O^{18}/O^{16} ratio of the preserved planktonic foraminifera has not been changed (which is incorrect in the deeper parts of the holes), it is possible to estimate the amount of isotopic reequilibration of nannofossils by comparing their isotopic ratio with that of the foraminifera. Another approach is to compare oxygen isotope ratios in the solid phase and the interstitial fluids (Lawrence, 1973).

The isotopic composition of planktonic foraminifera from Site 305 has not yet been measured. However, we can compare the oxygen isotope values of Figure 3 with the oxygen isotope profiles from Site 167 (Magellan Rise) and Site 47 (Shatsky Rise). The values from the nannofossils are the same in the Plio-Pleistocene or only slightly higher in the early Tertiary than those from planktonic foraminifera, but in the mid-Tertiary the nannofossils give values which are consistently higher by about $1^{o}/_{oo}$.

Textural observations indicate that these higher values are due to overgrowth cement precipitated preferentially on the nannofossils during early diagenesis. The lack of overgrowth cement in the overlying Plio-Pleistocene oozes explains the identical oxygen isotope compositions of nannofossils and planktonic foraminifera from this unit. The textural data and strontium values both indicate that below 50 meters the ratio of overgrowth cement to primary biogenic calcite increases with depth. This trend is inconsistent with cementation at the sea floor. In addition, the sediments have none of the textural characteristics of submarine lithified crusts (Milliman, 1974), nor have they been exposed for extended periods at the sea floor. They apparently accumulated continuously and at normal rates from the Early Cretaceous unt.. the Paleocene (Figure 3).

Our data are consistent with progressive lithification under a normal geothermal gradient. The sediments lithify by dissolution of thermodynamically instable biogenic particles and precipitation of overgrowth cement on the more stable ones. Lithification starts in the lower shallow burial realm (early diagenesis) and continues, with increasing overburden and temperature, in the deeper burial environment (late diagenesis). This is confirmed by the trend of strontium values which matches the oxygen isotope profile (Figure 3). A positive correlation is found between these parameters (Figure 5). The logarithm of mSr^{2+}/mCa^{2+} is related to δO^{18} with a correlation coefficient of 0.92. If we exclude the values from Site 306 the correlation coefficient is still 0.91. This correlation has a geologic meaning and is not merely the result of the independent correlation of each parameter with depth.

Dissolution and reprecipitation involve a considerable ion exchange as well as an isotopic exchange between solids and interstitial waters (Lawrence, 1973). Because we do not know the amount of isotope fractionation nor the isotopic composition of the pore waters, it is impossible to quantitatively estimate the isotopic composition of the cement.

The difference in the isotopic composition of benthonic and planktonic foraminifera from the same samples (Douglas and Savin, this volume) suggests that they did not re-equilibrate with the pore fluids. Thus, there is only partial isotopic adjustment in pelagic carbonates, namely in the reprecipitated calcite cement preferentially deposited on nannofossils. Therefore, the δO^{18} value of nannofossil calcite is an average of the isotopic composition of the original nannofossil and the overgrowth cement. Because geothermal temperature increases downhole, the isotopic composition of the calcite cement should approach the primary oxygen isotope composition of the nannofossils. In addition, the tests of planktonic foraminifera recrystallize and accrete calcite cement in the deeper depths of burial, which brings them isotopically closer to the overgrown nannofossils.

Although lithification of carbonate oozes may occur during periods of nondeposition, our data from Sites 305 and 306 indicate that this is a minor process. There-

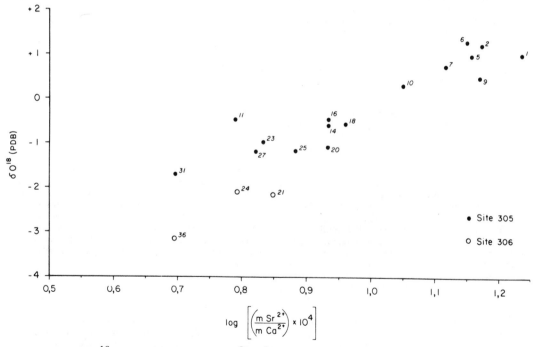

Figure 5. *Plot of δO^{18} (nannofossils) against $log(mSr^{2+}/Ca^{2+})$ of the bulk sediment.*

231

fore, layers such as the *Braarudosphaera* chalk (Wise and Hsü, 1971) are atypical. Our data show a systematic increase in cementation with depth which is accompanied by an equally systematic decrease of δO^{18}. The oxygen "shift" in the middle Tertiary, however, reflects mainly rapid cooling of the water masses and less the isotopically re-equilibrated calcite overgrowth effect of diagenesis. Because the C^{13}/C^{12} ratios in shells and in the water in which they live show only a very small temperature-dependent difference, we believe that the observed trend reversal at Site 305 reflects an isotopic change in the water mass rather than diagenetic alteration. Increasing C^{13} concentrations from the Pleistocene to Eocene were also noted at Site 147, drilled in the Cariaco Trench off Venezuela. These high concentrations are attributed to precipitation of small amounts of authigenic calcite cement, assuming derivation of the C^{12} from organic matter (Lawrence, 1973).

EVOLUTION OF TEXTURES

The texture of ancient sedimentary rocks is composed of a depositional and a diagenetic component. The depositional texture of a freshly deposited sediment is a function of composition, particle size, sorting, and packing of the grains. These properties are altered by diagenetic processes which include dissolution of particles, compaction, and deposition of cement.

Dissolution of planktonic foraminifera and calcareous nannoplankton starts as they settle through the water column and continues after burial. Apparently dissolution of planktonic foraminifera is minor during settling as compared to their dissolution rate on the sea floor (Adelseck and Berger, 1974). Although the mass of nannofossils settles as aggregates in fecal pellets produced by planktonic predators and not individually, the settling time of these pellets is still an order of magnitude less than that of foraminiferal tests (Smayda, 1971; Honjo, 1974). Nannofossils are therefore more affected by dissolution during their transit from the photic layer to the deep-sea floor.

The evolution of the diagenetic texture with increasing overburden and age is governed by in situ progressive dissolution of the more unstable components and reprecipitation of the carbonate as overgrowth cement on the more robust constituents. Adelseck et al. (1973) have experimentally demonstrated that dissolution and reprecipitation take place in the same sample. Dissolution and overgrowth features were observed in many samples from Sites 305 and 306, even on the same nannofossil (Plate 3, Figure 13; Plate 4, Figures 7, 10, etc.). This does not prove, however, the simultaneity of both processes, because the dissolution feature might be a preburial phenomenon. A time relationship can be established if partly dissolved elements have been overgrown (Plate 4, Figure 7; Matter, 1974, pl. 7, fig. 6), or if overgrowths have been dissolved. Only in the latter case can we be sure that both processes have taken place in the diagenetic realm.

Several authors have shown that overgrowth cement is preferentially deposited on discoasters, braarudosphaerid plates, and on the proximal side of the distal shield of the more dissolution-resistant coccoliths. It

appears that accretion of overgrowth cement is dependent on (1) crystal size, the larger seed crystals have less free energy and grow at the expense of smaller ones whose free energy is greater (Bathurst, 1971; Berner, 1971; (2) crystal shape (Neugebauer, 1974) and; (3) the orientation of the crystals' c-axis. In the case of placoliths this occurs if each element of the distal and proximal shield is part of the same single crystal. Apparently, overgrowth is species selective and is open to a ranking analogous to the selective dissolution ranking (Berger, 1968; Parker, 1967; McIntyre and McIntyre, 1971; Bukry, 1971; and Schneidermann, 1973).

Black (1963, 1972) has published excellent studies on the crystal structure of calcareous nannofossils. The discoasters, braarudosphaerids, and most coccoliths are composed of rhombohedral crystals. The trigonal c-axis is oriented parallel to the line of vision in the microscope when viewing discoasters and the distal shields of some coccoliths, such as *E. ovalis*, *C. leptopora*, etc. It has been shown that diagenetic growth of crinoid ossicles by syntaxial overgrowth is fastest in the direction of the c-axis and conversely dissolution is slowest on the faces perpendicular to the c-axis (Bain, 1940, cited in Adelseck et al., 1973). The faster growth rate, parallel to the c-axis, of the rhomb-shaped coccoliths' elements explains why the intershield space of placoliths is bridged early during coccolith diagenesis.

PROGRESSION OF DIAGENESIS

The transitions from ooze to chalk to limestone are gradual and, although lithification increases with depth and age, inversions occur (i.e., less lithified layers may be found below harder ones). According to Schlanger and Douglas (1974) this is a function of the diagenetic potential of the sediment: beds with a higher potential will lithify earlier than those with a lower one. An example is the Upper Cretaceous to Paleocene sequence at Site 305 where oozes alternate with chalk layers. The transitions appear to be gradual at Sites 305 and 306, yet several distinct diagenetic zones can be distinguished which correspond more or less with the lithologic units.

Zone 1

Soft oozes (0 to 50 m) are characterized by abundant nannofossils which show few signs of dissolution and overgrowth and a strongly dissolved temperate planktonic foraminiferal fauna with low planktonic to benthonic ratios. The concentration of Ca^{2+} in the interstitial water is identical to that of average seawater, which suggests that dissolution of the foraminifera took place during the preburial stage. Gravitational compaction is the dominant process in this zone. This process leads to expulsion of pore waters and to a porosity reduction of 10%.

Zone 2

Zone 2 is composed of stiff, pure carbonate oozes (50 to 140 m). In the upper two-thirds of this zone, bracketed by unconformities (Figure 2), marked changes in all the measured parameters are noticed. The calcareous nannoplankton assemblage shows a fair proportion of fragmented and partly dissolved specimens.

Most delicate central structures are missing and the proximal shields show signs of strong dissolution. Poorly preserved proximal shields of the dissolution-resistant *C. pelagicus* occur together with well-preserved ones.

Thick overgrowths of calcite cement are found on discoasters, sphenoliths, and elements which bridge the intershield space of Prinsiaceae. Considerable dissolution of planktonic foraminifera is also indicated by the low planktonic to benthonic ratios. These textural changes are accompanied by a large decrease of strontium and magnesium in the bulk sample and large shifts to lower δO^{18} and higher δC^{13} values. The porosity decreases by another 10% in this interval to 50%.

Benthonic foraminifera indicate that bottom temperatures cooled during the mid-Tertiary, increasing the dissolution rate of skeletal particles on the sea floor. Particularly the planktonic foraminiferal species richer in magnesium, which are also the shallower dwelling ("warmer") ones, were preferentially dissolved. This caused a reduction of the magnesium content of the accumulating sediment and also can explain the low accumulation rate of 2 m/m.y.

After burial, dissolution of biogenic particles, especially of planktonic foraminifera, delicate parts of coccoliths, and tiny supersoluble grains continued, and strontium-depleted calcite cement was precipitated on nannofossils. This cement, precipitated in equilibrium with the pore waters, has a higher δO^{18} concentration than the biogenic particles. The post-Paleocene δC^{13} decrease is probably attributed to a different water mass with much lower temperatures as indicated by the large shift in the oxygen isotope composition of the calcareous nannoplankton.

Zones 1 and 2 represent the shallow burial realm of Schlanger and Douglas (1974).

Zone 3

Below the lower Eocene unconformity the stiff oozes described above gradually pass into Late Cretaceous chalks and then into radiolarian limestones and porcellanites. As indicated by the high planktonic to benthonic ratios, dissolution of planktonic foraminifera was less than in the post-Paleocene. However, dissolution of calcareous nannofossils during the two periods was similar. Dissolution of nannofossils is severe below 250 meters down to the bottom of the hole. In the porcellanites and lowermost chalks only the most robust nannofossils, such as *Watznaueria* and some benthonic foraminifera, are preserved. In the absence of discoasters and sphenoliths, calcite is precipitated onto zygoliths' walls and micritic grains. *Watznaueria* and other placoliths have welded shields, and their central structures are masked by overgrowth cement.

Many coccoliths and micritic grains are replaced by silica in the radiolarian limestones and porcellanites. Silicification enhanced lithification and the reduction of porosity. The nannofossil chalk below the porcellanites has retained more pore space and has a better preserved low-diversity flora than do the silicified sediments. Many particles have accreted large amounts of syntaxial calcite cement which frequently envelops parts of or entire coccoliths and occludes much of the pore space and fills chambers of foraminifera. Authigenic dolomite takes up most or all of the magnesium which is released by dissolution of biogenic calcite. However, some magnesium is eventually incorporated into calcite cement.

Increasing lithification with depth is achieved by dissolution of planktonic foraminifera and all but the most robust calcareous nannofossils and the precipitation of strontium-poor cement onto the remaining calcareous particles. Progressive lithification and stabilization of pelagic carbonates starts during early diagenesis in the shallow burial realm and continues in the deeper burial environment with increased overburden and age (late diagenesis). This is clearly indicated by the decrease of strontium with increasing lithification.

SOME QUANTITATIVE ASPECTS OF LITHIFICATION OF PELAGIC CARBONATES FROM SITES 305 AND 306

To fully understand the mechanisms of lithification, it is important to get a quantitative estimate of how much of the original constituents have been dissolved and how much cement has been reprecipitated. If benthonic foraminifera comprise about 0.02% of a well-preserved tropical assemblage and they are considered insoluble, we can calculate the carbonate loss caused by dissolution of planktonic foraminifera using schemes proposed by Berger (1971) and Douglas (1973a, b). The average foraminiferal carbonate loss is 94% in the 52 to 100 meter interval at Site 305. But if a large proportion of the planktonic foraminifera was dissolved at the sea floor, the porosity reduction caused by dissolution of planktonic foraminifera, as advocated by Schlanger et al. (1973) and Schlanger and Douglas (1974), can only be accounted for if sea-floor dissolution is ruled out. Because the amount of nannofossil carbonate loss is almost impossible to determine, the total carbonate loss due to dissolution cannot be estimated by visual examination of the samples.

However, investigation of the oxygen and carbon isotopes (Lawrence, 1973) and the strontium abundance in the sediments and pore waters may provide the key to quantification of the diagenetic processes. We shall only consider strontium because we have no isotope data on the interstitial fluids. Both strontium and magnesium concentrations in recent deep-sea carbonate oozes are much lower than in shallow-water carbonate sediments. However, from the sparse published data (Manheim and Sayles, 1971) it is suspected that deep-sea carbonates, which obviously have been in continuous contact with pore fluids derived from seawater, are being depleted of strontium with progressive diagenesis. It was suggested by Kinsman (1969) that the strontium concentrations of diagenetically altered limestones could be used to follow the diagenetic pathways. We must know the strontium content of the unaltered sediment and of the interstitial water, as well as the strontium partition coefficient for calcium in order to estimate the amount of calcite cement. The accuracy of these estimates depends largely on how close the experimentally determined partition coefficient of 0.05 (Katz et al., 1972) reflects the true coefficient. If we simplify the calculations by assuming the total cement in a sample was precipitated in equilibrium with the same pore water found today in this sample, the carbonate at 53 meters would consist of 87% biogenic calcite and 13% cement containing 510 ppm strontium. The cement would amount to 64% of the solid carbonate phase at 100 meters, and to 76% at 240 meters (Table 3). The low strontium concentration of the carbonates at the bottom of the holes (350 to 550 ppm) can only be reached via the dissolution-reprecipitation process provided the strontium concentration of the pore waters is low and decreases with depth. This would be the case if a high amount of previously precipitated low strontium cement is dissolved. Our studies of textures show that this is the case.

If the sediments are depleted of strontium during shallow and deep burial, and not during dissolution at the sea floor, we must have a mechanism by which the excess strontium is moved out of the sediment column. The ratio of mSr^{2+} to mCa^{2+} would otherwise increase, because of the dissolution of the more soluble magnesium-bearing biogenic particles, to values which would be much higher than the values observed now in the pore waters (Table 3).

The total quantity of pore water expelled by compaction has only carried back to the sea a small amount of the strontium which is missing in the sediments. The only effective way to remove the dissolved strontium is by diffusion. This seems to be an unsatisfactory explanation because there is barely any concentration gradient. The problem could be solved by assuming (a) that the Sr^{2+} which was dissolved from the carbonates was incorporated into another solid phase, e.g., clay minerals in which case the bulk composition of the samples should remain constant or (b) that the studied sediments do not reflect a steady-state process.

Manheim and Sayles (1971) reported increased strontium concentrations with depth, reaching values of 40 to 119 ppm, from several holes. Using the partition coefficient of 0.05 and their calcium data, the cements precipitated from such solutions should contain between 1500 to 2200 ppm. Certainly deposition of these cements would not decrease the Sr^{2+} concentration of pelagic carbonates during lithification.

Thus, strontium is certainly an excellent quantitative measure of the diagenetic grade of pelagic carbonates, and it is a valuable tool for estimating the amount of precipitated cement. However, several critical aspects, such as the mode of removal of strontium from the sediment column, have to be solved before we fully comprehend the lithification process.

TABLE 3
Estimates of the Amount of Biogenic Particles
and Calcite Cement, Calculated on the Basis of
Pore-water Composition and a Primary Concentration
of 1350 ppm for Strontium in the Unaltered Sediment,
Sr = 0.05

Depth (m)	Composition of bulk carbonate Biogenic (%)	Cement (%)	Sr^{2+} in cement (ppm)
53	87	13	510
100	37	64	480
240	24	76	440

ACKNOWLEDGMENTS

The senior author would like to thank the scientific and technical teams aboard *Glomar Challenger* for the excellent cooperation during the 2-month cruise. The study has benefited from discussions with E. Jäger (Bern), D. Imboden, T. Li, and S. Emerson (Zürich). The cooperation and helpful assistance of F. Zweili with the SEM and of A. Egger and B. Wieland with atomic absorption and X-ray fluorescence analyses are kindly acknowledged. Mrs. Ivy Yeh and Miss Diane Eskenasy assisted in the isotopic analyses. The illustrations were prepared by U. Ernst, U. Furrer, H. Ischi, and A. Breitschmid and their help is greatly appreciated. Miss S. Sahli typed the manuscript. R. Herb (Bern) and K.J. Hsü (Zürich) critically reviewed and improved the manuscript. Finally, we would like to express our gratitude to J.V. Gardner of the Deep Sea Drilling Project for editing this article. The isotopic and foraminiferal analyses were supported by NSF Grants GA16827 and GA31622.

REFERENCES

Adelseck, C.G., Geehan, G.W., and Roth, P.H., 1973. Experimental evidence for the selective dissolution and overgrowth of calcareous nannofossils during diagenesis: Geol. Soc. Am. Bull., v. 84, p. 2755-2762.

Adelseck, C.G. and Berger, W.H., 1974. Dissolution of foraminifera from the sediment-seawater interface: Symp. "Marine plankton and sediments" and Plankt. Conf. 3rd, Kiel, p. 5.

Anderson, T.F. and Schneidermann, N., 1973. Stable isotope relationships in pelagic limestones from the Central Caribbean: Leg 15, Deep Sea Drilling Project. *In* Edgar, N.T., Saunders, J.B., et al., Initial Reports of the Deep Sea Drilling Project, Volume 15: Washington (U.S. Government Printing Office), p. 795-803.

Bathurst, R.G.C., 1971. Carbonate sediments and their diagenesis: Amsterdam (Elsevier).

Berger, W.H., 1968. Planktonic foraminifera: selective solution and paleoclimatic interpretation: Deep-Sea Res., v. 15, p. 31-43.

———, 1971. Sedimentation of planktonic foraminifera: Marine Geol., v. 11, p. 325-358.

Berger, W.H. and von Rad, U., 1972. Cretaceous and Cenozoic sediments from the Atlantic Ocean. *In* Hayes, D.E., Pimm, A.C. et al., Initial Reports of the Deep Sea Drilling Project, Volume 14: Washington (U.S. Government Printing Office), p. 787-954.

Berner, R.A., 1971. Principles of chemical sedimentology: New York (McGraw Hill).

Black, M., 1963. The fine structure of the mineral parts of Coccolithophoridae: Linnean Soc. London, Proc., v. 174, p. 41-46.

———, 1972. Crystal development in Discoasteraceae and Braarudosphaeraceae (planktonic algae): Paleontology, v. 15, p. 476-489.

Broecker, W.S., 1974. Chemical oceanography: New York (Harcourt Brace Jovanovich, Inc.).

Bukry, D., 1971. Cenozoic calcareous nannofossils from the Pacific Ocean: San Diego Soc. Nat. Hist. Trans, v. 16, p. 303-328.

Cook, F.M. and Cook, H.E., 1972. Physical properties synthesis. *In* Hayes, J.D. et al., Initial Reports of the Deep Sea Drilling Project, Volume 9: Washington (U.S. Government Printing Office), p. 645-646.

Coplen, T.B. and Schlanger, S.O., 1973. Oxygen and carbon isotope studies of carbonate sediments from Site 167, Magellan Rise, Leg 17. *In* Winterer, E.L., Ewing, J.I. et al., Initial Reports of the Deep Sea Drilling Project, Volume

17: Washington (U.S. Government Printing Office), p. 505-509.

Davies, T.A. and Supko, P.R., 1973. Oceanic sediments and their diagenesis: some examples from deep-sea drilling: J. Sediment. Petrol., v. 43, p. 381-390.

Douglas, R.G., 1973a. Benthonic foraminiferal biostratigraphy in the Central North Pacific, Leg 17, Deep Sea Drilling Project. In Winterer, E.L., Ewing, J.I., et al., Initial Reports of the Deep Sea Drilling Project, Volume 17: Washington (U.S. Government Printing Office), p. 607-671.

Douglas, R.G., 1973b. Planktonic foraminiferal biostratigraphy in the Central North Pacific Ocean. In Winterer, E.L., Ewing, J.I. et al., Initial Reports of the Deep Sea Drilling Project, Volume 17: Washington (U.S. Government Printing Office), p. 673-694.

Douglas, R.G. and Savin, S.M., 1971. Isotopic analyses of planktonic foraminifera from the Cenozoic of the Northwest Pacific, Leg 6. In Fischer, A.G. et al., Initial Reports of the Deep Sea Drilling Project, Volume 6: Washington (U.S. Government Printing Office), p. 1123-1127.

Goldsmith, J.R., Graf, D.L., and Heard, H.C., 1961. Lattice constants of the calcium-magnesium carbonates: Am. Mineralogist, v. 46, p. 453-457.

Honjo, S., 1969. Study of the fine grained carbonate matrix: sedimentation and diagenesis of "micrite." In Matsumoto, T. (Ed.), Litho- and bio-facies of carbonate sedimentary rocks—a symposium: Pal. Soc. Japan, Spec. Papers 14, p. 67-82.

————, 1974. Sedimentation of coccoliths in deep-sea, procedure of replication from the surface to the sediment assemblage (abst). Symposium "Marine plankton and sediments" and Plankt. Conf. 3rd, Kiel.

Katz, A., Sass, E., Starinsky, A., and Holland, H.D., 1972. Strontium behavior in the aragonite-calcite transformation: an experimental study at 40-98°C. Geochim. Cosmochim. Acta, v. 36, p. 481-496.

Kennedy, W.J. and Garrison, R.E., in press. Morphology and genesis of nodular chalks and hardgrounds in the Upper Cretaceous of Southern England: Sedimentology, v. 22.

Kinsman, D.J.J., 1969. Interpretation of Sr^{2+} concentrations in carbonate minerals and rocks: J. Sediment. Petrol., v. 39, p. 486-508.

Lawrence, J.R., 1973. Interstitial water studies, Leg 15—Stable oxygen and carbon isotope variations in water, carbonates, and silicates from the Venezuela Basin (Site 149) and the Aves Rise (Site 148). In Heezen, B.C., MacGregor, Ian D., et al., Initial Reports of the Deep Sea Drilling Program, Volume 20: Washington (U.S. Government Printing Office), p. 891-899.

Lloyd, R.M. and Hsü, K.J., 1972. Stable isotope investigations of sediments from the DSDP III cruise to South Atlantic: Sedimentology, v. 19, p. 45-58.

Manheim, F.T. and Sayles, F.L., 1971. Interstitial water studies on small core samples, Deep Sea Drilling Project, Leg 8. In Tracey, J.I., Jr., et al., Initial Reports of the Deep Sea Drilling Project, Volume 8: Washington (U.S. Government Printing Office), p. 857-869.

Matter, A. 1974. Burial diagenesis of pelitic and carbonate deep-sea sediments from the Arabian Sea. In Whitmarsh, R.B., Weser, O.E., and Ross, D.A., et al., Initial Reports of the Deep Sea Drilling Project, Volume 23: Washington (U.S. Government Printing Office), p. 421-469.

McIntyre, A. and McIntyre, R., 1971. Coccolith concentrations and differential solution in oceanic sediments. In Funnel, B.M. and Riedel, W.R. (Eds.), The micropaleontology of oceans: Cambridge (Cambridge University Press).

Milliman, J.D., 1966. Submarine lithification of carbonate sediments: Science, v. 153, p. 994-997.

————, 1971. Examples of submarine lithification. In Bricker, O.P. (Ed.), Carbonate cements: The Johns Hopkins Univ. Studies in Geol., no. 19, p. 95-102.

————, 1974. Marine carbonates, Part 1 of Recent sedimentary carbonates. In Milliman, J.D., Müller, G., and Förstner, U., (Eds.), Berlin, New York (Springer).

Neugebauer, J., 1973. The diagenetic problem of chalk: N. Jb. Geol. Paläont. Abh., v. 143, p. 223-245.

————, 1974. Some aspects of cementation in chalk. In Hsü, K.J. and Jenkyns, H.C. (Eds.), Pelagic sediments: on land and under the sea: Spec. Publ. 1, Int. Assoc. Sedimentologists.

Okada, H. and Honjo, S., 1970. Coccolithophoridae distribution in southwest Pacific: Pacific Geol., v. 2, p. 11-21.

Parker, F.L., 1967. Distribution of planktonic foraminifera in Recent deep-sea sediments. In Funnell, B.M. and Riedel, W.R. (Eds.), The micropaleontology of the Oceans: Cambridge (Cambridge University Press), p. 289-307.

Perch-Nielsen, K., 1971. Elektronenmikroskopische Untersuchungen an Coccolithen und verwandten Formen aus dem Eozän von Dänemark: Kongelige Danske Videnskabernes Selskab, Biologiske Skrifter, v. 18 (3).

————, 1972. Remarks on late Cretaceous to Pleistocene coccoliths from the North Atlantic. In Langhton, A.S., Berggren, W.A., et al., Initial Reports of the Deep Sea Drilling Project, Volume 12: Washington (U.S. Government Printing Office), p. 1003-1070.

Pray, L.C., 1960. Compaction in calcilutites: Geol. Soc. Am. Bull., v. 71, p. 1946 (abstract).

Savin, S.M. and Douglas, R.G., 1973. Stable isotope and magnesium geochemistry of Recent planktonic foraminifera from the South Pacific: Geol. Soc. Am. Bull., v. 84, p. 2327-2342.

Sayles, F.L., Manheim, F.T., and Waterman, L.S., 1974. Interstitial water studies on small core samples, Leg 15. In Heezen, B.C., MacGregor, I.D., et al., Initial Reports of the Deep Sea Drilling Project, Volume 20: Washington (U.S. Government Printing Office), p. 783-804.

Schlanger, S.O. and Douglas, R.G., 1974. The pelagic ooze-chalk-limestone transition and its implications for marine stratigraphy. In Hsü, K.J. and Jenkyns, H.C. (Eds.), Pelagic sediments: on land and under the sea: Spec. Publ. 1, Int. Assoc. Sedimentologists, p. 403-434.

Schlanger, S.O., Douglas, R.G., Lancelot, Y., Moore, T.C., Jr., and Roth, P.H., 1973. Fossil preservation and diagenesis of pelagic carbonates from the Magellan Rise, Central North Pacific Ocean. In Winterer, E.L., Ewing, J.L. et al., Initial Reports of the Deep Sea Drilling Project, Volume 17: Washington (U.S. Government Printing Office), p. 407-427.

Schneidermann, N., 1973. Deposition of coccoliths in the compensation zone of the Atlantic Ocean. In Smith, L.A., and Hardenbol, J. (Eds.), Proc. Symp. Calcareous nannofossils, Gulf Coast Section Soc. Econ. Paleont. Min: Houston, Texas, p. 140-151.

Scholle, P., 1974. Diagenesis of Upper Cretaceous chalks from England, Northern Ireland, and the North Sea. In Hsü, K.J. and Jenkyns, H.C. (Eds.): Pelagic sediments: on land and under the sea. Spec. Publ. 1, Int. Assoc. Sedimentologists, p. 97-103.

Smayda, T.J., 1971. Normal and accelerated sinking of phytoplankton in the sea: Marine Geol., v. 11, p. 105-122.

Tracey, J.I. Jr. et al., 1971. Initial Reports of the Deep Sea Drilling Project, Volume 8: Washington (U.S. Government Printing Office).

Wise, S.W., 1973. Calcareous nannofossils from cores recovered during Leg 18, Deep Sea Drilling Project: Biostratigraphy and observations of diagenesis. In Kulm, L.D., von Huene, R., et al., Initial Reports of the Deep Sea Drilling Project, Volume 18: Washington (U.S. Government Printing Office), p. 569-615.

Wise, S.W. and Hsü, K.J., 1971. Genesis and lithification of a deep sea chalk: Ecolog. Geol. Helv., v. 64, p. 73-278.

ERRATA

Page 899, line 21 in the 1st column should read: "... broken coccoliths are observed, and to have ..."

Page 899, line 13 in the 2nd column should read: "... amounts. Figure 3 shows ..."

PLATE 1

Microfacies and texture of Site 305 carbonates. All figures are scanning electron photomicrographs of fracture surfaces.

Figure 1   Soft siliceous foram-rich nanno ooze from Unit 1 with many broken shields of *Coccolithus leptoporus* (Murray and Blackman) (arrows), abundant micritic carbonate particles formed by disaggregation of nannofossils and moderately preserved diatoms. 305-5-3, 50 cm (39.0 m).

Figures 2, 3   Soft foram nanno ooze from Unit 2. Figure 2 shows many isolated distal shields with dissolved central structures of Prinsiaceae and heavy overgrowth on discoasters. In Figure 3 dissolution has affected discoaster (center left) prior to overgrowth. Pressure dissolution is recognizable between coccoliths (arrow). Notice much denser packing in Figure 3 than in Figure 2.
2. 305-10-3, 130 cm (86.3 m).
3. 305-13-4, 90 cm (115.9 m).

Figure 4   Foram-rich nanno chalk from upper part of Unit 3. Abundant small carbonate particles with heavy overgrowths on the calcareous nannofossils which show euhedral crystal faces, and a remnant of a foraminiferal wall (arrow) are seen. 305-23-5, 120 cm (212.2 m).

Figure 5   Silicified radiolarian nannofossil limestone from basal part of Unit 3 showing few dissolution-resistant coccoliths or their replicas or molds in a dense silicified groundmass. Notice larger euhedral carbonate crystals. 305-47-1, 128 cm (439.28 m).

Figure 6   Weakly silicified radiolarian nanno limestone from upper part of Unit 4 showing many tightly packed and overgrown dissolution-resistant *Watznaueria barnesae* (Black) and small carbonate particles. 305-64-1, 100 cm (599.0 m), ×3600.

PLATE 1

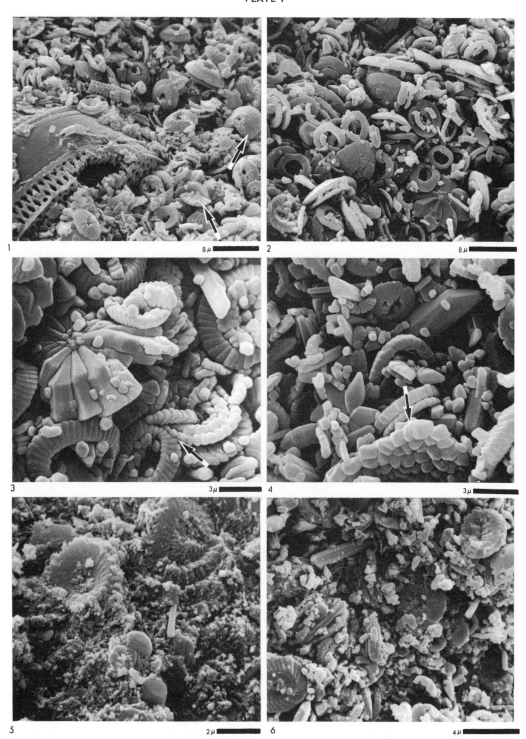

PLATE 2

Microfacies and texture of Site 306 carbonates. Figures 1 to 5 are
scanning electron photomicrographs of fracture surfaces, Figure 6
is a photomicrograph.

Figures 1, 2   Calcareous porcellanite. Heavily overgrown coc-
                   coliths or their molds are seen in a dense silicified
                   groundmass in Figure 1. The acid-treated speci-
                   men in Figure 2 shows that most of the carbonate
                   particles except the sturdiest coccoliths have been
                   replaced by silica. Compare also Plate 7. 306-8-1,
                   38 cm (113.38 m).

Figures 3, 4   Chalk samples from Unit 5 showing dissolution
                   and overgrowth features on coccoliths and dis-
                   aggregated coccoliths. Notice larger calcite
                   crystals which have grown by deposition of
                   overgrowth cement on coccolith fragments and
                   which now occlude some of the pore space (Figure
                   4).
                   3. 306-21-1, 62 cm (281.62 m).
                   4. 306-26, CC (337.0 m).

Figure 5   Twinned dolomite crystals from 306-26, CC (337.0
                   m). Average composition of many crystals is
                   $Ca_{54}Mg_{46}(CO_3)_2$. For detail of crystal surface see
                   Plate 7, Figure 11.

Figure 6   Limestone with *Lenticulina* sp. showing recrystal-
                   lized walls and chambers filled with calcite spar
                   and bipyramidal prisms of authigenic baryte. 305-
                   64-1, 100 cm (559.0 m).

PLATE 2

PLATE 3
Scanning electron photomicrographs showing preservation of Tertiary calcareous
nannofossils from
Site 305 (0 to ca 52 m subbottom).

Figure 1   Proximal view of *Coccolithus pelagicus* (Wallich) with most of the
                proximal shield and minor parts of the distal shield removed by dis-
                solution. 305-1-2, 20 cm (1.70 m), ×4000.

Figure 2   Distal view of well-preserved proximal shield of *Cyclococcolithus
                leptoporus* (Murray and Blackman) whose distal shield has broken
                off, probably because of dissolution. 305-1-2, 20 cm (1.70 m),
                ×8250.

Figure 3   Proximal view of dissociated distal shield of *Cyclococcolithus lepto-
                porus* (Murray and Blackman). Dissolution has attacked the cocco-
                lith along sutures destroying a large part of the elements. 305-4-3,
                40 cm (29.9 m), ×7050.

Figure 4   Distal view of *Cyclococcolithus leptoporus* (Murray and Blackman)
                showing minor overgrowth on a few strongly overlapping elements
                of the distal shield. 305-4-3, 40 cm (29.9 m), ×6700.

Figure 5   *Discoaster* sp. which has accreted a large amount of secondary cal-
                cite forming thick overgrowths with euhedral crystal faces along the
                rays. Original bifurcation at the tip of the rays is still recognizable
                (arrow). 305-6-4, 92 cm (50.92 m), ×5200.

Figure 6   *Discoaster* sp. with one ray removed completely and others partly
                by dissolution. 305-6-4, 92 cm (50.92 m), ×5200.

Figure 7   *Sphenolithus heteromorphus* (?) Deflandre. The proximal shield has
                been removed entirely by dissolution. Faint overgrowth is visible on
                apical segments. 305-6-4, 92 cm (50.92 m), ×7700.

Figure 8   *Sphenolithus distentus* (Martini) whose branches have broken off.
                The proximal column and the rest of branch show overgrowths.
                305-6-5, 135 cm (52.35 m), ×7550.

Figure 9   A relatively well-preserved specimen of *Reticulofenestra* sp. Notice
                absence of any welded elements. 305-6-4, 92 cm (50.92 m), ×7750.

Figure 10   *Discoaster* sp. showing heavy overgrowth and perfect euhedral crys-
                tal faces. Dimples (arrow) are seen where overgrowth cement
                envelopes micrite particles (see also Wise, 1973, pl. 6, fig. 2). 305-6-
                5, 135 cm (52.35 m), ×7050.

Figure 11   *Sphenolithus moriformis* (Brönnimann and Stradner) in oblique
                proximal view showing secondary calcite overgrowth and euhedral
                crystal faces on all elements. 305-6-5, 135 cm (52.35 m), ×9600.

Figures 12, 13 Proximal views of *Ericsonia ovalis* Black with a complete proximal
                shield in Figure 12 and a partly dissolved proximal shield in Figure
                13. Also notice secondary calcite overgrowths, particularly on the
                proximal side of the distal shield and the central area in Figure 13.
                305-6-5, 135 cm (52.35 m), ×4500 and ×8000.

240

PLATE 3

PLATE 4

Scanning electron photomicrographs showing preservation of Paleogene calcareous
nannofossils from
Site 305 (from 77 to 116 m subbottom).

Figure 1   *Isthmolithus recurvus* Deflandre, heavily coated by secondary calcite. Partly dissolved shields of coccoliths are also seen. 305-9-3, 100
cm (77.0 m), ×6150.

Figure 2   *Discoaster* sp. which has developed crystal faces along the rays due
to secondary calcite deposition. The elements of a coccolith which
are partly incorporated into the discoaster are good evidence for
diagenetic overgrowth on the rays. 305-9-3, 100 cm (77.0 m),
×2950.

Figure 3   Heavily overgrown *Discoaster kuepperi* Stradner, in proximal view
showing considerably thickened rays with discrete euhedral crystal
faces. 305-11-4, 90 cm (96.9 m), ×6400.

Figure 4   Distal view of *Dictyococcites bisectus* (Hay et al.). Some of the
originally lath-shaped elements are blocky (arrow) due to accretion
of secondary calcite. In the advanced stage all the elements of the
central field show a blocky habit as shown by Wise (1973). 305-9-3,
100 cm, (77.0 m), ×3950.

Figure 5   A *Sphenolithus* sp. almost completely overgrown. 305-11-4, 90 cm
(96.9 m), ×3500.

Figure 6   *Sphenolithus* cf. *Sphenolithus radians* Deflandre with secondary calcite overgrowth on proximal, lateral and apical elements all of
which show more or less discrete euhedral crystal faces. Notice also
the isolated wall of *Campylosphaera dela* (Bramlette and Sullivan)
whose central structure has been removed by dissolution (upper
left). 305-11-4, 90 cm (96.9 m), ×7000.

Figure 7   *Discoaster lodoensis* Bramlette and Riedel usually has long and bent
rays. The thick and stubby rays with euhedral faces seen on the
specimen shown here indicate considerable dissolution prior to
overgrowth. 305-11-4, 90 cm (96.9 m), ×4300.

Figure 8   Distal view of *Ericsonia ovalis s.l.* showing heavy overgrowth on elements of the shield and the central area. 305-11-4, 90 cm (96.9 m),
×5900.

Figure 9   Fractured surface of nanno ooze showing a late Paleocene assemblage including a fasciculith in side view and *Ericsonia ovalis* Black
whose shields have been largely welded by deposition of calcite cement on most elements. 305-13-4, 90 cm (115.9 m), ×4600.

Figure 10   Proximal view of *Ericsonia ovalis s.l.* The distal shield is partly welded with the first cycle of the proximal shield. Both the first and second cycle are partly dissolved. 305-13-4, 90 cm (115.9 m), ×7300.

Figure 11   Proximal view of *Chiasmolithus* sp. with both shields intact. *Ericsonia ovalis* Black with partly dissolved proximal shield is seen in
lower left. The sample is taken from a fracture surface of ooze. 305-
13-4, 90 cm (115.9 m), ×2900.

Figure 12   *Discoaster multiradiatus* Bramlette and Riedel showing differential
growth of elements by accretion of secondary calcite. The sample is
taken from a fracture surface of ooze. 305-13-4, 90 cm (115.9 m),
×3350.

PLATE 4

PLATE 5

Scanning electron photomicrographs showing preservation of Cretaceous calcareous
nannofossils
from Site 305 (from 145 to 599 m subbottom).

Figure 1   *Stradneria crenulata* (Bramlette and Martini) with partly dissolved
                  central area (arrow) and welded shields (arrow). 305-16-5, 23 cm
                  (145.23 m), ×4100.

Figure 2   Central process of *Stradneria limbicrassa* Reinhardt. The laths are
                  fairly well preserved whereas the elements of the terminal structure
                  are heavily overgrown and show crystal faces. 305-19-2, 130 cm
                  (169.8 m), ×7500.

Figure 3   Proximal view of *Arkhangelskiella cymbiformis* Vekshina displaying
                  moderate overgrowth, mainly on the elements of the central struc-
                  ture. 305-19-2, 130 cm (169.8 m), ×7500.

Figure 4   *Prediscosphaera cretacea* (Arkhangelsky) in side view. The two
                  shields are almost completely welded by accretion of cement on seg-
                  ments. 305-19-2, 130 cm (169.8 m), ×7850.

Figure 5   *Micula staurophora* (Gardet), one of the most dissolution-resistant
                  calcareous nannofossils in the Upper Cretaceous, is here moderate-
                  ly overgrown but still easily identified. Note fragments of cocco-
                  liths which are being incorporated into *Micula*. 305-19-2, 130 cm
                  (169.8 m) ×6350.

Figure 6   *Broinsonia parca* (Stradner) in distal view showing blocky crystals
                  on distal central structure indicating some of the originally lath-
                  shaped elements have accreted secondary calcite cement. 305-27,
                  CC (251.5 m), ×4500.

Figure 7   Terminal structure of a rhabdolith which has accreted a large
                  amount of overgrowth calcite cement. Notice well-defined crystal
                  faces and partly "digested" micrite grains and coccoliths. 305-27,
                  CC (251.5 m), ×6800.

Figure 8   Terminal structure of *Prediscosphaera cretacea* (Arkhangelsky)
                  affected by overgrowth. 305-34, CC (317.5 m), ×6500.

Figure 9   *Watznaueria barnesae* (Black) (arrow) with minor dissolution and
                  major overgrowth features. Dissolution has caused disaggregation
                  of most coccoliths whose segments occur now as partly overgrown
                  euhedral micrite grains. 305-34, CC, ×5050.

Figure 10   Fracture surface of limestone showing pore filling neomorphic eu-
                  hedral calcite growing on and enveloping coccoliths (lower side)
                  and thereby partly incorporating other smaller coccoliths (arrow).
                  305-59-1, 140 cm (551.9 m), ×5250.

Figure 11   Coccolith affected by dissolution (center) and relatively large neo-
                  morphic euhedral calcite crystals which have grown into pore
                  space. The sample is taken from a fracture surface of limestone.
                  305-59-1, 140 cm (551.9 m), ×6350.

Figure 12   A neomorphic calcite crystal which has incorporated a coccolith. A
                  trace of sutures of coccolith are still seen on crystal surface. 305-59-
                  1, 140 cm (551.9 m), ×7900.

Figure 13   Central process of *Eiffellithus turriseiffeli*-type showing dissolution
                  of faintly overgrown laths. 305-64-1, 100 cm (559.0 m), ×8600.

PLATE 5

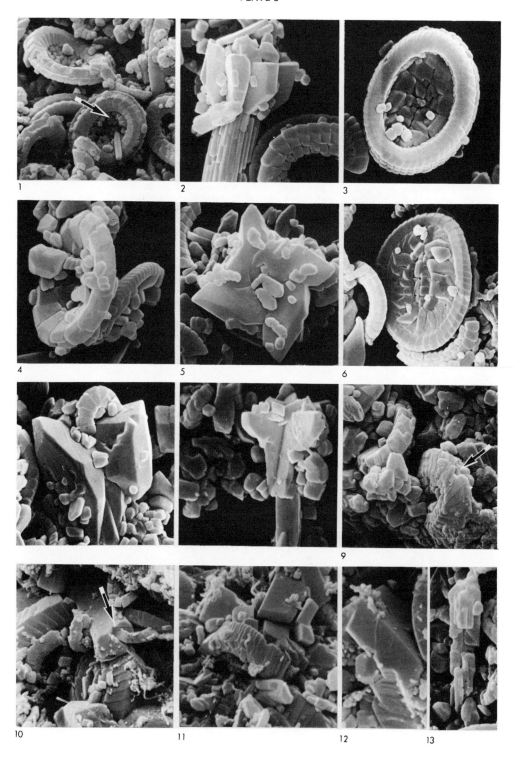

245

PLATE 6

Scanning electron photomicrographs showing preservation of
Cretaceous nannofossils from
Site 306.

Figures 1, 3   Siliceous framework of a calcareous porcellanite
which has been treated with hydrochloric acid to
dissolve away the carbonate. Calcitic fossils and
particles are therefore seen as molds. Notice in
Figure 1 the mold of *Watznaueria barnesae* (Black)
and partly silicified rhabdolith of the *Eiffellithus
turriseiffeli*-type. In Figure 3 a similar rhabdolith
is seen whose central canal is filled with silica.
Both figures from Sample 306-8-1, 38 cm (113.38
m), ×6900 and ×9450.

Figures 2, 4-8   Fracture surfaces of calcareous porcellanites
showing molds of coccoliths in the siliceous
groundmass. Some of these molds may have
formed during sample preparation, hence are
replicas, however, others are real. Molds of complete distal parts of *Watznaueria barnesae* (Black)
are seen in Figures 2, 4, and 8. Notice the mold of
an internal part of a coccolith in Figure 7 (arrow).
Apparently dissolution has affected the carbonate
particles both before and after cementation (see
text).
2, 6. 306-8-1, 38 cm (113.38 m), ×3450, ×6900.
4. 306-3, CC (28.0 m), ×7000.
5. 306-6-1, 114 cm (76.64 m), ×3450.
7, 8. 306-10-1, 69 cm (151.7 m), ×6940 and
×3470.

Figures 9, 10   A fracture surface of chalk. In Figure 9 large
crystal is seen which has grown by deposition of
overgrowth cement on a nannofossil and pressure
solution between a coccolith and a rhabdolith
(arrow). 306-24-1, 140 cm (310.4 m), ×8000.
Figure 10 also shows large pore-filling crystals
enveloping coccoliths as well as *Watznaueria
barnesae* (Black) in different preservation states.
306-29, CC (365.0 m), ×6940.

Figure 11   The surface of twinned dolomite crystals is shown
in Plate 2, Figure 6. Relatively well-preserved coccoliths appear like half-drowned flotsam on the
dolomite surface. 306-26, CC (337.0 m), ×426.

PLATE 6

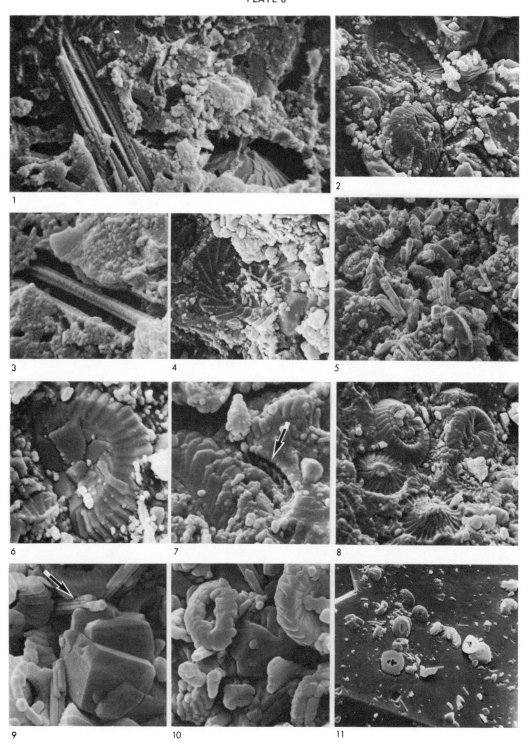

PLATE 7

Scanning electron photomicrographs showing different stages
of preservation of *Watznaueria barnesae* (Black) in samples
from Site 305.

Figures 1, 2   In Figure 1 elements of the shield and central areas
of Coccospheres have accreted secondary calcite
cement. ×7600. In Figure 2 the distal elements of
the central field have been removed by dissolution.
×11700. Both figures are from 305-27, CC (251.5
m).

Figures 3-5   Proximal views showing different steps of deposition of calcite cement leading to progressive
welding of shields and infilling of central area.
305-27, CC (251.5 m), ×8900; 305-46, CC (438.0
m). ×7250; and 305-64-1, 100 cm (599.0 m),
×8950; respectively.

Figure 6   Distal view showing moderate overgrowth on
elements of shield and central structure. 305-19-2,
130 cm (169.8 m), ×7300.

Figure 7   Distal view showing moderate overgrowth on
segments of shield whereas fine elements of central
structure have been dissolved. 305-34, CC (317.5
m), ×7850.

Figure 8   Distal view showing overgrowth on shield and
central structure. 305-64-1, 100 cm (599.0 m),
×7650.

PLATE 7

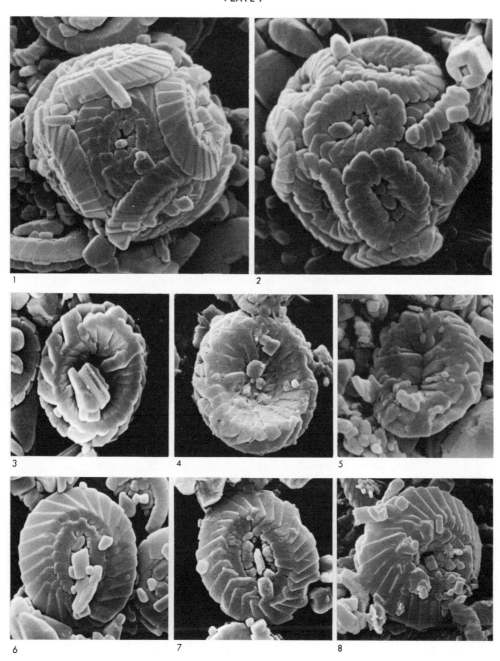

14

Some aspects of cementation in chalk*

JOACHIM NEUGEBAUER

*Geologisch-Paläontologisches Institut der Universität Tübingen,
Sigwartstrasse* 10, *D-74 Tübingen, West Germany*

ABSTRACT

Three aspects of chalk diagenesis are discussed: the production of cement by
pressure solution, the magnesium content in the pore fluid of deep-sea cores, and
the distribution of syntaxial cement.

The volume of cement produced by pressure solution of low-magnesium
calcite is calculated as a factor of overload. The production of cement by this
source is insignificant up to 300 m and volumes of 0·5–5% cement are not
generated before an overload corresponding to 1000 m is attained. Below this depth
there is an enhanced probability for consolidation of chalk. Only under favourable
conditions can chalk withstand a cover of 2000–4000 m sediment without total
hardening.

The magnesium content of the pore fluid, which is crucial for the diagenesis of
chalk, decreases only slowly with depth. The deficient magnesium is mainly consumed by substitution in carbonate minerals. Pressure solution is held to be an
important process for reduction of magnesium at greater depth. A concentration of
0·01 M magnesium seems to be critical for the hardening of chalk.

Chalk cement is normally a syntaxial cement. It avoids coccolith crystals,
which are handicapped because of their size and geometrical shape. The solubility
of coccolith overgrowths can differ from those of other fossils by several percent
due to the small size of the crystal plates. Layers rich in fossils other than coccoliths (inoceramids, *Braarudosphaera*-beds) are favoured in the hardening process.

Dissolution of Foraminifera, observed in shelf-sea chalks, is consistent with
an earlier hypothesis that the Late Cretaceous chalk sea floor was undersaturated
with respect to aragonite. Unlike Foraminifera, coccoliths are somewhat protected against dissolution by the nature of their crystal faces.

INTRODUCTION

Chalk is an extraordinary deposit. It is the only carbonate sediment which, under
certain circumstances, can endure an overload of 1500 m, or more, without hardening
to limestone.

This is usually attributed to a primary deficiency of metastable carbonate minerals,
of aragonite and high-magnesium calcite. As Neugebauer (1973) demonstrated, there

---

* Publication No. 7 of the Research Project 'Fossil-Diagenese' supported by the Special Research
Programme (Sonderforschungsbereich) 53—Palökologie, at the University of Tübingen.

is, however, one further factor: the pore fluid must be oversaturated with respect to low-magnesium calcite; if not, pressure solution indurates the chalk. A certain magnesium content of the pore fluid suffices to maintain an adequate oversaturation.

The present paper continues to discuss the diagenesis of marine low-magnesium calcite sediments. Using the model of chalk diagenesis advanced by the author in his 1973 paper, one can calculate approximately both the amount of cement which results from pressure solution under increasing overload, and the maximal overload which chalk can withstand in contact with magnesium-bearing fluids before turning into a compact limestone.

Pore solutions usually become poor in magnesium with increasing depth and the point of issue is whether chalk sediments (nannofossils + Foraminifera) can actually bear the maximal overload without hardening at lesser depths due to magnesium reduction. Pore fluid analyses from deep-sea bore holes give information about the actual alteration of the magnesium content and the dependent operative processes.

For our purposes the important question is to know what magnesium concentration in the pore fluid suffices to buffer pressure solution. Up to now there are but few and contradictory results, which we shall compare with those of pore fluid analyses from deep-sea cores.

Finally, we shall observe where in (soft) chalk the cement was deposited. The (biogenous) calcite crystals of the constituent fossils influence precipitation of the small volume of cement present in chalk. Operative factors are discussed, especially differences in free energy of the biogenous crystals.

CALCULATION OF CEMENT QUANTITIES

Our intention is to predict, by calculation, what quantity of cement originates with increasing depth as a result of pressure solution, and what overload chalk can withstand in contact with magnesium-bearing fluids before being solidified.

It is a drawback that the calculations must be simplified in a number of respects, which means that the results are subject to a large margin of error.

The method of calculation is as follows.

(1) We estimate the stress which obtains at points of contact in the chalk. This we can do in two independent ways: (a) by means of oversaturation—with the aid of the function for pressure solution; (b) by means of overload, density and texture.

(2) From the stress values we obtain the size of the areas of contact, as well as the approximate volume, which must be dissolved in order to produce the contact areas.

Oversaturation

As is well known, pressure increases the solubility of calcite. The following processes normally take place under overload: (a) solution of calcite at points of contact; (b) supersaturation of the pore fluid; (c) deposition of cement as a result of supersaturation. For low-magnesium calcite, stage (c), the deposition of cement, is kinetically hampered in those cases where the pore fluid is characterized by a certain magnesium content: numerous papers have acquainted us with the fact that magnesium specifically impedes the growth and consequently the precipitation of low-magnesium calcite. According to Lippmann (1960, 1973, p. 114) magnesium ions occupying atomic sites on the crystal surface of low-magnesium calcite block growth due to their high hydration energy.

The magnesium content results in an increase in CO_3^{2-} and Ca^{2+} until the precipitation of aragonite or high-magnesium calcite commences, which means that the solution is oversaturated for low-magnesium calcite. The pore fluids of chalk are characterized by a magnesium content sufficient for kinetic inhibition: in the case of deep-sea chalks this has been shown by the results of the Deep Sea Drilling Project (see page 158) and for chalks of shelf areas their limited induration provides indirect evidence (Neugebauer, 1973).

The extent of oversaturation in the pore fluid of chalk is therefore known: the concentration of Ca^{2+} and CO_3^{2-} corresponds to the solubility of aragonite or high-magnesium calcite. It cannot be lower, since during subsidence a sufficient volume of $CaCO_3$ is dissolved as a result of pressure solution, and it cannot be higher, as otherwise one of the phases mentioned would be deposited.

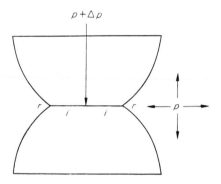

Fig. 1. Relation between stress $\triangle p$, hydrostatic pressure p, and total pressure at contact of particles $p + \triangle p$. r,i : rim and inner surface of the contact.

The effect of pressure solution

The effect of pressure on the solubility of calcite can be evaluated by the two functions (1) and (2) (cf. Neugebauer, 1973). Function (1), derived by Correns & Steinborn (1939), is valid for the rim (r) of contact and function (2), given by Owen & Brinkley (1941), for the inner surface (i) of contact (Fig. 1).

$$\ln \frac{K_{p+\triangle p}}{K_p} = \Delta p \frac{V}{RT} \tag{1}$$

$K_{p+\triangle p}$ = solubility product at the pressure $p + \Delta p$;
K_p = solubility product at the pressure p;
Δp = stress;
V = molar volume of calcite;
R = gas-law constant;
T = absolute temperature.

Joachim Neugebauer

$$\ln \frac{K_{p+\triangle p}}{K_p} = -\Delta\bar{V}\frac{(\Delta p - \Delta p^2 \Delta\bar{K}/2\Delta\bar{V}}{RT} \qquad (2)$$

$\Delta\bar{V}$ = partial molal volume change;
$\Delta\bar{K}$ = partial molal compressibility change;
other symbols as in equation (1).

The stress Δp is the crucial factor for pressure solution and is the difference between pressure on contact surfaces $(p+\Delta p)$ and the hydrostatic pressure p in the pore fluid (Fig. 1). Variations in the temperature have little effect and in Fig. 2a T is selected as $T_o = 298°K$. For other temperatures T the abscissa (stress Δp) in Fig. 2a must be amended by the factor T/T_o.

The reason for the pressure solution effect is the fact that the solution of calcite causes a volume reduction V and $\Delta\bar{V}$ respectively. The solution process reaches

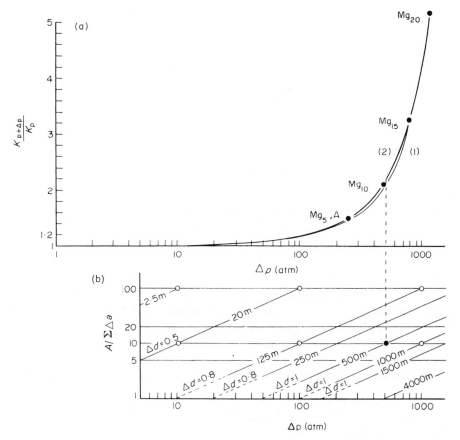

Fig. 2. Pressure solution of low-magnesium calcite (from Neugebauer, 1973). (a) Relation between the stress $\triangle p$ and the ratio 'solubility product under stress $(K_{p+\triangle p})$ / solubility product without stress (K_p)'. Curve (1) after equation (1) and curve (2) after equation (2). A, Mg_{5-20}: solubility product of aragonite and high-magnesium calcite with 5–20 mol % magnesium compared to that of low-magnesium calcite. (b) Dependence of the stress $\triangle p$ on superposition (2·5–4000 m), difference of density $)\triangle d$), and the ratio 'total area/contact area $(A/\Sigma\triangle a)$'. Compare equation (3).

equilibrium when the concentrations of the solution rise so far that the solubility product attains the value $K_{p+\Delta p}$ of the equations (1) and (2).

The reduction in volume for the two equations (systems) is of a different type. In system (1), where the stressed calcite is in direct contact with the pore fluid, the dissolved calcite disappears from the zone of stress and the volume reduction of equation (1) represents the (molar) volume of the calcite. Apart from an extremely small variation of the molar volume with variation of pressure, system (1) is not influenced by other factors.

In the system (2)—in the inner sphere of contact—the dissolved calcite passes into a 'solution film' (Weyl, 1959) (or into a surficial layer with marked transport characteristics), which is under the same high pressure $p+\Delta p$ as the solid calcite. In this case the volume reduction arises from the fact that the dissolved $CaCO_3$ ($+H_2O$) has a lesser volume than calcite ($+H_2O$). Equation (2) thus depends on the partial molal volume change. By chance it is of approximately the same order of magnitude as the molar volume of calcite. Both equations lead to almost the same result.*

Equation (2) has been corrected for compressibility, which is significant above about 500 atm Δp.

Both functions are shown in Fig. 2a. The abscissa of the diagram represents stress Δp (in atm); the ordinate indicates by what factor the solubility product under stress exceeds the solubility product with no stress obtaining; i.e. factor 2 thus means that the solubility product doubles itself.

The course of curve (2) continues to be influenced by the following variables, which were not taken into account in Fig. 2a: by the salt content of the pore fluid, which affects the partial molal volume and by the influence of stress on the dissociation constants of carbonic acid; the latter shifts curve (2) somewhat to the left (according to data from Culberson & Pytkowicz, 1968, p. 409). If one takes into account these variables (T included), then the same solubility product values (Fig. 2a) might emerge for stress values altered by some 10–30%.†

The above extensive discussion of sources of possible errors demonstrates that, strictly speaking, curve (2) should be understood not as a single curve, but rather as a group of curves. Calculation data are to be found in Neugebauer (1973).

Calculation of the stress Δp

With the aid of the diagram given in Fig. 2a we can calculate the stress obtaining at points of contact of the low-magnesium calcite particles. The stress results from the degree of oversaturation in chalk. In Fig. 2a the oversaturation is indicated by the points A and Mg_{5-20}, which correspond to the solubility product of aragonite and high-magnesium calcite (see p. 151).

The contact surfaces can only dissolve when the solution is undersaturated with

---

* The application of equation (2) to the solution film presupposes that the dissolved ions are hydrated as in a solution. Were equation (2) to be invalid, pressure solution alone would have to proceed in accordance with equation (1) at the edge of areas of contact and would advance from there inwards ('Bathurst's model', Neugebauer, 1973). Since functions (1) and (2) lead to the same results, our calculations of the pressure solution effect must, in any event, be of the correct order of magnitude.

† It is uncertain to what extent pressure affects the dissociation of $MgCO_3^\circ$ and $CaCO_3^\circ$. Pytkowicz, Disteche & Disteche (1967, p. 432) interpret discrepancies of some experimental data as due to this source of error (compare also Bathurst, 1971, p. 271).

respect to low-magnesium calcite. But since the solution is oversaturated, the stress must increase the solubility product of low-magnesium calcite by such an amount, that the oversaturation is compensated and exceeded. The 'compensation' stress necessary for this to happen can be read off the abscissa in Fig. 2a. The points A and Mg_{5-20} are attained by a stress of 250–1000 atm. Between these two limits, with a mean value of 500 atm, lies the compensation stress, which obtains at all points of particle contact in the chalk (during subsidence).

A higher stress value cannot maintain itself over long periods. Should, for example, the stress grow due to increased overload, the points of contact will dissolve in accordance with the increased solubility product until the stress has subsided to its compensation value.

Apart from the theoretical approach, the order of magnitude of the stress Δp can be estimated from equation (3):

$$\Delta p = \frac{t}{10}\, \Delta d\, \frac{A}{\Sigma \Delta a}\ \text{(atm)} \tag{3}$$

t = thickness of overload in metres;

Δd = difference between mean density of the sediment and density of pore fluid;

$\dfrac{A}{\Sigma \Delta a}$ = ratio of total area/contact area.

Besides the overload t and the difference of density between sediment and pore fluid Δd, the ratio $A/\Sigma \Delta a$ is also of great importance: A is the total area of a horizontal section and $\Sigma \Delta a$ the sum value of the contact areas of all particles bearing the overload in the section in question. Unfortunately in chalk $\Sigma \Delta a$ cannot be measured. From scanning electron micrographs we can deduce that $A/\Sigma \Delta a$ must be very large. The areas of contact may only rarely exceed 10–20% of the total horizontal section ($A/\Sigma \Delta a = 10$–5), although most chalk occurrences were once overlain by many hundreds of metres. By this means we obtain a stress value in the order of several hundred atmospheres (Neugebauer, 1973).

Von Engelhardt (1960) provides us with a minimum value of stress Δp. 30–40% porosity was measured in chalk under an overload of 1500 m. The theoretical minimum value of $A/\Sigma \Delta a$ can be derived if we assume that the area of contact of each grain corresponds to its maximum horizontal cross-section (cf. columns). At 30–40% porosity (or 0·7–0·6 proportion of solid matter) $A/\Sigma \Delta a$ is $1/0·7$–$1/0·6 = 1·4$–$1·7$ and under 1500 m overload and with $\Delta d = 1$ this results in $\Delta p = 210$–255 atm. The above-assumed 'columnar' mode of thought is, however, so naïve that the true value was certainly far in excess of this.

In Fig. 2b the stress Δp (abscissa) is plotted as a factor of $A/\Sigma \Delta a$ (ordinate) and various overloads t. For example: $t = 500$ and $A/\Sigma \Delta a = 10$ result in $\Delta p = 500$ atm (black dot).

Theory and observation both point to a compensation stress of about 500 atm (with variation by a factor of 2), a value which serves as a basis for further estimates.

Areas of contact

The areas of contact grow under increasing overload. By how much they grow may be deduced from equation (3) (Fig. 2b).

In Fig. 2b the (mean) compensation stress of 500 atm is hatched. This vertical line denotes correlated pair-values for overload t and ratio $A/\Sigma\Delta a$ (e.g. $t=500$ m and $A/\Sigma\Delta a=10$). The reciprocal values of $A/\Sigma\Delta a$ is the sum value of the contact areas related to the total cross-section.

For the sake of clarity the correlated pair-values for overload and contact area are assembled in columns 1 and 2 in Fig. 3. Column 1 represents a section starting with 70 m overload and continuing to 4000 m overload. In this section the area of contact increases from 1% at 70 m to 100% at 4000 m. This implies that under an overload of 4000 m the chalk is totally hardened due to pressure solution; the correlation presupposes that pore fluids containing sufficient magnesium are present. We see furthermore that chalk under an overload of 250–1000 m—values which commonly occur in nature—should have a contact area of 4–20%.

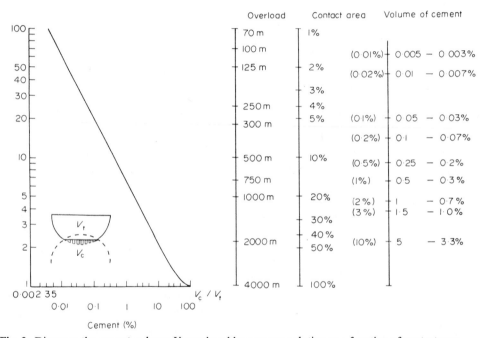

Fig. 3. Diagram: the cement volume V_c produced by pressure solution as a function of contact area. $V_t=$total volume of the solid matter. $A/\Sigma\Delta a=$ratio total area/contact area. Columns: relation between overload, contact area, and volume of cement at a compensation stress of 500 atm.

Let us also look at the results of a variation in values for stress Δp from 250–1000 atm (compare Fig. 2). At a compensation stress of 250 atm the values of overload (first column in Fig. 3) are approximately halved; at a compensation stress of 1000 atm they are nearly doubled (assuming $\Delta d=1\cdot25$ and $1\cdot5$). Total hardening then occurs at 2000 m and 7000 m respectively.

Cement volume

A much simplified model provides us with an approximation for the amount of cement resulting from pressure solution at various depths in the section.

We assume first of all that the sediment consists of spheres, which being densely

Joachim Neugebauer

packed touch each other in the manner shown in Fig. 3 (porosity is 48%). In Fig. 3 V_t is the volume of the hemisphere (the total volume of solid matter) and V_c represents the volume of the calotte (the volume of cement; hatched area). The abscissa V_c/V_t thus denotes the amount of cement given as a percentage of the total volume of solid matter. The cement volume increases nearly linearly with the area of contact (or with the decreasing ratio $A/\Sigma \Delta a$). The results are given in column 3 with the numerical values in brackets.

The use of this model exaggerates the amount of cement: each hemisphere has, instead of one surface of contact, on average about three upper or lower surfaces.* As the calculation shows, the volume of cement alters virtually in inverse proportion to the number of contact points; in the present case by one-third. On the other hand, the contact surfaces are no longer horizontal, but rather inclined, which increases the amount of cement somewhat. As a rough approximation, and bearing in mind the irregular grain shape and contact surfaces—some of which were large from the outset— we have recorded on the right in Fig. 3 one-half to one-third of the originally calculated amount as the 'volume of cement'.

In spite of considerable inexactitudes we can deduce the following conclusions from the results.

(1) In the upper 300 m of chalk, pressure solution of low-magnesium calcite produces minimal amounts of cement. If really greater amounts of cement are present, these must come from other sources.

(2) Noticeable amounts (0·5–1%) do not originate from pressure solution of low-magnesium calcite before an overload of 1000 m is present. If we presuppose a compensation stress of 250 atm, we arrive at 3–5% volume of cement. Under greater overload the amount of cement increases rapidly.

Some of the conclusions which can be drawn from these results are dealt with in the next section.

MAGNESIUM IN THE PORE FLUID OF CHALK

The amount of magnesium required for the inhibition effect

In estimating the area of contact and the volume of cement we presupposed a pore fluid containing 'sufficient' magnesium. For our purposes, 'sufficient' magnesium refers to a solution where the precipitation of low-magnesium calcite is inhibited to such an extent that aragonite or high-magnesium calcite appear. What is the lowest permissible magnesium content at which this effect can still occur?

Very few observations and experiments are available in this field.

(1) Müller, Irion & Förstener (1972) have investigated the conditions of formation of low-magnesium calcite, high-magnesium calcite, and aragonite in lakes. According to these authors low-magnesium calcite is formed when the Mg/Ca ratio of the solution is less than 0·8, whereas high-magnesium calcite is found where the Mg/Ca ratio exceeds 2·4. The minimum content permissible for the 'inhibition effect' lies within these limits; the above authors assume a value of about 2.

(2) In Lippmann's experiments (1960, 1973) low-magnesium calcite and aragonite

* According to von Engelhardt (1960, p. 5) the average number of contact points at 45% porosity is about 7, or 3·5 per hemisphere.

Table 1. Precipitation of $CaCO_3$ in the presence of varying amounts of Mg^{2+} (Lippmann, 1960)

Mg^{2+}/Ca^{2+} molar ratio	Calcite*	Aragonite	Mg^{2+} M
Less than 1·09	+	−	0·003
1·45	+/−	+	0·004
2·91	(−)/+	+	0·008
4·36 and above	−	+	0·012

* With maximally 5 mol % magnesium.

were precipitated from homogeneous solutions ($T = 20°C$), in which the minerals did not appear until weeks or months had passed. Slow precipitation from homogeneous solution prevents the occurrence of aragonite solely as a consequence of momentary or local oversaturation. The results of two test series are given in Table 1.

Aragonite always appeared suddenly and in substantial amounts, indicating a considerable oversaturation with respect to aragonite (Lippmann, 1973, p. 110). When both minerals appear jointly, it is probable that the calcite was formed before the aragonite.

A third test series of similar duration, characterized by lower Ca^{2+} content and higher CO_3^{2-} rate of production, yielded the results presented in Table 2.

Table 2. Precipitation of $CaCO_3$ in the presence of varying amounts of Mg^{2+} (Lippmann, 1973)

Mg^{2+}/Ca^{2+} molar ratio	Calcite*	Aragonite	Mg^{2+} M
Less than 3·4	+	−	0·0024
5·7	+	+	0·004
11·5 and above	−	+	0·008

* With maximally 5 mol % magnesium.

Judging from the Mg/Ca ratio, it would appear that only the first two test series are compatible with our knowledge of the precipitation of aragonite and low-magnesium calcite in sea water and lakes; the third test series seem to be contradictory. They are only compatible with the first two when, instead of the Mg/Ca ratio, the absolute concentration of magnesium is taken as crucial. Lippmann suggests the alternative explanation that a magnesium concentration of '0·01 M appears to be the critical order of magnitude above which aragonite forms as the only phase at normal temperature'.

(3) Möller & Rajagopalan (1972) investigated the relationship between the Mg/Ca ratio of the surface layer of low-magnesium calcite and the Mg/Ca ratio of the solution. According to these authors, the Mg/Ca ratio at the surface (which is crucial for the inhibition effect) changes by (less than) 10% as the Mg/Ca ratio of the solution decreases from 5·5 (sea water) to 2. Down to the latter value we can thus expect almost the same inhibition effect as in sea water. Below 2 the magnesium of the surface layer rapidly becomes impoverished.

Although these data do not provide us with a clear answer to our question, they do at least define the limits within which the 'minimum content' of magnesium lies, this level determining that aragonite and high-magnesium calcite occur in place of low-magnesium calcite.

Two solutions appear feasible.

(1) The 'minimum content' occurs approximately at a Mg/Ca ratio of 1·0–2·0 (in sea water Mg/Ca = 5·0–5·5).

(2) The 'minimum content' occurs approximately at 0·01 M magnesium (in sea water Mg = 0·05–0·055 M).

Certainly, some inhibition should also occur below this limit. Thus Bischoff & Fyfe (1968) observed that concentrations of magnesium as low as 0·0001 M (Mg/Ca = 0·1) can delay the growth of low-magnesium calcite. Similar experiments of Taft (1967) indicated that magnesium concentrations from 0·001 to 0·002 M upwards (Mg/Ca = ?) can prevent the recrystallization of aragonite to low-magnesium calcite, at least within the period of observation, which was one year.

The diagenetic significance of these latter results is that they suggest that chalk does not harden abruptly, if the concentration of magnesium subsides below the discussed 'minimum content'. The compensation stress (see page 154) should only decrease rapidly below this level and the points of contact withstand less and less stress, without being dissolved. However, even very small magnesium concentrations should cause a noticeable reduction in the effect of stress.

Magnesium in the pore fluid of deep-sea chalks

The invaluable investigations of Manheim, Sayles, Chan, Waterman and other associated workers (1969–73) have yielded a large number of analyses of interstitial water from deep-sea boreholes. Here we can examine changes in the magnesium concentration and of the Mg/Ca ratio, and check whether the two alternative 'minimum contents' referred to above are exceeded in deep-sea chalks. Of special interest are those (few) cases, where chalk has hardened to limestone.

To begin with we shall consider all published geochemical studies of deep-sea boreholes in which primarily nannofossil oozes and chalks were encountered.

Pore-water composition and processes reducing the magnesium concentration

Figures 4 and 5 show those constituents of the interstitial water of chalk which are important for our purposes.* These are the molar concentrations of Mg^{2+}, Ca^{2+}, HCO_3^- and SO_4^{2-} as well as the Mg/Ca ratio, plotted against borehole depth. The concentrations of Na^+ and Cl^- and thus the salinity as such remain practically constant.

For the purpose of the following discussion it is appropriate to arrange the pore fluids of chalk into two groups according to calcium content:

(a) Fig. 4: the calcium content scarcely increases with depth.

(b) Fig. 5: the calcium content markedly increases with depth, comparable to the calcium increase of argillaceous sediments.

Between these two groups there is no definite division; in Fig. 4 a few boreholes, which are intermediate between the two groups, are represented hatched.

The interstitial waters of all deep-sea chalks (Figs 4, 5) are characterized by a relatively slight reduction of the magnesium content with increasing depth. This distinguishes them from the interstitial waters of argillaceous and volcanic successions,

* A small group of pore fluids influenced by evaporites has been excluded. Furthermore, for the sake of clarity, those series of pore-water samples which extended to depths of less than 100 m have been omitted, since their analyses show no further variation.

where the magnesium reduction is usually more pronounced (Manheim, Chan & Sayles, 1970b; Manheim & Sayles, 1971a). The bicarbonate content of the interstitial water of chalks varies remarkably little down to a depth of 500 m; even the sulphate level is scarcely reduced. This is yet another dissimilarity between the interstitial water of chalk and that of argillaceous sediments, the latter being characterized by marked changes in the SO_4^{2-} and HCO_3^- level. The interstitial water of chalk is thus seen to be relatively unreactive.

Correlations between some of the ionic displacements lead to certain inferences about the processes which alter the magnesium and calcium content of interstitial water in chalk, as shown by Manheim, Sayles, Chan, Waterman and other co-workers (1969–73).

The increase in calcium content with depth of the group in Fig. 5 cannot be attributed to increased solubility of $CaCO_3$, since otherwise the HCO_3^- content would have to show a similar increase. There are two possible sources for the additional calcium.

(1) Calcium is liberated from silicates (i.e. plagioclases, Sayles, Manheim & Waterman, 1971; Manheim, Chan & Sayles, 1970b).

(2) Calcium is released by substitution of magnesium in calcite (Manheim & Sayles, 1971b).

Processes concerning the magnesium concentration are of special interest in our context. As far as we know at present, the relatively slight magnesium reduction observed in chalks (Figs 4, 5) cannot be correlated with the formation of dolomite (Sayles, Manheim & Waterman, 1971). This is exemplified by borehole 10 (marked '*dol*' in Fig. 4), which is perhaps even characterized by magnesium increase in the dolomite-bearing section.

Silicate reactions can use up magnesium. A process in which magnesium replaces the iron of clay minerals (Drever, 1971) would appear to be important; this process is thought to take place during the formation of pyrite. It must manifest itself in the pore solution in such a way that the magnesium reduction is accompanied by an equivalent (double) sulphate reduction and bicarbonate increase.

This process can be seen in the SO_4^{2-} curve from borehole 94 (Fig. 4) which displays an unusual shape. Logically, this is accompanied by increased bicarbonate content and a slight calcium decrease. However, the decrease in magnesium concentration corresponding to the decrease of sulphate is not sufficient to explain fully the actual magnesium decrease, and is only adequate to explain the 'hump' in the magnesium curve.

In other boreholes the sulphate decrease is so minute that it is difficult to demonstrate the operation of this process. In boreholes 116 (Fig. 4) and 135 (Fig. 5) it possibly plays a supplementary role and is perhaps one of the reasons for the width of the parallel band of magnesium curves. However, we cannot in general attribute a role of decisive importance to this process in the diagenesis of chalks.

A third possible explanation for magnesium reduction is the substitution of calcium by magnesium in calcite (see above). Should calcite dissolve and reprecipitate in the form of a carbonate mineral with a higher magnesium content, we would expect an increase in calcium concentration in the pore water corresponding to the magnesium consumed.

Analyses support the hypothesis that this process takes place. The boreholes of Fig. 5—but also those of Fig. 4—show reductions in magnesium which, in the

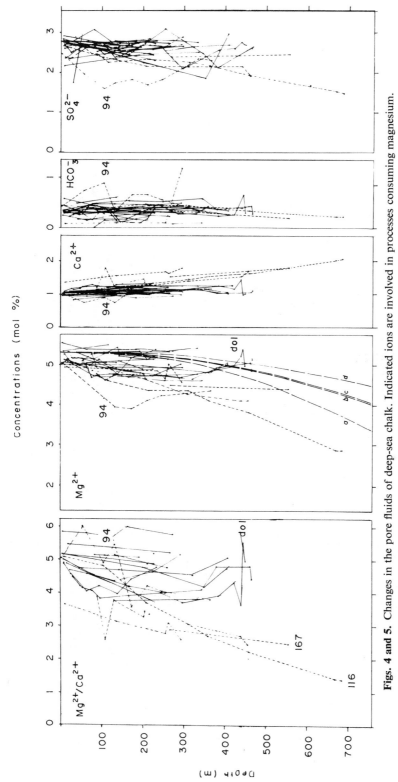

Figs. 4 and 5. Changes in the pore fluids of deep-sea chalk. Indicated ions are involved in processes consuming magnesium.

Fig. 4. Main group of pore fluids, showing small decrease of magnesium and small increase of calcium with depth.

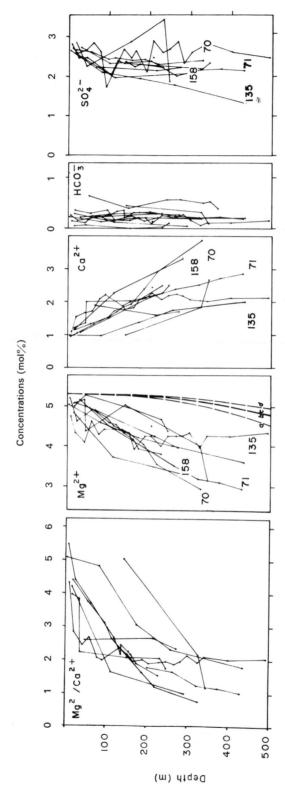

Fig. 5. More reactive pore fluids with respect to magnesium and calcium. Curves a, b, c, d: calculated decrease of magnesium caused by pressure solution. dol = dolomite occurrences; numbers refer to Deep Sea Drilling boreholes. Data in Figs 4 and 5 from Chan & Manheim (1970), Manheim *et al.* (1970a, b), Manheim & Sayles (1971a, b), Manheim, Sayles & Friedman (1969), Manheim, Sayles & Waterman (1972, 1973), Sayles & Manheim (1971), Sayles, Manheim & Chan (1970), Sayles, Manheim & Waterman (1971), Sayles, Waterman & Manheim (1972, 1973a, b) and Waterman, Sayles & Manheim (1972, 1973).

gradients and to a great extent also in absolute values, are mirror-images of the calcium increases.* The behaviour of magnesium and calcium in the interstitial fluids can thus be well explained if we postulate the formation of carbonates containing more magnesium in solid solution; dolomite at the expense of the available calcites, or high-magnesium calcite at the expense of low-magnesium calcite.

The first possibility, the formation of dolomite, is believed to be of no importance (see page 159). The changes in ionic concentrations should therefore be related to the formation of high-magnesium calcite at the expense of low-magnesium calcite.

This result confronts us with a problem which has already been dealt with by Manheim & Sayles (1971b). What process can lead to the transformation of stable low-magnesium calcite to a calcite containing more magnesium in solid solution? A possible process would be that of pressure solution, the role of which we will discuss in detail. By using the calculated cement volumes of Fig. 3 we are in a position to estimate the effect of pressure solution. Assuming that the dissolved low-magnesium calcite is precipitated as high-magnesium calcite with 10 mol % magnesium (see footnote † below) and assuming that the porosity is 40–50%, we attain the magnesium reduction curves a–d (long dashes), given in Fig. 4 or 5. A porosity of 40% combined separately with the two columns for cement volume in Fig. 3 gives the curves a + b, whereas 50% porosity results in the curves c + d.

If one compares the calculated curves with those for the boreholes, it is evident that the major calcium and magnesium shifts in Fig. 5 cannot be explained by pressure solution, even considering the larger margin of error of the calculated curves (see p. 155). Other processes must be involved, perhaps such as replacement of calcite by other growing minerals (e.g. SiO_2 modifications, pyrite) or recrystallization (see page 165).

In Fig. 4 the gradients of the calculated and the borehole curves are parallel; since this is mainly due to the minimal shifts in magnesium content in either case, this is no argument either for or against the magnesium reduction by pressure solution.

For greater depths the calculated curves suggest a pronounced change in magnesium content due to pressure solution. At the bottom of Fig. 4 the curves a–d already begin to flatten. If we extend Fig. 4 downwards they meet the 0·01 M Mg concentration line between 1100 and 1650 m. This would mean that sea water should be considerably depleted in magnesium at those depths.† Accordingly, there is a strong possibility that extensive cementation of chalk through pressure solution should occur even at these depths. In combination with other magnesium-consuming processes the critical level may be even slightly higher and far above the calculated depth of burial of 4000 m.

So far, few boreholes of the Deep Sea Drilling Project have reached carbonate sediments under such depths of burial. Soft 'chalk' was still encountered at site 192 at a depth of 930–1050 m below the sea floor (Scholl *et al.*, 1971). On the other hand, a sequence of chalks has been found to change into limestone at 827 m at site 167;

---

* Comparing the individual boreholes, the possible sulphate corrections (drill sites 116, 135) should be borne in mind.

† If calcite is precipitated in accordance with the partition coefficient from a pore fluid containing magnesium, it does not normally contain just 10 mol % magnesium as supposed for the curve a–d. But, if calcite with a lower magnesium content is formed, the compensation stress is also lower, resulting in a greater volume of cement. In every case the total magnesium consumption attains nearly the same level. This conclusion is only invalid in the case of aragonite formation.

this limestone remained very porous as deep as 1172 m, where it overlies basalt (Schlanger *et al.*, 1973). In boreholes 288 and 289 'limestone with interbeds of chert' and 'Radiolaria-bearing limestone, siliceous limestone, nannochalk' underlie chalks at a depth of 840–988 m and 969–1262 m respectively. Unfortunately, no analyses of the interstitial water are as yet available from such depths.

The effect of strong decrease of magnesium concentration

As a next step, we shall examine published pore-fluid data with the following questions in mind. (a) To what concentrations does the magnesium content actually decrease? (b) Is there a change from chalk to limestone when the magnesium content falls below 0·01 M or below Mg/Ca = 1–2?

All known pore fluids from sedimentary sequences dominated by chalk (Figs 4, 5) contain about 0·03 M magnesium or more. This is three times the concentration suggested by Lippmann (1973) for the inhibition effect.

Intercalations of chalk in argillaceous and volcanic sequences sometimes contain less than 0·03 M magnesium in the pore fluid.

The magnesium content was found to drop below 0·01 M at three sites.

Site 1. Borehole 2 at 140 m: 'calcite caprock' of a salt dome. In this case a sharp drop in salinity was also observed (Manheim, Sayles & Friedman, 1969; Ewing *et al.*, 1969). Measured magnesium content = 0·008 M (Mg/Ca = 0·18).

Site 2. Borehole 53 at 174 m and 193–200 m: 'altered' (partly recrystallized) chalk ooze and limestones (Pimm, Garrison & Boyce, 1971). Andesite and basalt have been found associated with the carbonate rocks below 195·4 m.* Magnesium content = 0·007 M (Mg/Ca = 0·09).

Site 3. Borehole 155 at 490 m: 'dolomitic (chalk), massive and *well* indurated' (van Andel *et al.*, 1973, p. 23). Magnesium content = 0·008 (3) M (Mg/Ca = 0·27).

These cases indicate a transition of chalk into limestone at a concentration below 0·01 M Mg.

The Mg/Ca ratio drops at a number of sites into the 1–2 range (site 116 and most of the boreholes in Fig. 5), but excessive induration of chalk was not observed. Lithification or excessive induration of chalk was not reported even from the few cores in which the Mg/Ca ratio drops below 1.

(1) Site 155 at 515 m; chalk ooze. Mg/Ca = 0·34 (Mg = 0·012 M). Thus the magnesium content increases again below the 'dolomitic well indurated chalk' mentioned above. At the same time the sediment 'is again softer' (van Andel *et al.*, 1973).

(2) Site 137 at 265–382 m; 'nanno marl to chalk ooze'; Mg/Ca = 0·63–0·35 (Mg = 0·035–0·028 M).

The very limited available data thus suggests that the magnesium content must decrease to a very low level for the change to limestone to occur with less than a few thousand metres of overlying sediments. The order of magnitude of 0·01 M magnesium, given by Lippmann (1973), seems to be critical for the premature lithification of chalk. Connate waters often contain such small amounts of magnesium (von Engelhardt, 1973).

---

* On the basis of the oxygen isotope composition of altered carbonates Anderson (1973) states that the partial recrystallization observed in this borehole is not the result of thermal metamorphism, but 'probably a consequence of chemical changes in ambient pore waters resulting from the submarine weathering of volcanic material'.

So far we have disregarded chalk pore fluids influenced by evaporites. These are characterized not only by increased sodium and chloride concentrations, but also by a sometimes drastic increase of calcium and magnesium. The Mg/Ca ratio frequency conforms to Figs 4 and 5. In principle, additional magnesium from the evaporites may keep the magnesium content at a sufficiently high level to allow the chalk to reach great depths in a soft state. This mechanism may be active, for instance, in the deep subsurface of the North Sea, under the influence of Permian salts.

We cannot close this discussion without mentioning that a number of other limestones have been found in the Deep Sea Drilling Project: (1) limestones derived from redeposited shallow-water sediments; (2) siliceous limestones; (3) limestones in direct contact with basalt.

They are not considered here because their pore fluids have not been analysed. However, these limestones suggest other ways in which deep-sea carbonate sediments can be lithified. (1) In the case of redeposited shallow-water sediments we can assume considerable amounts of aragonite and high-magnesium calcite in the sediment, which behave totally differently during diagenesis from low-magnesium calcite (e.g. in the case of pressure solution; cf. Neugebauer, 1973). (2) A larger proportion of siliceous skeletons presumably contributes in various ways to lithification (cf. for example page 162). (3) Near to basalt the magnesium content of the pore solution may drop markedly as indicated in borehole 53 (see page 163 and footnote).

Apart from these limestones there is a fourth group of thin lithified or excessively hardened seams intercalated in soft chalk, which occur in the deep-sea as well as in the shelf environment. The consolidation of these beds, characterized by a high content of low-magnesium calcite fossils (other than coccoliths), is discussed in the next section.

DEPOSITION OF CEMENT IN CHALK

In this third section we will discuss the problem of where the deposition of cement in chalk takes place. The amount of carbonate cement present in chalk is small as long as a sufficient magnesium concentration in the pore fluid prevents the rapid transition to limestone. It must be borne in mind that this small amount of cement must not exclusively have been produced by the pressure solution of low-magnesium calcite. In the first few hundred metres of burial depth, where the cement production by this process is very small (Fig. 3), it is imaginable and sometimes probable (see page 162 and Fig. 5) that more cement originates from other sources, for example, the highly efficient pressure solution of high-magnesium calcite particles, or perhaps the replacements of low-magnesium calcite by non-carbonate minerals (see page 162). In contrast to the first part (Fig. 3) we include here all sources and discuss the total (small) amount of cement present.

Cement in typical chalk originally containing no or very small amounts of high-magnesium calcite and aragonite exhibits two features.

(1) Normally most cement appears to be syntaxial that is, it grows in optical continuity with the available biogenic calcites.*

---

* Isolated idiomorphic calcite crystals are sometimes interspersed between the coccolith plates. Their origin is open to question. They are possibly entirely of biogenic origin or they may possess a biogenic core (cf. Wise & Kelts, 1972, p. 183).

(2) Different groups of fossils (coccoliths, discoasters, Foraminifera, etc.) attract different amounts of cement.

These peculiarities can be observed in deep-sea chalk (see below), in experiments with samples of deep-sea chalk subjected to elevated temperatures and pressures (Adelseck, Geehan & Roth, 1973) and will be demonstrated by the following figures of fossils from shelf-sea chalks.

The first group we will look at are the coccoliths. Cement deposition is rather unimportant on coccolith crystals (Fig. 6) and it is difficult to find crystals with demonstrable overgrowth formation. The excrescences, marked by arrows (Fig. 6b), should be interpreted as one type of diagenetic overgrowth on coccoliths. They appear on certain crystal faces, whereas adjacent faces are smooth. Similar formations are found on larger coccolith crystals of the size of one micron or more. From deep-sea chalks, and from experiments, some overgrowth on coccolith crystals was reported by Berger & von Rad (1972), Wise & Kelts (1972) and Schlanger *et al.* (1973) and by Adelseck *et al.* (1973).

The same authors noted also the common occurrence of overgrowth on the bigger crystals of discoasters.

Foraminifera regularly show a certain amount of cement, particularly on the inner surface of their chambers (Fig. 7; Pimm *et al.*, 1971; Schlanger *et al.*, 1973). These inner surfaces are smooth in the living animal and are now studded with centripetal crystals of diagenetic origin. It should be noted that the crystals in Fig. 7 chiefly exhibit the cleavage rhombohedron as a growing face and that, by and large, the 'teeth' of overgrowth on foraminiferal walls are larger than on coccolith plates (Figs 6 and 7). Cement is preferentially deposited on Foraminifera rather than on coccoliths.

Disturbing in this context is that an additional process plays a role in the diagenesis of Foraminifera. As shown in Fig. 7, the foraminiferal walls no longer have their primary structure. Originally most of the Foraminifera occurring in chalk (all of the prevalent planktonic and some of the benthonic forms) were built up of 'prisms', which are probably in turn composed of minute crystals (compare Towe & Cifelli, 1967; thickness of the crystal units scarcely 0.2 μm). The originally compact walls of the foraminifers are altered to a porous state by dissolution and 'recrystallization', possibly because of the high solubility of small crystals (Fig. 10). The dissolution should predominantly have taken place at the bottom of the chalk sea, comparable to the solution processes in the present deep-sea environment (Berger, 1967, 1972; Thiede, 1971 and others).

This process screens the starting form and size of the crystals effected by overgrowth. It appears that the top surfaces of the prisms served as the basis for the overgrowth.

Here and there larger crystals grew vertically on the wall of Foraminifera, probably via enlargement of some of the syntaxial 'teeth'.* It should be noted that, with increasing size of the cement crystals, the number of crystal forms increases also.

On prisms of inoceramids and ossicles of crinoids relatively large epitaxial crystals are found, resulting in a large amount of cement. Figure 8 shows cement crystals in the pore space between two prisms of a disintegrated *Inoceramus* shell, whilst in Fig. 9 cement can be seen to have grown into the cavity between two articulating

---

* The composition of the large crystal in the centre of Fig. 7b was verified as $CaCO_3$ with an EDAX microprobe.

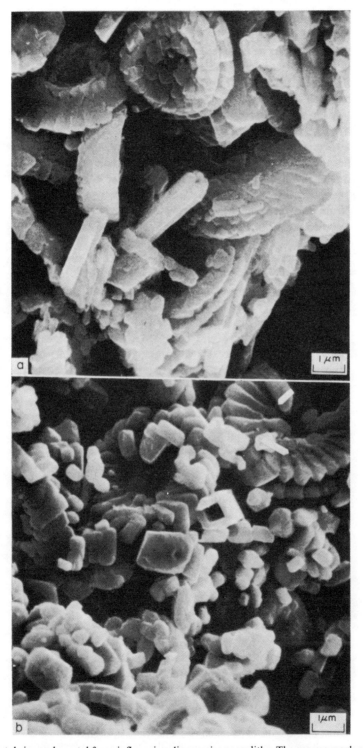

Fig. 6. Crystal size and crystal form influencing diagenesis: coccoliths. The arrows mark overgrowth. (a) Chalk of Kansas, *Uintacrinus-zone*. Scanning electron micrograph. (b) Chalk from Calais, France. Scanning electron micrograph.

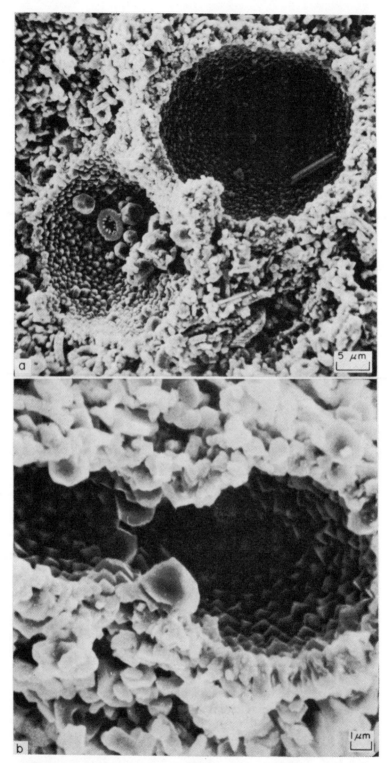

Fig. 7. Dissolution of the walls of Foraminifera and syntaxial overgrowth on the surface of their chambers. The small cement crystals chiefly exhibit the cleavage rhombohedron as a growing face. Chalk from Calais, France; shelf environment. Scanning electron micrographs. Numbers of the stereoscan electron micrographs of Figs 6–9: SEM Geol. Paläont. Inst. Tübingen Nos. 1447/31964: Fig. 6a; 1447/32963: Fig. 6b; 1447/33002: Fig. 7a; 1447/37741: Fig. 7b; 1447/33728: Fig. 8; 1447/ 33144: Fig. 9. The samples for the stereoscan studies (Figs 6–9) were simply broken and covered with Au/Pd or carbon.

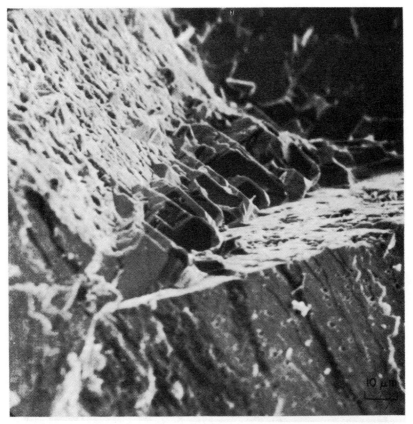

Fig. 8. Larger cement crystals on big prisms of *Inoceramus*. Chalk of Kansas. Scanning electron micrograph.

crinoid ossicles. The whole photomicrograph shows a part of a complicated single crystal of cement covering a biogenic single crystal of a crinoid ossicle.

The question now arises: what controls the selective deposition of cement on the biogenic crystals? Why is cementation on coccolith crystals discriminated against in comparison to Foraminifera, discoasters, inoceramids and echinoderms?

There are two factors of special importance: (1) the crystal size and (2) the crystal shape.

The crystal size

In Figs 6–9 we observed an increasing amount of syntaxial cement with increasing crystal size. The important point here is whether or not this is truly a causal relationship.

For the last 50 years the differences in solubility and free energy resulting from differences in crystal size have been known theoretically and experimentally. Differences in crystal size are only important for very small crystals, whilst they can be neglected if the crystals are larger (cf. Spangenberg, 1935). The size of calcite crystals in nannofossils and microfossils of chalk are just in between these two categories and we must therefore establish whether or not the differences in the free energy (solubility

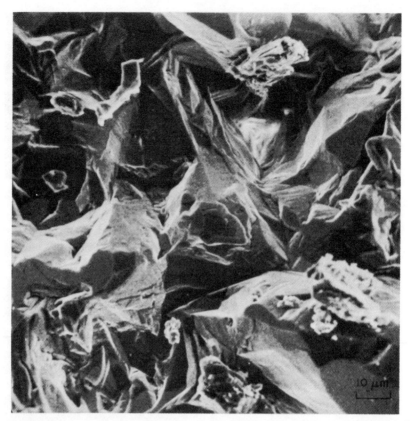

Fig. 9. Cementation on a crinoid ossicle from chalk of Rügen, Germany; the whole complicated overgrowth is optically a single crystal.

product) of the different-sized calcite crystals are great enough to effect the deposition of cement.

To begin with we will consider the theoretical principles for the dependence of the solubility product on crystal size (cf. Fig. 10). The theoretical principles for Fig. 10 were already given by Gibbs (in Spangenberg, 1935):

$$\Delta G = RT \ln \frac{K_d}{K_\infty} = \frac{F}{d} \,. \tag{4}$$

ΔG = difference in free energy; R = gas-law constant; T = absolute temperature; K_d and K_∞ = solubility products of crystals of the dimension d and ∞ respectively; F = factor and d = linear dimension of the crystal. According to Chave & Schmalz (1966) the factor F depends on the surface energy

$$F = s \sigma V \tag{5}$$

s = shape factor; σ = surface energy of the exposed faces and V = molar volume.

For small calcite crystallites of unknown form Chave & Schmalz determined (via ΔG) F values between $1{\cdot}0$ and $1{\cdot}6 \cdot 10^{-3}$ cal cm, the average being $1{\cdot}3 \cdot 10^{-3}$ cal cm. A similar value $1{\cdot}22 \cdot 10^{-3}$ cal cm results from the free specific surface energy of the

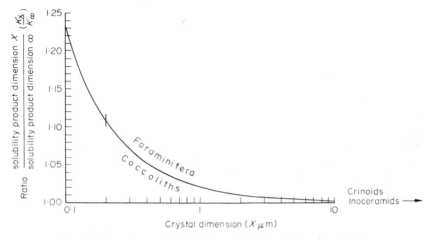

Fig. 10. Dependence of solubility product of calcite on crystal size.

cleavage plane of calcite ($\sigma = 230$ ergs/cm^2), as determined by Gilman (1960). Because the smallest cement crystals should chiefly exhibit cleavage rhombohedra as growing faces (see below) the latter value was used as basis for the calculation ($T = 298°$ K).

Figure 10 shows the resulting curve for the dependence of the solubility product on the crystal dimension d of calcite (bounded by the cleavage rhombohedron). Furthermore, the crystal sizes regarded as roughly comparable to linear dimension d of formula (4) are shown for the fossil groups under discussion.

The crystal size of coccoliths and Foraminifera (after dissolution) ranges from about 0·2 to more than 1 μm (or even 2–3 μm). Accordingly, the differences in the solubility product of the overgrowth on coccoliths and Foraminifera on one hand, and on crinoids, inoceramids and on discoasters on the other hand, are about 1–10%.* The solubility product of the overgrowth on the larger coccolith crystals approaches that of the overgrowth on crinoids and inoceramids far more closely than that of the overgrowth on the small coccolith crystals.

Once again we may ask whether these differences in the solubility product caused by differences of crystal sizes are sufficiently large to affect the amount of cement deposited. The differences are small compared to differences in solubility product caused by stress (Fig. 2), but they should still have an effect: it is known from growth experiments with NaCl, KCl and other substances that supersaturations of a small fraction of a percent determine growth and growth velocities (Honigmann, 1958; Schüz, 1969 and others).

Though the substances employed in the experiments are easily soluble compared with calcite, the differences in the solubility product are most probably great enough to affect the deposition of cement. Compared to coccolith and foraminiferal crystals the larger units of crinoids and inoceramids should therefore preferentially accrete cement.

Above a crystal dimension of a few microns, grain-size differences have negligible effect on the solubility product (Fig. 10). This could be the reason why many fine-grained carbonate rocks do not recrystallize beyond a grain size of a few microns (micrite, microsparite, Folk, 1962).

* The real difference in the solubility product should be somewhat less than given in Fig. 10 since, to simplify matters, a constant (rhombohedral) form as well as a pure solution were considered.

The small crystals of coccoliths (0·2 µm) in chalk did not dissolve and form larger crystals a few microns across because the pore fluid was supersaturated with respect to low-magnesium calcite.

The crystal shape

The effect of crystal size does not explain the differences of overgrowth between equal-sized crystals of coccoliths and Foraminifera or of inoceramids and crinoids. There is another determining factor: the shape of biogenic crystals. Many biogenic crystals have a shape which is only partially, or not at all, bounded by crystal faces. For this reason these biogenic crystals are favoured in growth until crystal faces of the so-called 'equilibrium form' (Gleichgewichtsform, Honigmann, 1958) are reached. Subsequent growth is slow.

The influence of the biogenic shape on cement deposition is schematically depicted in Fig. 11. The cleavage rhombohedron $\{10\bar{1}1\}$ is used as the equilibrium form because, theoretically, and in a series of experiments with and without impurities, it has the lowest (or a very low) specific surface energy (Honigmann, 1958). Figure 7 shows that we find this form predominantly in the form of the small 'teeth' on the foraminiferal walls.

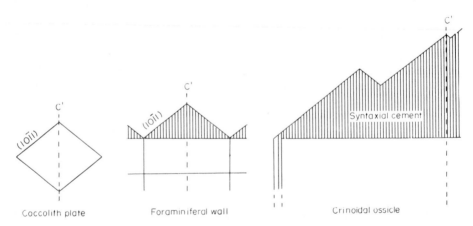

Fig. 11. Influence of the shape of biogenic crystals on the deposition of syntaxial cement (schematically).

The influence of surface energy decreases with increasing crystal size (Spangenberg, 1935); compare equations (4) and (5). This results in a larger number of crystal forms occurring on bigger crystals. In this respect the crinoids in Fig. 11 are not quite correctly represented, because the cleavage rhombohedron, though present on crinoids (Fig. 9), is no longer the dominant crystal form.

Figure 11 illustrates that the coccoliths are handicapped by the shape factor. The coccolith plates appear to be largely bounded by slowly growing crystal faces; according to Black (1963) the cleavage rhombohedron is frequently used as the habit of coccolith crystals. The shape of the coccolith plates therefore impedes the deposition of cement. This is consistent with Fig. 6, where cement-like excrescences appear only on certain faces of big crystals.

On the other hand, due to their stable form, the coccolith plates are more difficult to dissolve in comparison with other biogenic crystals of the same size. Together with the ultrastructure of the prisms (see p. 165) this could contribute to the fact that on the deep-sea bottom Foraminifera are dissolved easier than coccoliths (cf. Berger, 1972). The selective dissolution of Foraminifera and coccolith species (Bukry *et al.*, 1971; Douglas, 1971; McIntyre & McIntyre, 1971) should be examined from the aspect of the participant crystals (cf. Adelseck *et al.*, 1973).

Summarizing, we can state that the quantity of the deposited cement (and dissolved biogenic calcite?) is influenced by the size and shape of the biogenic crystals.

What are the geological implications of this statement? Cementation in chalk avoids coccoliths and concentrates on all other carbonate fossils, but favours big crystals. Layers within chalk that are rich in carbonate fossils other than coccoliths can therefore draw larger amounts of cement. This should cause partial or total hardening of such layers.

This is our interpretation of the hardening of *Inoceramus* layers in the chalk of Kansas (Fig. 8 and Neugebauer, 1973). More or less consolidated, Foraminifera-rich layers in England (Black, 1953) and the *Braarudosphaera* horizon in the South Atlantic (Wise & Hsü, 1971; Wise & Kelts, 1972) should be examined from this point of view.

RESULTS

The present study specifies some conclusions which result from the diagenetic model of chalk diagenesis formulated by Neugebauer (1973).

Part I–II. The magnesium content of the pore fluid is crucial for the diagenesis of chalk. It causes a supersaturation of the pore fluid with respect to low-magnesium calcite. Consequently, pressure solution can only be effective when a certain compensating stress is exceeded (Neugebauer, 1973). The compensation stress present at all grain contacts of compacting chalk, irrespective of burial depth, amounts to about 250–1000 atm.

The volume of cement produced by pressure solution of low-magnesium calcite can be approximately evaluated. The cement production is negligible up to 300 m overload. Depending upon the strength of the compensation stress, volumes of 0·5–5% cement are generated from about 1000 m overload onwards. The production of cement increases considerably with further burial.

Chalk can remain highly porous down to great depths. Under favourable conditions complete lithification requires an overload of 2000–4000 m or more.

Pressure solution lithifies chalk at shallower depths when the magnesium content of the pore fluid is exhausted. As evidenced by pore fluid analyses of the deep-sea cores, processes consuming magnesium are not very important in the first 500–700 m of chalk deposits. At greater depths pressure solution is held to become an important process for reduction of magnesium through the formation of magnesium-bearing cement at the cost of biogenic low-magnesium calcite. For this reason the magnesium content of the pore fluids should often be exhausted at a depth of about 1000–1600 m. A few changes from chalk to limestones have been observed at these sub-bottom depths. Where additional magnesium is supplied by evaporites, one should expect soft chalk below 1600 m burial depth.

The growth of low-magnesium calcite is influenced by very low concentrations of magnesium (0·0001–0·001 M). So far, the lowest magnesium concentration that leads to precipitation of aragonite or high-magnesium calcite instead of low-magnesium calcite is unknown. However, various lines of evidence indicate that an absolute concentration of about 0·01 M Mg or a Mg/Ca ratio from 1–2 might represent the concentrations below which low-magnesium calcite is formed. Pore fluids of chalk normally contain more than 0·01 M magnesium. In exceptional cases, where the concentration of magnesium lies below 0·01 M, stronger induration or limestones are observed. In some chalks the Mg/Ca ratio reaches values of 1–2. No special induration is observed in these cases.

Part III. Chalk is characterized by its dearth of cement. The small amounts of cement which are produced by pressure solution and by other processes are usually precipitated as syntaxial cement.

The biogenic calcite crystals of the constituent fossil groups attract different amounts of cement. Shape and size of the skeletal crystals are of crucial importance. Cement precipitation on coccoliths is handicapped by these two factors. Due to their small crystal size the supersaturation level is lowered by several percent for the overgrowth on coccoliths. Coccoliths are also discriminated against with respect to cementation by the shape factor. Consequently, cementation in chalk avoids coccoliths and concentrates on all other carbonate fossils, preferentially fossils composed of large crystals. Layers within the chalk which are rich in fossils other than coccoliths can thus draw larger amounts of cement and will be selectively lithified.

Foraminifera of chalk from the shelf environment show the same dissolution characteristics as Foraminifera from the deep-sea bottom. The dissolution of small-sized low-magnesium calcite particles indicates a very low concentration of $CaCO_3$ on the bottom of the Chalk Sea, a concentration which was below the solubility product of aragonite. This is consistent with the hypothesis of Hudson (1967).

ACKNOWLEDGMENTS

For fruitful discussions and help I am indebted to Dr W. Bay, Dr Chr. Hemleben, Dr F. Lippmann, Professor A. Seilacher, Dr N. Shrivastava and Professor K. M. Towe. I thank Professor A. Seilacher for samples of chalk and inoceramite from Kansas. Dipl.-Geol. F. Fürsich and Dr U. von Rad kindly helped to obtain copies of the papers on interstitial water analyses of the Deep Sea Drilling Project. Miss R. Freund instructed me during stereoscan studies. The impulse for my interest for chalk diagenesis was given by my friend Dr G. Ruhrmann. The translation of this manuscript was accomplished especially with the help of Dipl.-Geol. S. Chrulew, and further of Professor A. Seilacher, Dr N. Shrivastava and Professor R. D. K. Thomas. Dr A. Matter and Professor K. J. Hsü kindly reviewed the manuscript and made many useful suggestions. I would like to thank all who helped in this work.

REFERENCES

ADELSECK, C.G. JR, GEEHAN, G.W. & ROTH, P.H. (1973) Experimental evidence for the selective dissolution and overgrowth of calcareous nannofossils during diagenesis. *Bull. geol. Soc. Am.* **84**, 2755–2762.

174 *Joachim Neugebauer*

ANDERSON, T.F. (1973) Oxygen and carbon isotope compositions of altered carbonates from the western Pacific, core 53.0, Deep Sea Drilling Project. *Mar. Geol.* **15**, 169–180.

BATHURST, R.G.C. (1971) *Carbonate Sediments and their Diagenesis*, pp. 620. Elsevier, Amsterdam.

BERGER, W.H. (1967) Foraminiferal ooze: solution at depths. *Science*, **156**, 383–385.

BERGER, W.H. (1972) Deep-sea carbonates: dissolution facies and age–depth constancy. *Nature, Lond.* **236**, 392–395.

BERGER, W.H. & VON RAD, U. (1972) Cretaceous and Cenozoic sediments from the Atlantic Ocean. In: *Initial Reports of the Deep Sea Drilling Project*, Vol. XIV (D. E. Hayes, A. C. Pimm *et al.*), pp. 787–954. U.S. Government Printing Office, Washington.

BISCHOFF, J.L. & FYFE, W.S. (1968) Catalysis, inhibition, and the calcite–aragonite problem. I. The calcite–aragonite transformation. *Am. J. Sci.* **266**, 65–79.

BLACK, M. (1953) The constitution of the Chalk. *Proc. geol. Soc. Lond.* no. **1499**, 81–86.

BLACK, M. (1963) The fine structure of the mineral parts of Coccolithophoridae. *Proc. Linn. Soc. Lond.* **174**, 41–46.

BUKRY, D., DOUGLAS, R.G., KLING, S.A. & KRASHENINNIKOV, V. (1971) Planktonic microfossil biostratigraphy of the northwestern Pacific Ocean. In: *Initial Reports of the Deep Sea Drilling Project*, Vol. VI (A. G. Fischer *et al.*), pp. 1253–1300. U.S. Government Printing Office, Washington.

CHAN, K.M. & MANHEIM, F.T. (1970) Interstitial water studies on small core samples, Deep Sea Drilling Project, leg. 2. In: *Initial Reports of the Deep Sea Drilling Project*, Vol. II (M. N. A. Peterson *et al.*), pp. 367–371. U.S. Government Printing Office, Washington.

CHAVE, K.E. & SCHMALZ, R.F. (1966) Carbonate–seawater interactions. *Geochim. cosmochim. Acta*, **30**, 1037–1048.

CORRENS, C.W. & STEINBORN, W. (1939) Experimente zur Messung und Erklärung der sogenannten Kristallisationskraft. *Z. Kristallogr.* **101**, 117–133.

CULBERSON, C. & PYTKOWICZ, R.M. (1968) Effect of pressure on carbonic acid, boric acid, and the pH in seawater. *Limnol. Oceanogr.* **13**, 403–417.

DOUGLAS, R.G. (1971) Cretaceous Foraminifera from the northwestern Pacific Ocean: leg 6, Deep Sea Drilling Project. In: *Initial Reports of the Deep Sea Drilling Project*, Vol. VI (A. G. Fischer *et al.*), pp. 1027–1046. U.S. Government Printing Office, Washington.

DREVER, J.I. (1971) Magnesium-iron replacement in clay minerals in anoxic marine sediments. *Science*, **172**, 1334–1336.

EWING, M. *et al.* (1969) *Initial Reports of the Deep Sea Drilling Project*, Vol. I, pp. 672. U.S. Government Printing Office, Washington.

FOLK, R.L. (1962) Spectral subdivision of limestone types. In: *Classification of Carbonate Rocks* (Ed. by W. E. Ham). *Mem. Am. Ass. Petrol. Geol.* **1**, 62–84.

GILMAN, J.J. (1960) Direct measurements of the surface energies of crystals. *J. appl. Phys.* **31**, 2208–2218.

HAUSSÜHL, S. (1964) Das Wachstum großer Einkristalle. *Neues. Jb. Miner. Abh.* **101**, 343–366.

HONIGMANN, B. (1958) *Gleichgewichts- und Wachstumsformen von Kristallen*, pp. 161. Steinkopf, Darmstadt.

HUDSON, J.D. (1967) Speculations on the depth relations of calcium carbonate solution in Recent and ancient seas. *Mar. Geol.* **5**, 473–480.

LIPPMANN, F. (1960) Versuche zur Aufklärung der Bildungsbedingungen von Kalzit und Aragonit. *Fortschr. Miner.* **38**, 156–161.

LIPPMANN, F. (1973) *Sedimentary Carbonate Minerals*, pp. 228. Springer, Berlin.

MANHEIM, F.T., CHAN, K.M., KERR, D. & SUNDA, W. (1970a) Interstitial water studies on small core samples, Deep Sea Drilling Project, leg 3. In: *Initial Reports of the Deep Sea Drilling Project*, Vol. III (A. E. Maxwell *et al.*), pp. 663–666. U.S. Government Printing Office, Washington.

MANHEIM, F.T., CHAN, K.M. & SAYLES, F.L. (1970b) Interstitial water studies on small core samples, Deep Sea Drilling Project, leg 5. In: *Initial Reports of the Deep Sea Drilling Project*, Vol. V (D. A. McManus *et al.*), pp. 501–511. U.S. Government Printing Office, Washington.

MANHEIM, F.T. & SAYLES, F.L. (1971a) Interstitial water studies on small core samples, Deep Sea Drilling Project, leg 6. In: *Initial Reports of the Deep Sea Drilling Project*, Vol. VI (A. G. Fischer *et al.*), pp. 811–821. U.S. Government Printing Office, Washington.

MANHEIM, F.T. & SAYLES, F.L. (1971b) Interstitial water studies on small core samples, Deep Sea Drilling Project, leg 8. In: *Initial Reports of the Deep Sea Drilling Project*, Vol. VIII (J. I. Tracey, Jr, *et al.*), pp. 857–869. U.S. Government Printing Office, Washington.

MANHEIM, F.T., SAYLES, F.L. & FRIEDMAN, I. (1969) Interstitial water studies on small core samples, Deep Sea Drilling Project, leg 1. In: *Initial Reports of the Deep Sea Drilling Project*, Vol. I (M. Ewing *et al.*), pp. 403–410. U.S. Government Printing Office, Washington.

MANHEIM, F.T., SAYLES, F.L. & WATERMAN, L.S. (1972) Interstitial water studies on small core samples, Deep Sea Drilling Project, leg 12. In: *Initial Reports of the Deep Sea Drilling Project*, Vol. XII (A. S. Laughton, W. A. Berggren *et al.*), pp. 1193–1200. U.S. Government Printing Office, Washington.

MANHEIM, F.T., SAYLES, F.L. & WATERMAN, L.S. (1973) Interstitial water studies on small core samples, Deep Sea Drilling Project, leg 10. In: *Initial Reports of the Deep Sea Drilling Project*, Vol. X (J. L. Worzel, W. Bryant *et al.*), pp. 615–623. U.S. Government Printing Office, Washington.

MCINTYRE, A. & MCINTYRE, R. (1971) Coccolith concentrations and differential solution in oceanic sediments. In: *The Micropalaeontology of Oceans* (Ed. by B. M. Funnell and W. R. Riedel), pp. 253–261. Cambridge University Press, Cambridge.

MÖLLER, P. & RAJAGOPALAN, G. (1972) Cationic distribution and structural changes of mixed Mg–Ca layers on calcite crystals. *Z. phys. Chem.* N.F. **81**, 47–56.

MÜLLER, G., IRION, G. & FÖRSTNER, U. (1972) Formation and diagenesis of inorganic Ca–Mg carbonates in the lacustrine environment. *Naturwissenschaften*, **59**, 158–164.

NEUGEBAUER, J. (1973) The diagenetic problem of chalk: the role of pressure solution and pore fluid. *Neues. Jb. Geol. Palaont. Abh.* **143**, 223–245.

OWEN, B.B. & BRINKLEY, S.R. (1941) Calculation of the effect of pressure upon ionic equilibria in pure water and in salt solutions. *Chem. Rev.* **29**, 461–474.

PIMM, A.C., GARRISON, R.E. & BOYCE, R.E. (1971) Sedimentology synthesis: lithology, chemistry and physical properties of sediments in the northwestern Pacific Ocean. In: *Initial Reports of the Deep Sea Drilling Project*, Vol. VI (A. G. Fischer *et al.*), pp. 1131–1252. U.S. Government Printing Office, Washington.

PYTKOWICZ, R.M., DISTECHE, A. & DISTECHE, S. (1967) Calcium carbonate solubility in sea water at *in situ* pressures. *Earth Plan. Sci. Letts*, **2**, 430–432.

SAYLES, F.L. & MANHEIM, F.T. (1971) Interstitial water studies on small core samples, Deep Sea Drilling Project, leg 7. In: *Initial Reports of the Deep Sea Drilling Project*, Vol. VII (E. L. Winterer *et al.*), pp. 871–881. U.S. Government Printing Office, Washington.

SAYLES, F.L., MANHEIM, F.T. & CHAN, K.M. (1970) Interstitial water studies on small core samples, leg 4. In: *Initial Reports of the Deep Sea Drilling Project*, Vol. IV (R. G. Bader *et al.*), pp. 401–414. U.S. Government Printing Office, Washington.

SAYLES, F.L., MANHEIM, F.T. & WATERMAN, L.S. (1971) Interstitial water studies on small core samples, leg 11. In: *Initial Reports of the Deep Sea Drilling Project*, Vol. XI (C. D. Hollister, J. I. Ewing *et al.*), pp. 997–1008. U.S. Government Printing Office, Washington.

SAYLES, F.L., WATERMAN, L.S. & MANHEIM, F.T. (1972) Interstitial water studies on small core samples, leg 9. In: *Initial Reports of the Deep Sea Drilling Project*, Vol. IX (J. D. Hays *et al.*), pp. 845–855. U.S. Government Printing Office, Washington.

SAYLES, F.L., WATERMAN, L.S. & MANHEIM, F.T. (1973a) Interstitial water studies on small core samples from the Mediterranean Sea. In: *Initial Reports of the Deep Sea Drilling Project*, Vol. XIII (W. B. F. Ryan, K. J. Hsü *et al.*), pp. 801–808. U.S. Government Printing Office, Washington.

SAYLES, F.L., WATERMAN, L.S. & MANHEIM, F.T. (1937b) Interstitial water studies on small core samples, leg 19. In: *Initial Reports of the Deep Sea Drilling Project*, Vol. XIX (J. S. Creager, D. W. Scholl *et al.*), pp. 871–874. U.S. Government Printing Office, Washington.

SCHLANGER, S.O., DOUGLAS, R.G., LANCELOT, Y., MOORE, T.C. JR, & ROTH, P.H. (1973) Fossil preservation and diagenesis of pelagic carbonates from the Magellan Rise, central North Pacific Ocean. In: *Initial Reports of the Deep Sea Drilling Project*, Vol. XVII (E. L. Winterer, J. I. Ewing *et al.*), pp. 407–427. U.S. Government Printing Office, Washington.

SCHOLL, D.W., CREAGER, J.S., BOYCE, R.E., ECHOLS, R.J., FULLAM, T.J., GROW, J.A., KOIZUMI, I., LEE, J.H., LING, H-Y., SUPKO, P.R., STEWART, R.J., WORSLEY, T.R., ERICSON, A., HESS, J., BRYAN, G. & STOLL, R. (1971) Deep Sea Drilling Project, leg 19. *Geotimes*, **16**, (11) 12–15.

SCHÜZ, W. (1969) Über den Einfluß der Übersättigung auf das Wachstum von KCl-Einkristallen in reinen wässerigen Lösungen. *Z. Kristallogr.* **128**, 36–54.

SPANGENBERG, K. (1935) Wachstum und Auflösung der Kristalle. In: *Handwörterbuch der Naturwiss.* Vol. X, 362–401. Fischer, Jena.

TAFT, W.H. (1967) Physical chemistry of formation of carbonates. In: *Carbonate Rocks, Physical and Chemical Aspects* (Ed. by G. V. Chilingar, H. J. Bissel and R. W. Fairbridge), pp. 151–167. Elsevier, Amsterdam.

THIEDE, J. (1971) Planktonische Foraminiferen in Sedimenten vom ibero-marokkanischen Kontinentalrand. *Meteor Forsch.-Ergebnisse,* R.C., **7**, 15–102.

TOWE, K.M. & CIFELLI, R. (1967) Wall ultrastructure in the calcareous Foraminifera: crystallographic aspects and a model for calcification. *J. Paleontol.* **41**, 742–762.

VAN ANDEL, T.H. *et al.* (1973) *Initial Reports of the Deep Sea Drilling Project*, Vol. XVI, pp. 1037. U.S. Government Printing Office, Washington.

VON ENGELHARDT, W. (1960) *Der Porenraum der Sedimente*, pp. 207. Springer, Berlin.

VON ENGELHARDT, W. (1973) *Die Bildung von Sedimenten und Sedimentgesteinen*, pp. 378. Schweizerbart, Stuttgart.

WATERMAN, L.S., SAYLES, F.L. & MANHEIM, F.T. (1972) Interstitial water studies on small core samples, leg 14. In: *Initial Reports of the Deep Sea Drilling Project*, Vol. XIV (D. E. Hayes, A. C. Pimm *et al.*), pp. 753–762. U.S. Government Printing Office, Washington.

WATERMAN, L.S., SAYLES, F.L. & MANHEIM, F.T. (1973) Interstitial water studies on small core samples, leg 16, 17 and 18. In: *Initial Reports of the Deep Sea Drilling Project*, Vol. XVIII (L. D. Kulm, R. von Huene *et al.*), pp. 1001–1012. U.S. Government Printing Office, Washington.

WEYL, P.K. (1959) Pressure solution and the force of crystallization—a phenomenological theory. *J. geophys. Res.* **64**, 2001–2025.

WISE, S.W., JR & HSÜ, K.J. (1971) Genesis and lithification of a deep sea chalk. *Eclog. geol. Helv.* **64**, 273–278.

WISE, S.W., JR & KELTS, K.R. (1972) Inferred diagenetic history of a weakly silicified deep sea chalk. *Trans. Gulf-Cst Ass. geol. Socs,* **22**, 177–203.

15

Reprinted from *Earth and Planetary Sci. Letters* **25**:1–10 (1975)

HYDROGEN AND OXYGEN ISOTOPE RATIOS IN SILICA FROM THE JOIDES DEEP SEA DRILLING PROJECT*

L. PAUL KNAUTH[1] and SAMUEL EPSTEIN

*Division of Geological and Planetary Sciences, California Institute of Technology
Pasadena, Calif. (U.S.A.)*

Received September 23, 1974
Revised version received November 19, 1974

Water extracted from opal-CT ("porcellanite", "cristobalite"), granular microcrystalline quartz (chert), and pure fibrous quartz (chalcedony) in cherts from the JOIDES Deep Sea Drilling Project is 56‰ to 87‰ depleted in deuterium relative to the water in which the silica formed. This large fractionation is similar in magnitude and sign to that observed for hydroxyl in clay minerals and suggests that water extracted from these forms of silica has been derived from hydroxyl groups within the silica.

$\delta^{18}O$-values for opal-CT at sites 61, 64, 70B and 149 vary from 34.3‰ to 37.2‰ and show no direct correlation with depth of burial. Granular microcrystalline quartz in these cores is 0.5‰ depleted in ^{18}O relative to coexisting opal-CT at sediment depths of 100 m and the depletion increases to 2‰ for sediments buried below 384 m. These relationships suggest that opal-CT forms before significant burial while granular microcrystalline quartz forms during deeper burial at warmer temperatures. The temperature at which opal-CT forms is thus probably approximately equal to the temperature of the overlying bottom water. Isotopic temperatures deduced for opal-CT formation are preliminary and very approximate, but yield Eocene deep-water temperatures of 5–13°C, and 6°C for the upper Cretaceous sample.

Pure euhedral quartz crystals lining a cavity in opal-CT at 388 m in core 8-70B-4cc have a $\delta^{18}O$ value of +29.8‰ and probably formed near maximum burial. The isotopic temperature is approximately 32°C.

1. Introduction

Silica samples from deep-sea sediment cores contain many of the known phases of silica and probably have not been subject to alteration in the presence of meteoric waters which are usually depleted in deuterium and ^{18}O relative to seawater. It is likely that the silica phases formed in water isotopically similar to ocean water and that the isotopic composition of the silica is primarily related to temperature and mineralogy. In this paper, one of a series of three on the isotopic composition of authigenic silica, we attempt to determine the geochemical significance of the D/H

ratio of water extracted from deep-sea silica and the $^{18}O/^{16}O$ ratio of the total silica for the various deep-sea silica phases.

2. Paragenesis of deep-sea silica

Although silica in deep-sea sediments is usually labeled "chert", it actually exists in a variety of forms, both amorphous and crystalline. The paragenesis of these forms is briefly summarized below to provide a framework for interpreting the isotope data.

Most marine silica is initially deposited as biogenic opal, mainly in the form of radiolarians, sponge spicules, and diatoms. This opal phase is amorphous and is called opal-A according to the X-ray classification scheme of Jones and Segnit [1]. After deposition and burial the opal-A is often dissolved by interstitial

* Contribution No. 2533. Publications of the Division of Geological and Planetary Sciences, California Institute of Technology, Pasadena, California 91109.

[1] Now at the Department of Geology, Louisiana State University, Baton Rouge, Louisiana 70803.

waters. The time and depth of sediment burial at which this mobilization occurs as well as its cause are not well understood. Some investigators (e.g. [2]) have suggested that additional dissolved silica may originate from volcanic ash.

The dissolved silica is then precipitated as a nearly-amorphous silica phase characterized by broad, weak X-ray diffraction peaks at approximately 4.05 Å and 2.48 Å. This silica phase is referred to as opal-CT [1] and can be identified optically by its index of refraction and its extremely weak birefringence. Most of the hard silica with conchoidal fracture called "chert", "porcellanite", or "cristobalite" in JOIDES DSDP reports is actually composed of opal-CT.

With time and/or increased temperature due to burial the opal-CT crystallizes or recrystallizes to form "micro quartz". Ernst and Calvert [3] and Heath and Moberly [4] have suggested that this crystallization occurs in the solid state without an intervening solution-precipitation step, but the actual mechanism of this transformation is not known. Petrographically, the "micro quartz" in deep-sea sediments is very similar to the granular microcrystalline quartz found as chert nodules and beds in sediments of all ages. This stable silica phase will be referred to hereafter in this paper as "granular microcrystalline quartz" and rock composed of this silica phase we shall call "chert".

Fibrous quartz, or chalcedony, is normally thought to form by growth of microcrystalline quartz in cavities [5, 6], but is probably not part of the typical deep-sea silica paragenetic sequence. Where found in deep-sea sediments it has developed in fractures and cavities in the opal-CT deposits and consequently has probably grown directly from solution during burial.

Drusy quartz in DSDP core 8-70B-4cc consists of subhedral quartz crystals protruding into a cavity. The crystals have grown on a cavity-lining composed of fibrous quartz, a configuration common in cherts of all ages. Folk and Pittman [6] have argued that silica in cavities is initially deposited as fibrous quartz and that the drusy quartz represents a final growth stage in waters of lower silica concentration. In deep-sea sediments drusy quartz is uncommon, and as a cavity-filling, is unrelated to the typical sequence of silica transformations.

In summary, in the case of DSDP samples, biogenic silica refers to silica formed in seawater; opal-CT is a phase formed from silica solutions in the sediments;

granular microcrystalline quartz is a product of crystallization of opal-CT; and fibrous and drusy quartz result from the direct precipitation of silica from solutions in fractures and cavities in the earlier-formed phases.

3. Notation

$^{18}O/^{16}O$ and D/H data are reported in the usual δ-value notation with Standard Mean Ocean Water as reference standard.

The fractionation factor for oxygen isotope distribution between silica and water, α, is defined by:

$$\alpha \, {}^{oxygen}_{silica\text{-}water} = ({}^{18}O/{}^{16}O)_{silica}/({}^{18}O/{}^{16}O)_{water}$$

Similarly,

$$\alpha \, {}^{hydrogen}_{silica\text{-}water} = (D/H)_{silica}/(D/H)_{water}$$

For all practical purposes, α depends only on temperature.

4. Samples studied

Radiolarian ooze, opal-CT, granular microcrystalline quartz, chalcedony (fibrous quartz), and drusy quartz were sampled from JOIDES DSDP cores. In addition, a sample of granular microcrystalline quartz from a 7-foot long chert nodule from the Horizon Guyot was analyzed. In Table 1 the DSDP sample numbers, the age of the sediments in which the silica occurs, the mineralogy, and the isotopic results are listed. Wherever possible, the different silica phases were separated for isotopic analyses. When separations were not made, estimates of the percentages of the different constituents were made. The silica phases were identified petrographically by examining grain mounts in immersion oils. All samples except the radiolarian ooze were crushed to pass 200 mesh prior to isotope analyses.

5. Hydrogen isotope analysis

Our intial effort to extract hydrogen for δD anal-

TABLE 1
Sediment depth, mineralogy, isotopic composition of oxygen, hydrogen and carbon, water content, and oxygen isotopic temperature of deep-sea silica and coexisting carbonates (samples 356 and 356-0 are from an Upper Cretaceous horizon; all others are from strata of Eocene age)

Sample No.	JOIDES No.	Sediment depth (m)	Mineralogy	$\delta^{18}O$ (‰)	$\delta^{13}C$ (‰)	wt.% H_2O	δD (‰)	Isotopic temp. (°C)
352	8-70B-4cc	388	95% gmc qtz, 5% opal-CT	33.2				18
353	8-70B-4cc	388	drusy quartz	29.8				32
353-F	8-70B-4cc	388	chalcedony	35.2		1.09	−85.6	11
354	7-64-1-11	984	>95% gmc qtz	35.2		0.92	−78.9	11
354-LA	7-64-1-11	984	67% calcite	30.8	+2.34			
354-LB	7-64-1-11	984	65% calcite	30.6	+2.34			
354-L	7-64-1-11	984	acid residue: 80% opal-CT	37.2				4
355	8-69-6-5	190	100% opal-A air dried	35.8				
355	8-69-6-5	190	100% opal-A dehydrated	42.4				
356	7-61-1cc	86	>98% gmc qtz	35.6	1.0		−79.1	9
356	7-61-1cc	86	>98% gmc qtz	35.9				8
356-0	7-61-1cc	86	>98% opal-CT	36.1	1.2		−56.5	7
356-0	7-61-1cc	86	98% opal-CT dehydrated	36.8				5
367	8-70B-1B-1	383	7% calcite	29.5	+2.35			
367	8-70B-1B-1	383	acid residue: 90% opal-CT	34.9				
368	8-70B-2B-1	385	>98% gmc qtz	32.6				20
368A	8-70B-2B-1	385	5% gmc qtz, 95% opal-CT	34.7				12
368B	8-70B-2B-1	385	98% gmc qtz (porous)	32.46				21
369	8-70B-1B-1	383	>98% opal-CT	34.7				12
370	15-149-43-1	390	acid residue: 90% opal-CT	34.3				14
370	15-149-43-1	390	25% calcite	29.1	+2.21			
370	15-149-43-1	390	27% calcite	28.9	+2.13			
343	Horizon Guyot	dredge haul	>98% gmc qtz	38.9				

Abbreviation: gmc qtz = granular microcrystalline quartz.

TABLE 2
δ D-values of successive increments of water extracted from 0.3 g of opal-CT upon heating the sample from 25°C to 1000°C

Fraction	Temp. interval (°C)	Time interval of heating	µm water / mg SiO₂	δD(‰)
1	25	18 hr 45 min	0.313	−17.6
2	25−126	3 hr	0.141	−13.1
3	126−228	4 hr 05 min	0.109	−36.4
4	228−308	3 hr 25 min	0.077	−60.8
5	308−416	2 hr 30 min	0.113	−74.3
6	416−1000	4 hr 30 min	0.257	−79.6

yses was done on 0.3 g of an opal-CT sample no. 356.0 (7-61-1cc), which contains less than 2% granular microcrystalline quartz. This sample was heated under vacuum in a stepwise fashion. The evolved water was collected in successive separate increments and analyzed for δD. The results are given in Table 2 and are shown in Fig. 1.

Fraction-1 water degassed from the opal-CT at 25°C. Its δD-value of −18‰ probably reflects, in part, exchange during sample-crushing, exposure to air, and isotopic fractionation during the intial evacuation of the experimental apparatus.

Water outgassed at progressively higher temperatures shows progressively lower δ-values. It is apparent

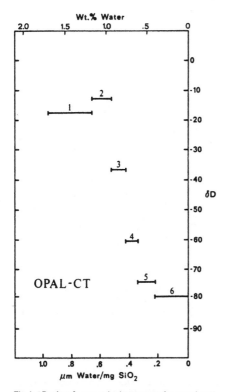

Fig. 1. δD values for successive increments of water released from opal-CT as 0.3 g of the sample is heated from 25° to 1000°C. Water sample 1 is released at 25°C whereas sample 6 is released between 416° and 1000°C. The combined δD-value of these increments is −57‰.

from Fig. 1 that most of the water within opal-CT is greatly depleted in deuterium relative to seawater.

The δD dehydration pattern shown in Fig. 1 cannot be attributed to simple isotopic fractionation due to dehydration of opal-CT because the δD of sample 1 is more negative than the δD of the successive sample 2 which is in turn more positive than the δD of sample 3. It is difficult to imagine a single fractionation process causing such a δD dehydration pattern. It is thus more likely that samples 2–5 are simply mixtures

of various proportions of low-temperature contaminant water of δD ≃ −10‰ represented by sample 2 and "high-temperature released water" of δD = −80‰ represented by sample 6. Although it is possible that the δD dehydration pattern in Fig. 1 including samples 2–6 can be due to an isotopic fractionation where D is preferentially removed from opal-CT upon heating, this is unlikely because isotopic fractionation of this type should follow a Raleigh type of pattern. In such a pattern the δD difference between successive fractions increases at later stages of the dehydration whereas the opposite appears to be true in this case.

The opal-CT sample used in the above experiment formed in the absence of low δD meteoric waters. The −80‰ δ-value for the water extracted at high temperature must reflect an isotopic fractionation of about −80‰ between ocean water and the high-temperature released in opal-CT. This fractionation is similar in magnitude and sign to that observed for hydrogen isotope fractionation in clay minerals [7], suggesting that the high-temperature water has been derived from hydroxide groups within the opal-CT.

6. Granular microcrystalline quartz and chalcedony

It has been demonstrated above that an unquestionably marine hydrous silica phase (opal-CT) contains probable hydroxide groups depleted in deuterium by about 80‰ relative to the δD value of the water in which they formed. The presence of such hydrogen in granular microcrystalline quartz would be geologically much more significant because hydroxide groups extracted from such samples may provide an isotope record extending as far back as the age of the oldest cherts, approximately 3×10^9 years B.P. Consequently, water was extracted from a powdered chert from a DSDP core by heating the sample at 1000°C for 3 hours after first pumping on the sample for about 12 hours at 25°C to remove surficial moisture. Pumping at 25°C was shown by Knauth and Epstein [8] to be sufficient to remove nearly all adsorbed and non-"hydroxide" water in cherts.

As shown in Table 1 δD-values for cherts from two different cores are both −79‰. Pure fibrous quartz from a third core, 8-70B-4cc, has a δD-value of −85.6‰. The δD values are very similar to those for the opal-CT "hydroxide" waters and again represent a fractionation factor similar in magnitude and sign to that observed for clay minerals. We therefore suggest that the D/H ratio of water in these cherts was acquired during isotopic equilibration of *hydroxide groups* with water isotopically similar to ocean water.

There is a possibility that a major portion of the water extracted exists in the chert in the form of mechanically trapped H_2O in microscopic to sub-microscopic inclusions and fractures, and that it has formed from hydroxide groups originally present in the precursor nearly-amorphous phase (D. White, personal communication, 1973). However, this latter possibility would be of no consequence in determining the δD of the water in the chert as long as no water is lost or gained during the transformation of the OH group to H_2O. Any partial loss of mechanically trapped H_2O could be accompanied by isotopic fractionation. However, it is unlikely that such water loss would be uniform enough to produce the consistent -79 to $-85‰$ fractionation observed.

7. Oxygen isotope analyses

7.1. Radiolarians

The $\delta^{18}O$-value of air-dried radiolarian ooze, no. 355, is $+35.8‰$. The same sample after being totally dehydrated under vacuum at $1000°C$ had a $\delta^{18}O$-value of $+42.4‰$. The $6.6‰$ difference between the $\delta^{18}O$-value of the air-dried and the completely dehydrated radiolarians indicates that the presence of water, and possibly the manner of dehydration, have profound effects on the $\delta^{18}O$-value of opaline silica. Because ^{18}O analysis always involves vacuum dehydration of the sample, and probably also variable degrees of dehydration, Mopper and Garlick [9], as well as ourselves, encountered difficulties in analyzing radiolarians. Although reproducibility of ^{18}O analyses of water-rich silica is good after total dehydration [10, 11], it is probable that non-equilibrium isotopic exchange occurs between H_2O and SiO_2 during dehydration at higher temperatures and thus makes the $\delta^{18}O$-values difficult to interpret. The significance of the reproducible $\delta^{18}O$-values cannot be assumed unless it can be demonstrated that the dehydration processes for a single source of biogenic silica are accompanied by reproducible isotopic effects, and, in addition, that

the isotopic composition of hydrous silica of different types and different initial H_2O contents will be affected to the same extent upon dehydration.

7.2. Opal-CT

Extractions of oxygen from pure air-dried opal-CT for $\delta^{18}O$ analyses were done by fluorination and without any prior dehydration. However, samples containing carbonates were first treated with 100% phosphoric acid to remove the carbonate. The CO_2 generated by the phosphoric acid was analyzed for its $\delta^{18}O$ and $\delta^{13}C$ values.

The total water content of opal-CT content of opal-CT sample 365-0 is about 1.2 wt.%. This is much lower than that of biogenic silica and suggest that analytical difficulties with this material may not be as severe as for radiolarians. As shown in Table 1 the $\delta^{18}O$-value for opal-CT sample 356-0 when air-dried is $+36.1‰$. When dehydrated, the sample yields a reproducible δ-value of $+36.8 \pm 0.1‰$. The difference between the $\delta^{18}O$ values of opal-CT and its dehydrated equivalent is thus less than $1‰$. Consequently, useful $\delta^{18}O$-values of opal-CT can probably be obtained either with or without high-temperature dehydration. $\delta^{18}O$-values of several air-dried opal-CT samples are given in Table 1.

7.3. Crystalline silica

The water contents of air-dried granular micro-crystalline quartz, chalcedony, and drusy quartz are less than that of opal-CT. Consequently $\delta^{18}O$ analyses of these samples were also performed without high-temperature dehydration. The $\delta^{18}O$-values are shown in Table 1.

8. Discussion

8.1. General

The $\delta^{18}O$ values of the various forms of silica are generally very positive and are thus in good agreement with values reported in the literature for supposedly marine cherts. The $\delta^{18}O$ and $\delta^{13}C$ values for the few carbonates are also quite normal marine values. The δD values of the water extracted from the opal-CT and cherts are the first of their kind and indicate that there is a hydrogen isotope fractionation between the water that is chemically bound in the cherts (probably as hydroxyls) and the ocean water in which the silica is formed. The existence of such a fractionation factor may add considerably to the significance of both oxygen and hydrogen isotope studies and to the understanding of the genesis and history of cherts.

8.2. Relationship of $\delta^{18}O$-values to depth of burial

Fig. 2 shows $\delta^{18}O$-values of the silica phases and coexisting carbonates plotted against the depth below the sediment/water interface at which the samples were cored. The samples come from 4 separate localities in the Pacific ocean and from one locality in the Caribbean Sea (sample 370). All were recovered from mid-Eocene horizons except for sample 356, which was taken from an upper Cretaceous horizon.

Estimated in situ sediment temperatures, assuming bottom-water temperatures of 0°C and a normal oceanic

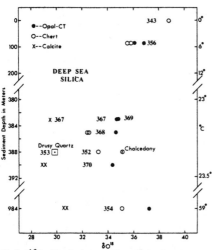

Fig. 2. $\delta^{18}O$-values for deep-sea silica samples plotted against their depth of burial. Temperatures on the right are the estimated temperatures of the sediment at the indicated depth of burial based on a geothermal gradient of 0.06°C/m.

geothermal gradient of 0.06°C/m [12], are shown in Fig. 2. An increase in temperature of about 60°C with depth of burial is apparent. However, the expected change of the $\delta^{18}O$-values with temperature for opal-CT does not take place. In particular, $\delta^{18}O$ values of the most deeply buried sample (no. 354, 984 m) and the least deeply buried sample (no. 356, 86 m) are almost identical. These data are strong evidence that opal-CT formation occurs before or during shallow burial (< 86 m) at near-ocean bottom temperatures and that the $\delta^{18}O$-values acquired at these temperatures are preserved to at least 1000 m depth. With the limited data available it is at present difficult to be more specific with respect to the actual temperature at which the opal-CT formed as compared to the actual temperature of the ocean bottom. Nevertheless this difference is apparently small, perhaps within several degrees, so that paleobottom-ocean temperature fluctuations might well be recorded in the $\delta^{18}O$ of opal-CT.

8.3. $\delta^{18}O$-values of opal-CT vs. $\delta^{18}O$-values of cherts

Coexisting opal-CT and chert occur in samples 356, 368, and 354. In every case the $\delta^{18}O$ of the opal-CT is more positive than that of the coexisting granular microcrystalline quartz in amounts varying from 0.5 to 2‰. This difference in $\delta^{18}O$ between the two phases could result from the transformation of opal-CT formed during shallow burial to granular microcrystalline quartz in the presence of water with $\delta^{18}O \simeq 0\%$ at the higher temperatures encountered during burial. Granular microcrystalline quartz which equilibrates or exchanges wholly or partially with this water at the warmer temperatures would have lower $\delta^{18}O$-values.

Sample 356, the shallowest of the samples listed (86 m), shows the smallest $\delta^{18}O$ difference between opal-CT and granular microcrystalline quartz. Because of the thermal gradient and shallow burial of sample 356 the two phases of silica in this sample had to have formed within no more than 6°C of each other. Sample 368 has been buried to greater depths and presumably was subjected to higher temperatures. The larger difference between $\delta^{18}O$-values of the two types of silica suggests that the granular microcrystalline quartz formed at these higher temperatures. Significantly, this $\delta^{18}O$ difference between the opal-CT and granular microcrystalline quartz is not larger for sample 354, in spite of the fact that the latter sample has been buried to twice the depth. The result indicates that the granular microcrystalline quartz in sample 354 formed at about the same depth as the granular microcrystalline quartz in sample 368 and has suffered little isotopic change with subsequent deepening of burial.

Lawrence [13] has reported a general decrease of the $^{18}O/^{16}O$ ratio with depth of burial in pore waters squeezed from sediments at DSDP sites 148, 149, 157, 225, and 228. Near partially silicified horizons the $\delta^{18}O$-value of pore water decreases to $-2.5‰$ relative to ocean water, and is even lower in other sediment types. It is thus possible that the lower $\delta^{18}O$-values for granular microcrystalline quartz forming at depth from precursor opal-CT reflects lower $\delta^{18}O$-values for the pore waters in addition to the increased temperatures of burial. However, it is difficult to evaluate this possibility with the limited amount of data that we have.

Drusy quartz crystals in a cavity in sample 353 have the least positive $\delta^{18}O$-values of deep sea silica. These crystals grew upon a layer of fibrous quartz which lined the cavity. The chalcedony has a more positive $\delta^{18}O$-value, similar to that of sample 356. The simplest explanation of the isotope data is that the cavity formed and was lined with chalcedony before it had reached a depth of 100 m. Subsequent growth of the drusy quartz probably occurred at greater sediment depths and higher temperatures.

8.4. Coexisting silica-carbonate

Carbonate coexisting with silica in these sediments consist of indurated calcareous oozes of pelagic origin which have been partially silicified. In thin-section, minor calcite overgrowths can be seen on foraminifera and nannofossils, indicating at least some diagenetic change in the conditions of the original calcite [14]. However, no correlation of $\delta^{18}O$-value with depth of burial is apparent for the three carbonate samples analyzed. Carbonate sample 354-L (mid-Eocene), buried to the greatest depth (984 m), has a $\delta^{18}O$-value of +30.7‰, almost identical to the $\delta^{18}O$-values of +30.3‰ obtained by Douglas and Savin [15] for well-preserved middle planktonic foraminifera from the western Pacific. At least in the case of this sample, major isotopic change has not resulted from indura-

tion of the calcareous ooze. The other two carbonate samples have lower $\delta^{18}O$-values, but it is not necessary to attribute these to diagenesis at higher temperatures during burial since the sediment temperatures of 20–25°C inferred from the thermal gradient are similar to the surface ocean temperatures (15–30°C) at which the pelagic calcite originally formed.

For the three cases examined the $\delta^{18}O$ for the calcareous pelagic oozes is lower than the coexisting opal-CT by 5.4 to 6.5‰. Since the variations in carbonate $\delta^{18}O$-values are probably related to the temperature variations of the surface ocean water in which they formed, this uniform difference suggests that the $\delta^{18}O$-values of opal-CT are also related to ocean temperatures.

8.5. Isotopic temperatures

In order to obtain isotopic temperatures for the formation of silica phases it is necessary to know the isotopic composition of the water in which the silica precipitated and the temperature dependence of the isotopic fractionation factors for the silica–water system. Unfortunately, the silica–water fractionation at temperatures below 400°C is poorly understood. In the case of oxygen isotopes, Becker [16] has calculated the fractionation for the quartz–water system, but Knauth and Epstein [17] have argued that Becker's results yield unreasonable temperatures when applied directly to cherts. The equilibrium isotopic fractionation factor is, in part, determined by the type of chemical and crystallographic bonds which exist in a crystal. Inasmuch as the crystallographic structure of chert is poorly understood and may be variable for different kinds of cherts it is not known if a single equilibrium isotope fractionation factor can be applied rigorously to all cherts.

For purposes of estimating the approximate range of temperature variations associated with the $\delta^{18}O$ variation we have assumed that the silica with the most positive $\delta^{18}O$-value obtained so far has formed at near-zero degrees C. We also assume that there is no equilibrium fractionation of oxygen isotopes between opal-CT and quartz.

Sample 343, with $\delta^{18}O = 39‰$, has the most positive $\delta^{18}O$-value of any deep-sea chert, and must represent formation at a lower temperature than the other samples. This sample is part of a 7-foot long

chert nodule dredged from an Eocene horizon exposed on the side of Horizon Guyot (W. Newman, personal
communication, 1972). Unlike the other samples, this
chert had been reexposed to the sediment/water interface after burial to an unknown depth (probably 150
m or less, the depth of Eocene sediments at DSDP
site 171 near the middle of the guyot). The formation
of this chert may thus have occurred in the presence
of interstitial waters which had isotopically exchanged
with ocean water at any time from the Eocene to the
Recent depending upon the history of submarine
erosion and sediment deposition along the flanks of
Horizon Guyot. The temperature at which the chert
formed could have been that of maximum burial or
that of Eocene to Recent ocean water. In spite of
these uncertainties it is apparent that unless this chert
formed from interstitial waters enriched in ^{18}O then
the isotopic fractionation factor at $0°C$ must be at
least 1.039. For discussion purposes we therefore use
1.039 as the value for α silica–water at $0°C$. An inter-
polation between this value and the α-values determined
by Clayton et al. [18] for completely equilibrated
quartz at high temperatures is shown in Fig. 3. The
curve is a standard $1000 \ln \alpha$ vs. $10^6/T^2$ plot. Temperatures deduced for silica formation using this curve and
a water $\delta^{18}O$-value of 0.0‰ are given in Table 1.

The calculated temperature of opal-CT formation is
between 5 and $13°C$ whereas the chert formed at
temperatures of about $2-9°$ warmer than the coexisting opal-CT, presumably depending upon the depth
at which the chert formed. The highest-temperature
silica is the drusy quartz with an isotopic temperature
of about $32°C$. This temperature exceeds that inferred from the geothermal gradient of $0.06°C/m$. If
accurate, this temperature implies either a larger geothermal gradient, formation of the quartz from pore
water depleted in $\delta^{18}O$ relative to seawater, or formation of the quartz in warmer waters derived from
deeper sediments. Chalcedony-filling fractures in
opal-CT yield a temperature of about $11°C$, similar
to the temperatures for adjacent opal-CT. It is apparent
that the fractures were opened and filled with
chalcedony before the opal-CT had been buried to
depths such that the temperature was significantly
increased. It is of course also possible that the fibrous
quartz simply formed at much different temperatures
from interstitial water which had been isotopically
altered.

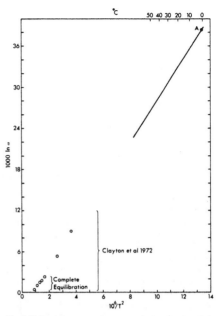

Fig. 3. Estimated quartz–water oxygen isotope fractionation.
Opal-CT is assumed to behave in an identical manner. Point
A is determined from the $\delta^{18}O$-value of sample 343, which,
for discussion purposes, is assumed to have formed at $0°C$.

Since isotopic arguments have been presented that
the opal-CT forms at temperatures similar to the
overlying bottom-water temperatures then the $\delta^{18}O$
for opal-CT may be used to infer ocean bottom temperatures. Temperatures so inferred range from $5°$ to
$13°C$ for Eocene opal-CT, and are in approximate
agreement with the deep-water temperatures deduced
by Douglas and Savin [15]. However, the $5-7°C$
value for Cretaceous opal-CT is significantly lower
than the $12-16°C$ value obtained by these authors
and the $17°C$ value obtained by Lowenstam and
Epstein [19].

In view of the considerable uncertainties associated
with assigning these temperatures, and with the small
number of samples we have studied, further discussion
of actual temperatures is unwarranted.

9. Conclusions

(1) In deep-sea sediments, granular microcrystalline quartz and fibrous quartz contain 0.9–1.2 wt% extractable water with δD values of $-57‰$ to $-86‰$ relative to ocean water. This fractionation of hydrogen by silica is similar in magnitude and sign to that observed for hydrogen isotope fractionation associated with the formation of hydroxyl groups in clay minerals, suggesting that the extractable water in the silica has been derived from hydroxyl groups within the silica.

(2) Opal-CT, the first silica phase to form during the diagenetic conversion of amorphous to crystalline silica, forms before or during shallow burial, within $5°C$ or less of ocean-bottom temperatures. $\delta^{18}O$-values acquired at these temperatures are then preserved during deep burial to at least 1000 m where the temperature is approximately $60°C$.

(3) Granular microcrystalline quartz (chert) has lower $\delta^{18}O$-values than coexisting opal-CT ("porcellanite", "cristobalite"). For the cases examined the granular microcrystalline quartz formed at temperatures about $2–9°C$ warmer than its precursor opal-CT, presumably depending upon the depth of burial at which it crystallized.

(4) Drusy quartz in a cavity in core 8-70B-4cc has a lower $\delta^{18}O$-value than coexisting opal-CT, and formed at depth at an estimated approximate temperature of about $32°C$.

(5) Isotopic temperatures deduced for opal-CT formations are preliminary and very approximate, but yield Eocene deep-water temperatures of $5–13°C$, and $5–7°C$ for one upper Cretaceous sample.

Acknowledgements

We are grateful to the JOIDES Deep Sea Drilling Project of the National Science Foundation for supplying samples. The chert sample from Horizon Guyot was obtained by Dr. W. Newman of the Scripps Institution of Oceanography and transmitted to us by Dr. Harmon Craig. Mrs. Jane Young performed the isotopic analyses of the carbonate. Drafting and typing services were provided by the Geology Department of the Louisiana State University. We thank Y. Kolodny for reading the manunscript. This investigation was initially presented as part of Knauth's Ph.D. Thesis at the California Institute of Technology and was funded by NSF grant GA-31325.

References

1 J.B. Jones and E.R. Segnit, The nature of opal: I. Nomenclature and consistent phases, Geol. Soc. Aust. 18 (1971) 57.

2 T.G. Gibson and K.M. Towe, Eocene volcanism and the origin of horizon A, Science 172 (1971) 152.

3 W.G. Ernst and S.E. Calvert, An experimental study of the recrystallization of porcellanite and its bearing on the origin of some bedded cherts, Am. J. Sci. 267-A (1969) 114–133.

4 G.R. Heath and R. Moberly, Cherts from the western Pacific, Leg 7, Deep Sea Drilling Project, in: Initial Reports Deep Sea Drilling Project, 7, E.L. Winterer et al., eds. (1971) 991.

5 R.L. Folk and E. Weaver, A study of the texture and composition of chert, Am. J. Sci. 250 (1952) 498–510.

6 R.L. Folk and J.S. Pittman, Length-slow chalcedony: A new testament for vanished evaporites, J. Sediment. Pet. 41 (1971) 1045–1058.

7 S.M. Savin and S. Epstein, The oxygen and hydrogen isotope geochemistry of clay minerals, Geochim. Cosmochim. Acta 34 (1970) 25–42.

8 L.P. Knauth and S. Epstein, The nature and isotope composition of water extracted from authigenic silica, in preparation (1974).

9 K. Mopper and G.D. Garlick, Oxygen isotope fractionation between biogenic silica and ocean water, Geochim. Cosmochim. Acta, 35 (1971) 1185–1187.

10 M.L. Labeyrie, Composition isotopique de l'oxygène de la silice biogénique, C.R. Acad. Sci. Paris 274 (1972) 1605–1608.

11 L.P. Knauth, Oxygen and hydrogen isotope ratios in cherts and related rocks, Ph. D. Thesis, Calif. Inst. Tech. (1973) 369 pp.

12 R.P. Von Herzen and M.G. Langseth, Present status of oceanic heat-flow measurements, in: Physics and Chemistry of the Earth, 6 (Pergamon Press, London, 1965) 365–407.

13 J.R. Lawrence, Stable oxygen and carbon isotope variations in the pore waters, carbonates, and silicates, sites 225 and 228, Res Sea, in: Initial Reports Deep Sea Drilling Project 23, R.B. Whitmarsh et al., eds. (1974) 939.

14 R. Moberly, Jr. and G.R. Heath, Carbonate sedimentary rocks from the western Pacific, in: Initial Report of the Deep Sea Drilling Project, 7, E.L. Winterer et al., eds. (1971) 977–989.

15 R.G. Douglas and S.M. Savin, Oxygen and carbon isotope analyses of Cretaceous and Tertiary foraminifera from the Central North Pacific – Leg 17, Initial Reports of the Deep Sea Drilling Project VII, (1973) 591–605.

16 R.H. Becker, Carbon and oxygen isotope ratios in ironformations and associated rocks from the Hamersley Range of Western Australia and their implications, Ph.D. Thesis, Univ. of Chicago (1971).

17 L.P. Knauth and S. Epstein, Oxygen and hydrogen isotope
   ratios in nodular and bedded cherts, in preparation (1974).
18 R.N. Clayton, J.R. O'Neil and T.K. Mayeda, Oxygen isotope
   exchange between quartz and water, J. Geophys. Res. 77
   (1972) 3057.

19 H.A. Lowenstam and S. Epstein, Cretaceous paleotemperatures as determined by the oxygen isotope
   method, their relations to the nature of rudist reefs,
   20th Int. Geol. Congr. Rep. (1956) 65–76.

16

This article has also been accepted for publication in *Geochimica et Cosmochimica Acta*.

DIAGENESIS OF SILICEOUS OOZES. I. CHEMICAL CONTROLS ON THE RATE OF OPAL-A TO OPAL-CT TRANSFORMATION—AN EXPERIMENTAL STUDY

M. Kastner, J. B. Keene, and J. M. Gieskes

Scripps Institution of Oceanography
University of California, San Diego

ABSTRACT

Evidence from deep-sea sediments supports the following diagenetic maturation sequence: opal-A (siliceous oozes) → opal-CT (porcelanite) → chalcedony or cryptocrystalline quartz (chert). A solution-redeposition mechanism is involved in the opal-A to opal-CT transformation. Exceptions to the overall maturation sequence are numerous, suggesting that temperature and time are not the only important factors controlling these mineralogical transformations. The rates of the opal-A to opal-CT to quartz transformations are strongly affected by the composition of the solution and of the host sediments; in Mesozoic clayey sediments, opal-CT predominates, while in carbonate sediments, quartz is most common.

Experiments at 25°C and 150°C over a period of one day to six months show that the transformation rate of opal-A to opal-CT is much higher in carbonate- than in clay-rich sediments and that opal-CT lepisphere formation is aided by the precipitation of nuclei with magnesium hydroxide as an important component. The role of carbonate is explained as follows: In carbonate sediments, the dissolution of carbonate provides the necessary alkalinity, and seawater provides the magnesium for the magnesium hydroxide in the nuclei. In contrast, in clay-rich sediments the clay minerals compete with opal-CT formation for the available alkalinity from seawater. As a result, the clays are enriched in magnesium, and the rate of opal-CT formation is strongly reduced. This mechanism also bears on the common observation of carbonate replacement by silica.

1. INTRODUCTION

In most of the pre-Cenozoic porcelanites[1] and cherts sampled on land, diagenetic processes have obliterated textural and fossil evidence, and the

---

[1]Many differences in the usage of the term *porcelanite* exist; in this manuscript, the term *porcelanite* (Taliaferro 1934) is used for rocks composed mainly of opal-CT (Jones and Segnit 1971), and the term *chert* for rocks composed mainly of nonclastic crypto to microcrystalline quartz and chalcedony.

controversy over their mode of origin dates back to the beginning of the century (Tarr 1917; Barton 1918; Davis 1918; Van Tuyl 1918). In siliceous oozes, porcelanites, and cherts recovered by the Deep Sea Drilling Project, however, ongoing diagenetic processes in upper Jurassic to Recent deep-sea sediments can be observed in their original depositional environment (for example: Calvert 1966, 1971; Heath and Moberly 1971; Berger and von Rad 1972; Wise et al. 1972; Lancelot 1973; Scholle and Creager 1973; Wise and Weaver 1973, 1974; Zemmels and Cook 1973; Heath 1974; Weaver and Wise 1974; Keene 1975).

Certainly since the early Mesozoic and most probably since the early Paleozoic, the major mechanism of silica precipitation at ordinary temperatures and pressures is biochemical. Siever (1962) suggested that even in the late Precambrian, silica sedimentation might have been primarily biogenic.

The initial silica of most deep-sea siliceous sediments is biogenic, a hydrous X-ray amorphous silica phase that shows a diffuse X-ray band centered at about 3.8–4.1 Å. It has been classified as opal-A by Jones and Segnit (1971). Subsequently, opal-A dissolves and opal-CT (Jones and Segnit 1971), a unidimensionally disordered 3-layered structure of low cristobalite with 2-layered low tridymite domains (Flörke 1955), precipitates, often in the form of lepispheres (Wise and Kelts 1972; Berger and von Rad 1972; Weaver and Wise 1972; and references therein). The stacking disorder of opal-CT is in the [111] direction (Flörke 1955), and the X-ray diffraction pattern shows the low cristobalite broad peaks at about 4.1–2.5 Å, with a low tridymite peak at 4.3 Å. Based on electron-diffraction patterns of the material that Jones and Segnit (1971) classified as opal-CT, Wilson et al. (1974) suggested that it should be called disordered low tridymite; but Jones and Segnit (1975) disagree with their conclusions.

Eventually, at higher temperatures (greater burial depth or higher heat flow) or with time, opal-CT transforms to chalcedony and/or microcrystalline quartz, the stable silica phase in diagenetic environments.

Bramlette (1946) introduced the concept of maturation of siliceous oozes as a function of burial depth (temperature) from a soft biogenous ooze through porcelanite and finally to chert. Evidence from deep-sea sediments and from the Monterey Formation, California, supports the following overall diagenetic sequence: opal-A (siliceous ooze) → opal-CT (porcelanite) → chalcedony or cryptocrystalline quartz (chert) → megaquartz (chert). The opal-A to opal-CT transformation is a solution-redeposition mechanism (Carr and Fyfe 1958; Mizutani 1966). A solid-state reaction was suggested for the transformation opal-CT → quartz by Ernst and Calvert (1969) and Heath and Moberly (1971). But on the basis of Carr and Fyfe's (1958) hydrothermal experiments, Murata and Larson's (1975) oxygen-isotope values of opal-CT versus quartz in the Monterey Formation, California, and Stein and Kirkpatrick's (1976) scanning and transmission microscopy studies of Ernst and Calvert's (1969) experimental products, a solution-redeposition mechanism has been suggested also for this phase transformation.

Bramlette (1946) already observed exceptions to the preceding overall diagenetic sequence in the Monterey Formation, California: Porcelanite and/or chert beds formed within the nonrecrystallized diatomite zone, and chert layers formed within the porcelanite zone. Similar observations were made, for example, by Winterer et al. (1971), Hays et al. (1972), Lancelot (1973), and Keene (1975) in deep-sea siliceous sediments of DSDP legs 7, 9, 17, and 32, respectively, suggesting that heat and time (Heath and Moberly 1971) are not the only decisive factors that promote the diagenesis of siliceous oozes to porcelanite and chert (Kastner and Keene 1975).

The diagenetic history of siliceous oozes is also strongly affected by the nature of the host sediments. According to Murata and Nakata (1974), in the Monterey Formation, California, the solution of opal-A and the precipitation of opal-CT occur first in pure diatomite layers and only later in diatomaceous mudstones. Millot (1964) suggested that metallic cations are responsible for the formation of the intermediate disordered cristobalite phase. And according to Lancelot (1973), in deep-sea sediments, "as a rule, chert occurring in clayey layers . . . is predominantly disordered cristobalite, while chert in carbonate sediments is predominantly quartz." They did not, however, specify mechanism(s) for these reactions. Keene (1975) made similar observations in nodular and bedded cherts recovered in the western Pacific Ocean, leg 32. Lancelot (1973) explained these observations by the difference in silica/ "metallic cations" ratio in the surrounding sediments; when the ratio is low, as in clayey sediments of low permeability, opal-CT (porcelanite) forms, but when the ratio is high, as in carbonate sediments, chalcedony and quartz (chert) form directly from opal-A.

In addition, Lancelot (1973) suggested that in permeable, almost clay-free carbonate sediments, quartz precipitates directly from opal-A without an intermediate opal-CT phase. In fact, Heath and Moberly (1971) found that in certain cherts, during the first stage in chert nodule formation in carbonate sediments, empty foraminifera chambers are filled by chalcedony, while the replacement of the carbonate groundmass by opal-CT occurs only during the second stage. Keene (1975), however, observed that the first silica phase to precipitate is opal-CT. Thus there is evidence for both to occur in contrast to Lancelot's (1973) suggestion. Wise and Kelts (1971) noted only opal-CT lepispheres in a permeable Oligocene nannofossil chalk with negligible amounts of clay minerals, and Wise and Weaver (1974) described the occurrence of opal-CT only, underlying a highly permeable clay-free diatom ooze (ELTANIN core 47-15).

The order of decreasing solubility of the common sedimentary silica phases is: amorphous silica (opal-A), cristobalite, chalcedony, and quartz (Figure 1). From Figure 1 the solubility of opal-CT can be inferred to lie between those of opal-A and α-cristobalite. The solubility of each of these phases increases with temperature (Kennedy 1950; Alexander et al. 1954; Iler 1955; Krauskopf 1956; Alexander 1957; van Lier 1959; Fournier 1960, 1973; Fournier and Rowe 1962; Morey et al. 1962, 1964; Siever 1962; Kato and Kitano 1968). The small positive effect of pressure on the solubility of quartz and amorphous silica was demonstrated by Kennedy (1950), Jones and Pytkowicz (1973), and Duedall et al. (1976). Both the internal structure and specific surface areas of amorphous silica and cristobalite affect their solubilities (Alexander 1957; Lewin 1961; Fournier 1973; Hurd and Theyer 1975).

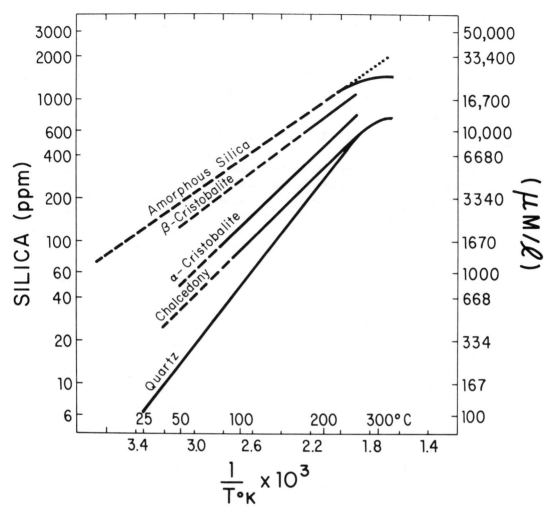

Fig. 1. Solubilities of the various silica phases along the two-phase curve, water plus vapor. Plotted as log concentration versus reciprocal of absolute temperature, after Fournier (1973, Figure 2).

It seems that quartz can crystallize directly from solution only when the silica concentration in solution is below the inferred equilibrium solubility of opal-CT. When silica in solution reaches concentrations above the equilibrium solubility of opal-CT as inferred from Figure 1, as in interstitial waters of most siliceous oozes, both in carbonate and clayey sediments (for example, Gieskes 1975), a disordered metastable silica phase, e.g., opal-CT, is likely to crystallize instead of the ordered stable phase quartz. Mackenzie and Gees (1971), for example, crystallized quartz directly from seawater at 20°C. In their experiments, silica in solution never exceeded $73 \pm 5 \ \mu M$. Harder (1965) and Harder and Flehmig (1970) synthesized quartz between 0°–80°C, at pH 7, from solutions saturated with respect to quartz, by ageing Al^{+3}, Fe^{+3}, Mg^{+2}, Mn^{+2}

hydroxide-silica precipitates in the presence of their aqueous solutions. Thus in most deep-sea sediments with high concentrations of siliceous skeletal remains, the diagenetic sequence opal-A $\rightarrow$ opal-CT $\rightarrow$ quartz will predominate, as suggested by Heath and Moberly (1971), Wise et al. (1972), Wise and Weaver (1974), and Keene (1975).

2. PREVIOUS EXPERIMENTS

Previous experiments on the transformation of amorphous silica to intermediate metastable phases such as disordered low-cristobalite, silica-X, keatite, and kenyaite, and to the stable-phase quartz, and on the epitaxial growth of quartz, emphasized mainly the effects of temperature, pressure, and pH on the rate and sequence of transformation. These experiments were conducted at low, moderate, and high temperatures, low to moderate pressures, and some in neutral but most in alkaline solutions (Corwin et al. 1953; Carr and Fyfe 1958; Campbell and Fyfe 1960; Fyfe and McKay 1962; Ernst and Blatt 1964; Harder 1965; Mizutani 1966; Ernst and Calvert 1969; Harder and Flehmig 1970; Mackenzie and Gees 1971; Oehler 1973; Betterman and Liebau 1975; Flörke et al. 1975). The following deductions are based on these experiments:

1. There is agreement that alkaline solutions affect the rate of each step of amorphous silica to quartz transformation and that the activation energy of opal-CT to quartz transformation is lower in KOH solutions (14.3 Kcal/mole, Mizutani 1966) than in H_2O (23.2 Kcal/mole, Ernst and Calvert 1969).

2. Reaction mechanisms appear to depend on temperature. For example, in dilute NaOH solutions at 245°C, the time required to form quartz from amorphous silica is a linear function of $1/(OH^-)$ (Campbell and Fyfe 1960), but at 330°C it is a linear function of $1/(OH^-)^2$ (Fyfe and McKay 1962).

3. The number of intermediate metastable phases and their sequence of formation depend on both temperature and pressure (Carr and Fyfe 1958; Betterman and Liebau 1975; Flörke et al. 1975) and on pH (Mizutani 1966).

4. There is disagreement about the order of the reactions and whether these transformations take place through solution and redeposition or in the solid state. Carr and Fyfe (1958) concluded that amorphous silica transforms to disordered low-cristobalite and quartz by solution and redeposition. Mizutani (1966), however, implied that both mechanisms are possible for the two transformation steps, which are first-order reactions, but leaned toward the solid-state mechanism. Similarly, Ernst and Calvert (1969) concluded that the opal-CT to quartz transformation is a zero-order reaction, and transformation takes place in the solid state. But recently Stein and Kirkpatrick (1976) examined with a scanning electron microscope Ernst and Calvert's (1969) experimental products and reinterpreted their results in terms of a dissolution-precipitation mechanism.

5. Based on X-ray and optical analyses, disordered low-cristobalite formed as an intermediate phase in most experiments; but the common habit of opal-CT lepispheres in deep-sea siliceous sediments, as described by Wise and Kelts (1971), has not been reproduced. Oehler's (1973) sphere-shaped crystals with tridymite habit blades are similar but not identical to natural opal-CT lepi-

spheres. Oehler's (1973) sphere-shaped crystals show bundles of parallel blades instead of the natural intersecting blades.

6. Quartz crystallizes from a highly disordered low-cristobalite only after the cristobalite attains a higher degree of ordering (Carr and Fyfe 1958; Mizutani 1966). Similar observations in the Monterey Formation, California, were described by Murata and Nakata (1974) and Murata and Larson (1975).

7. Quartz crystallizes from solutions with silica concentrations below the equilibrium solubility of cristobalite (Carr and Fyfe 1958; Mackenzie and Gees 1971) and, as suggested by Harder (1965) and Harder and Flehmig (1970) from solutions saturated with respect to quartz in the presence of Al^{+3}, Fe^{+3}, Mg^{+2}, and Mn^{+2} hydroxides.

Experiments concerning the influence of the carbonic acid system on the transformation rates of amorphous silica were conducted by Cox et al. (1916) and Lovering and Patten (1962). Cox et al. (1916) concluded that calcium bicarbonate is an effective precipitant of colloidal silica in the presence of calcite, whereas Lovering and Patten (1962), who conducted experiments at room temperature with silica in solution in the presence and absence of calcite and dolomite, disagreed with this conclusion and suggested instead that carbonic acid is the precipitating agent.

3. PRESENT EXPERIMENTS

a. Objectives

The deduction that quartz of all cherts in carbonate sediments had an opal-CT precursor, in conjunction with the observation that the quartz to opal-CT ratio is generally much higher in younger (Middle Eocene) pelagic carbonate sediments than in older (Cretaceous) clay-rich sediments implies either the presence of a "catalyst(s)" within the carbonate system and/or the presence of inhibitor(s) in clay-rich sediments. If we accept Lancelot's (1973) suggestion that "metallic cations" in clay-rich sediments retard the diagenesis of siliceous oozes, the following questions are pending: Do these cations affect (1) the rate of opal-A dissolution, as has been suggested by Iler (1955), Lewin (1961), and Hurd (1972, 1973); (2) the rate of nucleation and/or growth of opal-CT (Rieck and Stevels 1951; Millot 1964); or (3) the rate of opal-CT to quartz conversion as indicated by the high temperature experiments of Rieck and Stevels (1951), or the epitaxial growth of quartz (Mackenzie and Gees 1971).

The main objectives of our laboratory experiments were:

1. to test the observations that in deep-sea sediments the rate of diagenesis of siliceous oozes is enhanced in carbonate sediments relative to clay-rich sediments or to pure siliceous oozes;
2. to determine whether this increased rate of diagenesis in carbonate sediments is in the opal-A to opal-CT transformation step;
3. to determine the inorganic controls on the diagenetic conversion and rate of opal-A → opal-CT transformation; and

4. to compare the diagenetic susceptibility of diatom and radiolarian oozes in both marine and fresh water environments.

b. Methods

Experiments on opal-A to opal-CT transformation were conducted under hydrothermal conditions at 150°C and room temperature from one day to six months. The starting materials are described in Table 1. Practically all possible impurities were removed from the natural diatoms and radiolarians by size fractionation of ultrasonically treated suspensions and hand picking under a binocular microscope. The removal of all trapped impurities from diatoms is *much* more difficult than from radiolarians. Therefore, we are certain that our starting materials of Recent and Eocene radiolarians were very pure, but small amounts of trapped impurities, undetectable by surface chemistry techniques, may have been present in the Quarternary diatoms sample.

The cleaned fractions were checked for possible impurities, degree of preservation, and presence of organic matter by X-ray diffraction, transmission and scanning electron microscopes, and ESCA. When organic matter was detected, it was removed, either by low-temperature oxygen plasma ashing or by treatment with warm 10 percent H_2O_2 and cold 10 percent HCl. The samples were then washed with double-distilled H_2O and analyzed again by ESCA. The effect of the two different organic-matter removal methods on the experimental results was insignificant. The organic matter of the cultured diatoms was also removed by low-temperature oxygen plasma ashing, followed by washing with double-distilled H_2O, and ESCA analysis. The diatoms and radiolarians were analyzed by BET (Amico Adsorptonat) for specific surface areas.

Table 1. Starting materials for hydrothermal experiments.

Starting materials	*Sample description*
Diatom *Nitzschia thermalis*	Cultured marine diatom (Kates and Volcani 1966, p. 265)
Diatom *Navicula pelliculosa*	Cultured, fresh-water diatom (Kates and Volcani 1966, p. 265)
Quaternary diatoms	ELTANIN core 47-15, 57°17.3′S, 78°48.5′E, depth interval 525–550 cm
Quaternary radiolarians	SIO BNFC Expedition, box core 74, 14°35′N, 117°20′W, interface sediment/water
Eocene radiolarians	SIO LSDH Expedition, piston core 88, 8°33′N, 177°46′W, depth interval 289–291 cm
Quaternary foraminifera	SIO SOTW Expedition, 4°16′S, 164°24.6′W, interface sediment/water
$CaCO_3$	Reagent grade powder, Allied Chemical
$CaCO_3$	Iceland spar
Montmorillonite	A.P.I. #24, Otay, California (Kerr et al. 1950) $<2\mu$ fraction

The solutions used were: filtered seawater (0.45-μm millipore filter) from Scripps pier (Table 2), double-distilled H_2O, and several artificial solutions described in Tables 3 and 4. The solid/solution ratio in all experiments was 0.1 g solid/15 ml solution.

The room-temperature experiments were conducted in polyethylene bottles, and the 150°C experiments in sealed Pyrex tubes with some space left for water vapor and CO_2. Except for the distilled H_2O experiments, the pH of the starting solutions at 25°C was 8. To check for possible effects of impurities within the Pyrex glass on the experimental results, several duplicate experiments were conducted in sealed fused quartz tubes (highest purity grade); the experimental results were identical. Most experiments were run twice, and several experiments three to four times.

At the end of each experiment, the solution was immediately separated from the solids by filtering it through a 0.45-μm millipore filter. Because of the high dissolved silica values, it was imperative to analyze the solutions almost immediately at the end of each experiment. The quenching time was about 5 minutes. The solutions were analyzed for Si, Mg, and alkalinity, and where appropriate, for Ca and SO_4 as well. The methods used were essentially the same as those Gieskes (1974) described, with an accuracy of 2 percent for Si, Mg, and alkalinity. The solids were analyzed microscopically and by X-ray diffraction, electron diffraction, and scanning electron microscope with an X-ray energy dispersive attachment.

Due to the major difficulties in separating the opal-CT lepispheres from the siliceous tests and to the necessity to use electron diffraction to identify the newly formed silica phase, an attempt to quantify the amounts of opal-A dissolution and opal-CT formation was unsuccessful.

c. Description of the Results

Representative results of hydrothermal experiments at 150°C are presented in Tables 2, 3, and 4, and of those at room temperature in Table 5. In all experiments at 150°C, corrosion of siliceous tests was observed. After one month major differences were observed in both the solid and liquid phases, as shown in the 150°C seawater experiments of Table 2 and Figure 2. Embryonic lepispheres, as detected by the scanning electron microscope, were confirmed as highly disordered opal-CT by electron diffraction. They formed only in the experiments with $CaCO_3$ (exps. 3, 6, 9, and 12, Table 2); these samples also acquired a hydrophobic character. The opal-CT lepispheres preferentially developed on the foraminifera surfaces, while the siliceous tests were corroded, as shown in Figure 2a, b (exp. 3). After two months cementation of siliceous tests to foraminifera surfaces was observed (Figure 2c), and finally, after three months cementation between siliceous tests took place (Figure 2d); the sediment was partially cemented (porcelanite), and pseudomorphs of silica after the more soluble foraminifera species, e.g., *Globigerinoides ruber,* had formed. At the end of one month the experiments with $CaCO_3$ showed highly corroded siliceous tests, and many of the more delicate tests had dissolved completely. But as a result of opal-CT crystallization, silica concentrations in the solutions

Table 2. Solution chemistry of hydrothermal experiments in seawater, at 150° C.

Exp. number	Starting materials	Time (months)	$Si(OH)_4$ μM	Mg mM	Alkalinity meq/l	Observations in solids	Figure number
1	Eocene Radiolarians	1	2,550	50.6	0.2	Dissolution but no precipitation of silica	2e, f
2	Eocene Radiolarians	6	>10,000	50.4	0.1	Dissolution but no precipitation of silica	
3	Eocene Radiolarians + Foraminifera	1	1,980	19.2	3.8	Formation of embryonic opal-CT lepispheres	2a, b
4	Eocene Radiolarians + Foraminifera	6	7,850	7.3	2.7	Formation of embryonic opal-CT lepispheres, silica cementation	
5	Eocene Radiolarians + Montmorillonite	1	2,400	51.1	0.2	Formation of magnesium-rich clay	3a, b
6	Eocene Radiolarians + Foraminifera + Montmorillonite	1	2,060	18.5	3.7	Formation of embryonic opal-CT lepispheres	

7	Quaternary Diatoms	1	6,000	46.9	0.1	Dissolution but no precipitation of silica
8	Quaternary Diatoms	6	>10,000	50.1	0.1	Dissolution but no precipitation of silica
9	Quaternary Diatoms + Foraminifera	1	2,050	3.2	2.1	Formation of embryonic opal-CT lepispheres
10	Quaternary Diatoms + Foraminifera	6	7,350	5.0	1.4	Formation of embryonic opal-CT lepispheres, silica cementation
11	Quaternary Diatoms + Montmorillonite	1	8,000	46.6	0.1	Formation of magnesium-rich clay
12	Quaternary Diatoms + Foraminifera + Montmorillonite	1	1,950	26.6	0.7	Formation of embryonic opal-CT lepispheres
Average Surface Seawater			2–3	53.0	2.5	

Table 3. The effects of Mg, Ca, Na, K, and alkalinity on the rate of opal-A to opal-CT transformation; solution chemistry of hydrothermal experiments at pH 8, 150°C.

Exp. number	Starting materials	Time (months)	Si(OH)$_4$ μM	Mg mM	Alkalinity meq/l	Observations in solids	Figure number
1	Eocene Radiolarians + Seawater	1	2550	50.6	0.2	Dissolution but no precipitation of silica	2e, f
3	Eocene Radiolarians + Foraminifera + Seawater	1	1980	19.2	3.8	Formation of embryonic opal-CT lepispheres	2a, b
13	Eocene Radiolarians + Artificial Seawater without Magnesium*	1	2160	—	n.a.	Dissolution but no precipitation of silica	
14	Eocene Radiolarians + Foraminifera + Artificial Seawater without Magnesium*	1	2130	—	n.a.	Dissolution but no precipitation of silica	
15	Eocene Radiolarians + 0.5M NaCl, 0.03 M MgCl$_2$ sol. (l = 0.68)	1	6820	29.0	0.1	Dissolution but no precipitation of silica	
16	Eocene Radiolarians + Foraminifera + 0.5M NaCl, 0.03M MgCl$_2$ sol. (l = 0.68)	1	2550	2.2	2.8	Formation of embryonic opal-CT lepispheres	4a, b
17	Eocene Radiolarians + 0.027M MgCl$_2$, 0.028M NaHCO$_3$ sol. (l = 0.12)	1 day	4920	13.4	2.2	Formation of embryonic opal-CT lepispheres	
18	Eocene Radiolarians + 0.03M MgCl$_2$, 0.03M NaHCO$_3$ sol. (l = 0.12)	1	6550	16.6	1.1	Formation of well-developed opal-CT lepispheres	5a, c

No.	Composition	Time				Observation	Fig.
19	Eocene Radiolarians + 0.03M MgCl$_2$, 0.03M NaHCO$_3$ sol. (I = 0.12)	3	9750	14.4	0.8	Formation of well-developed and large opal-CT lepispheres	5d
20	Eocene Radiolarians + 0.03M MgCl$_2$, 0.006M NaHCO$_3$, 0.024M NaCl sol. (I = 0.12)**	1	9100	27.2	0.4	Formation of embryonic opal-CT lepispheres	5e
21	Eocene Radiolarians + 0.03M NaHCO$_3$, 0.1M NaCl sol. (I = 0.13)	1	3400	—	30.4	Dissolution but no precipitation of silica	5f
22	0.027M MgCl$_2$, 0.028 NaHCO$_3$ sol.	1 day	—	13.9	2.1	Formation of nuclei that contain Mg and (OH$^-$)	
23	Eocene Radiolarians + Montmorillonite + 0.03M MgCl$_2$, 0.03M NaHCO$_3$ sol. (I = 0.12)	1	7400	15.1	1.6	Formation of magnesium-rich clay	8
24	Montmorillonite + 0.03M MgCl$_2$, 0.03M NaHCO$_3$ sol. (I = 0.12)	1	3100	15.4	3.1	Formation of magnesium-rich clay	
25	Eocene Radiolarians + 0.03M MgCl$_2$, 0.03M KHCO$_3$ sol. (I = 0.12)	1	7900	16.0	1.6	Formation of well-developed opal-CT lepispheres	6
26	Eocene Radiolarians + 0.03M MgCl$_2$, 0.015M Na$_2$B$_4$O$_7$ · 10H$_2$O sol.**	1	7350	27.5	0.1	Formation of embryonic opal-CT lepispheres	7a, b

n.a. = not analyzed.
*Artificial seawater according to Lyman and Fleming (1940).
**The pH of the solution was adjusted to pH 8 at room temperature by acid titration. The alkalinity of the final solutions were 5.6 meq/l.

Table 4. The effect of ionic strength on the rate of opal-A to opal-CT transformation; solution chemistry of hydrothermal experiments at pH 8, 150°C, 1 month. Well-developed opal-CT lepispheres formed in all the experiments.

Exp. number	Starting materials	$Si(OH)_4$ μM	Mg mM	Alkalinity meq/l	Figure number
18	Eocene Radiolarians + 0.03M $MgCl_2$, 0.03M $NaHCO_3$ sol. (I = 0.12)	6550	16.6	1.1	5a, b
27	Eocene Radiolarians + 0.03M $MgCl_2$, 0.03M $NaHCO_3$, 0.58M NaCl sol. (I = 0.7)	6425	19.1	13.6	
28	Eocene Radiolarians + 0.03M $MgCl_2$, 0.03M $NaHCO_3$, 0.88M NaCl sol. (I = 1.0)	6225	16.6	3.2	
25	Eocene Radiolarians + 0.03M $MgCl_2$, 0.03M $KHCO_3$ sol. (I = 0.12)	7900	16.0	1.6	6
29	Eocene Radiolarians + 0.03M $MgCl_2$, 0.03M $KHCO_3$, 0.58M NaCl sol. (I = 0.7)	6975	17.4	2.8	9a, b

Table 5. Solution chemistry of room-temperature experiments.

Exp. number	Starting materials	$Si(OH)_4$ μM		
		1 week	2 weeks	3 weeks
a	Eocene Radiolarians + Distilled H_2O	n.d.	22	42
b	Eocene Radiolarians + Seawater	47	125	142
c	Eocene Radiolarians + Foraminifera + Seawater	<40	105	120
d	Quaternary Radiolarians + Distilled H_2O	1300	1430	1610
e	Quaternary Diatoms + Distilled H_2O	105	143	195
f	Quaternary Diatoms + Seawater	920	1030	1020
g	Quaternary Diatoms + Foraminifera + Seawater	840	935	
h	Quaternary Diatoms + Reagent grade $CaCO_3$ powder + Seawater	120	125	125
i	Quaternary Diatoms + Iceland Spar 150–250 μ + Seawater	910		
j	Quaternary Diatoms + Iceland Spar 63–105 μ + Seawater	860		
k	Quaternary Diatoms + Iceland Spar <44 μ + Seawater	840		
l	Cultured Diatom, *Navicula pelliculosa* + Seawater	480	950	1110
m	Cultured Diatom, *Nitzschia thermalis* + Seawater	350	960	1120

n.d. = not detected.

were lower than in the experiments without $CaCO_3$ (exps. 1, 5, 7, and 11, Table 2), in which siliceous tests only partially dissolved, many of the delicate tests were still present—although highly corroded, and no crystallization of opal-CT was detected, as shown in Figure 2e, f (exp. 1).

When montmorillonite was added to the radiolaria or diatom-seawater system (exps. 5, 11, Table 2), dissolution of siliceous tests but no opal-CT crystallization was observed. In addition, many of the pores of the siliceous tests were filled due to partial redistribution of the montmorillonite (Figure 3a, exp. 5), which most probably decreased the permeability of the system, and small

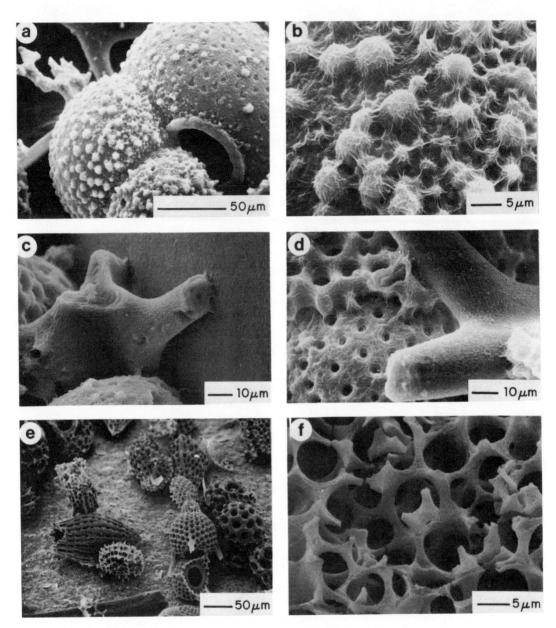

Fig. 2. Scanning electron microscope photographs showing dissolution and precipitation of silica during hydrothermal experiments.

a. Embryonic opal-CT lepispheres that formed on the surface of a foraminifera (exp. 3, Table 2).

b. Enlargement of (a), showing dense covering of the embryonic opal-CT lepispheres on the surface of a foraminifera.

c. A radiolarian fragment cemented by silica to the surface of a foraminifera test (exp. 3; 2 months).

d. Radiolarian fragments cemented by silica (exp. 3; 3 months).

e. Selective dissolution of the more delicate radiolarian species in an assemblage of more robust species (exp. 1, Table 2).

f. Severe dissolution of a delicate radiolarian test. The bars are attenuated and dissolved prior to the nodes of the test (exp. 1; 3 months).

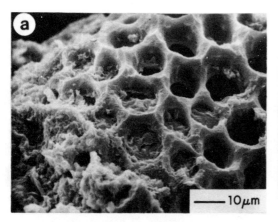 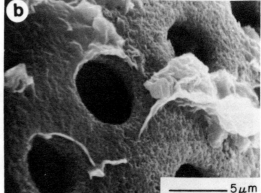

Fig. 3. Scanning electron microscope photographs of:
a. Filling of the pores of a radiolarian test by montmorillonite (exp. 5, Table 2; 2 months).
b. Neogenic clay attached to the surface of a radiolarian test (exp. 5; 3 months).

amounts of a magnesium-rich clay formed on the siliceous tests' surfaces (Figure 3b, exp. 5). This clay was analyzed semiquantitatively by the X-ray energy dispersive attachment to the SEM.

	A.P.I. Montmorillonite #24	*Recrystallized Clay Fig. 3b*
SiO_2/MgO	9.2	5.6
Al_2O_3/MgO	2.8	1.7

X-ray energy dispersive analyses of the embryonic opal-CT lepispheres that formed in the experiments with $CaCO_3$ revealed the presence of small amounts of magnesium (on the order of a few ppm) in addition to silica. The magnesium was extracted from seawater as indicated, for example, in exp. 3, Table 2, in which the magnesium concentration decreased by about 23 mM. Due to dissolution of foraminifera, the calcium concentration increased by about the same amount, but no corresponding increase in alkalinity was observed; the alkalinity increased by only 1 meq/l. On the other hand, in the experiments without $CaCO_3$ the magnesium values decreased only slightly, and the alkalinity decreased to almost zero.

The 150°C experimental results of Table 2 strongly suggest that both magnesium and $CaCO_3$ are the most important components responsible for the increased rate of opal-A diagenesis observed in the seawater experiments with carbonate, relative to the experiments without carbonate, but with montmorillonite. This conclusion is strengthened by the results of exps. 13 and 14, Table 3, in artificial seawater without magnesium, in which no opal-CT crystallization was observed and the silica values, with and without $CaCO_3$, are almost identical; the fact that exps. 15 and 16, Table 3, show the same trend as

Fig. 4. Scanning electron microscope photographs of:

a. Opal-CT embryonic lepispheres crystallized preferentially on a foraminifera surface. No similar crystallization of opal-CT is seen on the adjacent radiolarian tests (exp. 16, Table 3).

b. An enlargement of the opal-CT embryonic lepispheres on a foraminifera surface of Figure 4a.

exps. 1 and 3, Table 2, with seawater, points at the same direction. Indeed, Figure 4a, b (exp. 16) shows the same preferential precipitation of embryonic opal-CT lepispheres on foraminifera surfaces as Figure 2a, b (exp. 3). Thus it seems that the effect of all major elements in seawater, except magnesium and carbonate, on the rate of opal-A diagenesis is insignificant.

Exps. 17 and 18, Table 3, were conducted to demonstrate that bicarbonate, not the calcium of seawater and of the dissolved foraminifera, affects the rate of opal-CT crystallization. Even after one day (exp. 17), embryonic opal-CT lepispheres formed, and after one month (exp. 18), well-developed opal-CT lepispheres crystallized, as shown in Figure 5a. Their textural resemblance to pelagic opal-CT lepispheres is demonstrated in Figure 5b, which shows lepispheres from sediments of Early Cretaceous age, DSDP site 305 (Keene 1975). The degree of disorder of the synthetic opal-CT lepispheres was determined by electron diffraction to be similar to the ELTANIN core 47-15 lepispheres (Wise and Weaver 1974). Simultaneously with opal-CT crystallization, the delicate siliceous tests were completely dissolved, and the less delicate ones were highly corroded (Figure 5c, exp. 18). The solution chemistry of exps. 17 and 18 shows that the silica concentrations, although high, are below the equilibrium solubility of opal-A at 150°C (Figure 1) and that the magnesium concentration and alkalinity decreased by a ratio of about 1:2. After three months (exp. 19), a significant increase in the concentration of silica was observed, but no further significant decrease in the magnesium concentration and alkalinity was observed; the alkalinity had decreased to almost zero. In addition to the well-developed opal-CT lephispheres, radiolarian tests were also replaced by opal-CT lepispheres (Figure 5d). At a lower alkalinity of 5.6 meq/l (exp. 20) instead of 30 meq/l (exps. 17, 18, and 19), a small number of only embryonic opal-CT

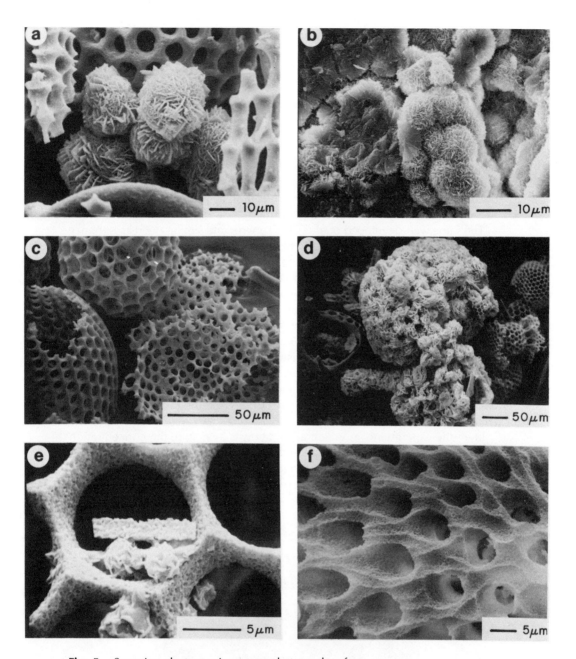

Fig. 5. Scanning electron microscope photographs of:

a. A cluster of well-formed opal-CT lepispheres that crystallized in exp. 18, Table 3.

b. Opal-CT lepispheres from a deep-sea porcelanite of Early Cretaceous age. The lepispheres line the inside of a radiolarian mold and texturally are very similar to the lepispheres of Figure 5a.

c. Several severely corroded radiolarian fragments and two less corroded robust species (exp. 18, Table 3).

d. Pseudomorphs of opal-CT after robust radiolarians (exp. 19, Table 3).

e. Small embryonic opal-CT lepispheres are attached to the surface of a radiolarian test and cement a radiolarian fragment to the test (exp. 20, Table 3).

f. Severe corrosion of a robust radiolarian test (exp. 21, Table 3).

lepispheres formed (Figure 5e), and the magnesium concentration decreased only by about 2.5 mM, while the alkalinity was consumed.

In exp. 21, Table 3, without magnesium but with 30 meq/l $NaHCO_3$ as in exp. 18, crystallization of opal-CT lepispheres was not detected, and no decrease in alkalinity was observed. The only reaction that took place was dissolution of siliceous tests (Figure 5f).

In a one-day blank experiment with a $MgCl_2$ and $NaHCO_3$ solution but without silica (exp. 22, Table 3), the magnesium and alkalinity values also decreased by a ratio of about 1:2, and nuclei of, as yet, an unidentified nature that contained magnesium and OH^- formed.

The same well-developed opal-CT lepispheres and similar decreases in magnesium and alkalinity values were obtained in exp. 25 (Figure 6, Table 3), which was identical to exp. 18 except for potassium instead of sodium.

Exps. 17, 18, 19, and 25, Table 3, in the presence of magnesium indicate the strong influence of OH^- on the rate of opal-CT precipitation. To further test and confirm this observation, exp. 26, Table 3, was conducted, which is identical with exp. 20, Table 3, except for the substitution of borate alkalinity of 5.6 meq/l for bicarbonate alkalinity of 5.6 meq/l. As the results of both experiments are the same, it follows that the decisive factor is the presence of alkalinity irrespective of its origin. In both experiments the magnesium concentration decreased by about 2.5 mM, the alkalinity was almost completely consumed, and only embryonic opal-CT lepispheres formed on the radiolarian test surfaces (Figures 7a, b, and 5e, respectively).

In all the experiments in which alkalinity was available, the concentrations of magnesium and alkalinity decreased in a ratio of about 1:2.

Except for the addition of montmorillonite, exp. 23, Table 3, is identical with exp. 18, in which well-developed opal-CT lepispheres formed (Figure 5a). Although the concentrations of magnesium and alkalinity decreased by about

Fig. 6. Scanning electron microscope photograph of well-formed opal-CT lepispheres (exp. 25, Table 3). The interpenetrating blades or plates of the lepispheres are clearly visible.

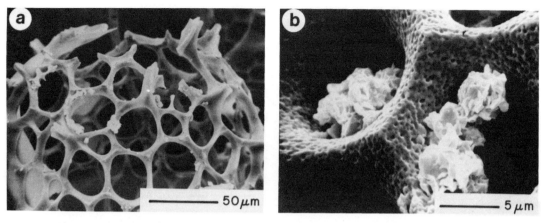

Fig. 7. Scanning electron microscope photographs of:
a. Embryonic opal-CT lepispheres in small clusters on the surface of a radiolarian test (exp. 26, Table 3).
b. An enlargement of Figure 7a, showing the embryonic opal-CT lepispheres attached to and growing into the open spaces of the pores of a radiolarian test. The pitting on the surface is due to dissolution.

15 mM and 28–29 meq/l, respectively, no well-developed opal-CT lepispheres crystallized. Only very few embryonic opal-CT lepispheres formed, siliceous tests partially dissolved (Figure 8), the montmorillonite was redistributed, and a magnesium-rich clay formed. Even in the absence of siliceous tests (exp. 24, Table 3), a similar major decrease in the concentrations of magnesium and alkalinity was observed. Semiquantitative X-ray energy dispersive analyses of the magnesium-rich clays of exps. 23 and 24 show that the composition of these clays differs from that of the A.P.I. #24 montmorillonite as follows:

	A.P.I. Montmorillonite #24	Clays Exp. 23	Exp. 24
SiO_2/MgO	9.2	8.1	6.1
Al_2O_3/MgO	2.8	1.0	1.2

In the presence of siliceous tests (exp. 23), the magnesium rich clay was more siliceous than in exp. 24 in which only montmorillonite was present.

Table 4 demonstrates that the effect of ionic strength on the extent of opal-CT crystallization appears to be very small; at $I = 0.7$ (exps. 27 and 29), the ionic strength of seawater, the extent of opal-CT crystallization was slightly reduced (higher magnesium content and higher alkalinity), but at $I = 1.0$ (exp. 28), it was similar to that at $I = 0.12$ (exps. 18 and 25). The same well-developed opal-CT lepispheres that formed at $I = 0.12$ (exps. 18 and 25, Figures 5a and 7) also formed at $I = 0.7$ and 1.0, as shown, for example, in Figure 9a (exp. 29, at $I = 0.7$). The small crystals indicated by an arrow on Figure 9a are shown at a higher magnification in Figure 9b. These crystals are the so-called embryonic

Fig. 8. Scanning electron microscope photograph showing a highly corroded radiolarian test and well-preserved robust species and neogenic clay attached to their surface. No opal-CT lepispheres are observed (exp. 23, Table 3).

opal-CT lepispheres that subsequently form the adjacent well-developed opal-CT lepispheres of Figure 9a (exp. 29).

Experiments with seawater, but at room temperature, also show the influence of magnesium and $CaCO_3$ on the rate of silica uptake, after only one week (exps. b, c, f, and g, Table 5). Because of the very small extent of reaction in the room-temperature experiments (Table 5), the changes in magnesium and alkalinity values were insignificant and are therefore not reported in Table 5.

The observed inverse relation between the available surface areas of $CaCO_3$

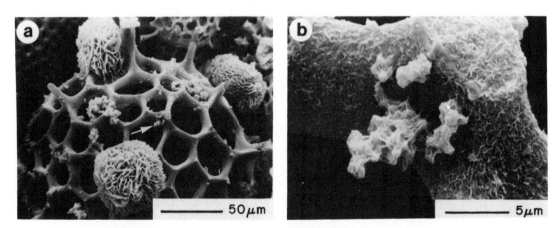

Fig. 9. Scanning electron microscope photographs of:
   a. Opal-CT lepispheres of two sizes: large, well-formed lepispheres and small clusters of embryonic lepispheres. Both are attached to a partially dissolved radiolarian test (exp. 29, Table 4).
   b. An enlargement of the area shown by an arrow on Figure 9a, showing the embryonic opal-CT lepispheres cemented to the surface of the radiolarian test.

and the amount of silica in solution, in exps. h, i, j, and k, Table 5, suggest that silica initially preferentially precipitates on the $CaCO_3$ surfaces, as was subsequently confirmed by X-ray energy dispersive analyses of the calcite crystal surfaces. The almost constant dissolved silica values of exp. h, during the first three weeks, most probably reflects the rapid coating of the surfaces of the finely powdered $CaCO_3$.

The dissolution rate of opal-A seems to be much faster in seawater than in distilled water, as indicated by the silica concentrations of exps. a and e, as compared with exps. b and f, respectively (Table 5). This observation agrees with Kato and Kitano's (1968) results but does not agree with Krauskopf's (1956) and Siever's (1962) results. Additional experiments are in progress on the effects of pH and ionic strength on the dissolution rate of opal-A.

Specific surface areas and the degree of disorder of the opal-A structure are two important factors that control the dissolution rates of the various diatom and radiolarian species. It is tempting to infer that these factors were responsible for the observed differences in silica concentrations in experiments with Eocene and Quaternary radiolarians, respectively (exps. a and d, Table 5), and in agreement with Moore's (1969) observations. These factors would also explain the generally higher dissolution rate of the more delicate diatoms as compared to that of the more robust radiolarians. Exps. a, d, and e (Table 5) show that the silica content of the solution increases with increasing surface area of the solids; for example, the specific surface area of the Eocene radiolarians is 7 m²/g and of the Quaternary radiolarians, 439 m²/g. Experimental work is in progress to verify these suggestions. Hurd and Theyer (1975) observed a trend of decreasing solubility of radiolarian and sponge spicule assemblages as a function of geologic age.

4. DISCUSSION

The main results of our experiments are: the presence of magnesium and a source of hydroxyl are important requirements in the conversion process of opal-A → opal-CT; in the experiments in which opal-CT lepispheres formed, the concentrations of magnesium and alkalinity decreased by a ratio of 1:2, ΔMg (in equivalents) $\simeq \Delta$ alkalinity; and nuclei of a compound containing magnesium and OH^- enhance the formation of opal-CT lepispheres. The nuclei appear to serve as sites on which opal-CT precipitation takes place. These nuclei form rapidly at 150°C, as indicated by the blank exp. 22, Table 3. In the presence of magnesium and hydroxyl ions, when montmorillonite was added, no opal-CT lepispheres formed within the time period of our experiments, but magnesium appeared to be taken up by the clays.

As mentioned already, in the blank exp. 22, Table 3, nuclei that contain magnesium and hydroxyl formed rapidly. In a similar experiment, but with radiolarians (exp. 17, Table 3), after only one day embryonic opal-CT lepispheres formed on such nuclei. As most of the available alkalinity was consumed in one day by the nuclei, no significant additional decreases in the magnesium and alkalinity concentrations were observed after one month (exp.

18, Table 3), but instead of embryonic lepispheres, well-developed opal-CT lepispheres crystallized (Figure 5a), and the silica concentration increased significantly. In an identical experiment (exp. 19, Table 3), but for three months, no major changes in magnesium concentration and alkalinity were observed, but the silica concentration increased by about 50 percent as a result of siliceous tests dissolution. The dimensions of the lepispheres that had formed after one month increased, the blades thickened, and several radiolarians were replaced by opal-CT (Figure 5d). But in exps. 20 and 26, Table 3, with initial bicarbonate and borate alkalinity, respectively, of 5.6 meq/l, merely a few embryonic opal-CT lepispheres formed (Figures 5e and 7a, b). Again magnesium and alkalinity decreased at a ratio of 1:2. The identical results of these last two experiments demonstrate that in the presence of magnesium, the OH^-, and not the bicarbonate or carbonic acid, is responsible for the increased rate of transformation of opal-A to opal-CT in both the experiments of high alkalinity and in pelagic carbonate sediments.

The formation of well-developed opal-CT lepispheres rather than just embryonic ones is the main difference between exps. 18 and 19, Table 3, of high initial alkalinity and the seawater experiments with $CaCO_3$ (exps. 3, 4, 9, and 10, Table 2), in which only about 2.5 meq/l alkalinity was initially available; additional alkalinity was continuously provided as a result of $CaCO_3$ dissolution. In these experiments the decrease in the concentration of magnesium equaled the increase in concentration of calcium, as a result of opal-CT formation and of foraminifera dissolution (and not due to dolomite formation).

The mechanism for the observed increased rate of the reaction opal-A → opal-CT, as deduced from our experiments, is as follows: Exp. 17, Table 3, in which the solution was depleted of most of the alkalinity and of an equivalent amount of magnesium after on day, indicates that at high alkalinity values and at 150°C, formation of nuclei with Mg:$OH^- \simeq$ 1:2 of critical size is rapid. These nuclei most probably attract silanol groups and thus act as sites of opal-CT nucleation and subsequent growth of lepispheres. As soon as the alkalinity is consumed, the rate of formation of new embryonic opal-CT lepispheres decreases considerably, and the already available embryonic lepispheres extend to well-developed opal-CT lepispheres as a result of continuous dissolution of opal-A. At low alkalinity values and at lower temperatures, the rate of nuclei formation of critical size and the subsequent formation of opal-CT lepispheres is reduced. In the seawater experiments of initially low alkalinity but with $CaCO_3$, due to continuous dissolution of $CaCO_3$, successive new nuclei that contain magnesium and OH^-, form, but only embryonic opal-CT lepispheres develop within one month. Due to the slower rate of nuclei formation, they might also be smaller than in the experiments with high initial alkalinity. As merely critical size nuclei are needed for the formation of embryonic opal-CT lepispheres, the amount of magnesium in the lepispheres could be negligible. (Kinetic experiments of rates of formation and dimensions of these nuclei as a function of temperature are in progress).

Iler (1955) has already observed that $Mg(OH)_2$ strongly affects the precipitation of silica from solution. A similar mechanism was also suggested by Harder (1965) and Harder and Flehmig (1970) for quartz crystallization by aging Al^{+3},

Mg^{+2}, Fe^{+3}, and Mn^{+2} hydroxide-silica precipitates in the presence of their aqueous solutions between 0° and 80°C. In addition to the preceding experiments with magnesium, we have also conducted similar preliminary hydrothermal experiments with Al^{+3}, Fe^{+3}, and Mn^{+2} at 150°C. The results suggest that the hydroxides of these elements also positively affect the rate of the reaction opal-A → opal-CT; but because of the high magnesium concentration in seawater compared with the very low concentrations of Al^{+3}, Fe^{+3}, and Mn^{+2}, their role in most deep-sea sediments is probably small, and magnesium was emphasized in our experiments.

Based on the above mechanism, in pelagic carbonate sediments the consumption of alkalinity as a result of nuclei formation with a $Mg : OH^- \simeq 1:2$ should promote the dissolution of $CaCO_3$ and thus make a continuous source of new alkalinity available. But in clay-rich sediments in which the $CaCO_3$ is not present, additional alkalinity can be provided by diffusion of bicarbonate.

While the generally observed slower rate of chert formation in clay-rich versus carbonate sediments can already be explained on the basis of these results and interpretations, exps. 23 and 24, Table 3, indicate that clay minerals play a complementary active role causing even further retardation in the rate of opal-A to opal-CT transformation, as they provide an additional sink for dissolved magnesium.

Except for the addition of montmorillonite, exp. 23 is identical with exp. 18, Table 3. But despite the high initial alkalinity value and the high silica value resulting from radiolarian dissolution (Figure 9), no opal-CT lepispheres formed. Instead, the montmorillonite recrystallized to a more magnesium-rich clay and thus almost depleted the solution of alkalinity, as indicated in exp. 24 in which only montmorillonite and no radiolarians were present. Consequently, not much alkalinity was available for the formation of nuclei that contain magnesium and OH^- and the subsequent crystallization of embryonic opal-CT lepispheres.

Despite the more extensive dissolution of the siliceous tests and embryonic opal-CT precipitation, silica concentrations in solution were lower in the seawater experiments than in exps. 5 and 11, Table 2, with montmorillonite but without $CaCO_3$, in which embryonic opal-CT lepispheres did not precipitate. In all these experiments, silica concentration values were below the equilibrium concentration for opal-A. Thus in calcareous pelagic sediments, embryonic opal-CT lepispheres could nucleate from solutions with silica concentrations in equilibrium with low cristobalite. In the absence of $CaCO_3$, however, and at silica concentrations in equilibrium with opal-A (exps. 2 and 8, Table 2), only very small amounts of silica precipitated even after six months, and no major reduction in magnesium concentration was observed because of the low alkalinity value of seawater.

In the experiments at 150°C in which silica values were above the equilibrium solubility value for quartz, silica precipitated in the form of disordered low-cristobalite. Under similar conditions in hydrothermal experiments of smectite to illite conversion (Kastner and Bada, in preparation), quartz and not low-cristobalite crystallized directly from solution, but the concentration of silica in solution was only 465 μM at 150°C. It is conceivable that in permeable

311

carbonate oozes with disseminated siliceous tests, in particular radiolaria tests, the concentration of silica in solution will not increase immediately to values as high as in pure siliceous oozes, especially diatom oozes (for example, Schink et al. 1974). During this early diagenetic stage, quartz (chalcedony) could crystallize directly from solution and, as Heath and Moberly (1971) observed, will crystallize preferentially within the calcareous tests. The rate of opal-A (siliceous tests) dissolution most probably rapidly exceeds the slow rate of quartz crystallization and silica diffusion. Consequently, the silica concentration in the interstitial waters will increase until it reaches concentration values above the equilibrium solubility value for cristobalite; under these conditions the commonly observed opal-CT phase will crystallize. This mechanism provides one possible explanation for the commonly observed quartz cores and opal-CT rims of chert nodules recovered by the DSDP from carbonate-rich horizons. Such chert nodules can, however, be explained by the overall maturation sequence of opal-A that proceeds from the center of the nodule outwards, as Wise and Weaver (1974) have suggested.

Primary quartz in veins and as drusy quartz in deeper sections of pelagic sediments was also described from DSDP samples (for example, Keene 1975; Knauth and Epstein 1975). These quartz occurrences that formed at a later diagenetic stage probably reflect a geochemical environment in which opal-A is no longer present and silica concentration in solution is relatively low.

5. GEOLOGICAL SIGNIFICANCE

The specific surface areas of both diatoms and radiolarians decrease as a function of geologic age, as Hurd and Theyer (1975) have shown. For example, the specific surface area of Quaternary radiolarians, Table 5, is 439 m²/g, but that of the Eocene radiolarians, Tables 2–5, is only 7 m²/g. Moore (1969) suggested that an evolutionary factor is responsible for a primary difference between the specific surface areas of Eocene and Quaternary radiolarians.

The results of exps. a and e versus b and f, respectively, of Table 5 suggest that in the marine environment the rate of diagenesis of pure siliceous oozes will be greater than in fresh-water environments such as subaerial weathering (for example, of the now exposed diatomites of the Monterey Formation, California) and fresh-water lake environments. The chemical controls responsible for the observed high rate of diagenesis of platform cherts has not been investigated, as yet.

In the experiments of Tables 2–5, it has also been shown that the transformation rate of opal-A to opal-CT is greatly enhanced by solutions high in magnesium and alkalinity in the absence of reactive clay minerals. In the marine environment, the most obvious source of magnesium and alkalinity is seawater with an average concentration of 53 mM magnesium and 2.5 meq/l alkalinity. In noncarbonate sediments, the low alkalinity value of seawater will then be the limiting factor for the reaction opal-A $\rightarrow$ opal-CT. In fresh-water environments with average concentrations of about 0.3 mM magnesium and 1 meq/l alkalinity (Stumm and Morgan 1970, Table 8–1), both magnesium and alkalinity are limiting factors. In subaerial weathering environments, dissolution of magnesium-carbonates would provide additional magnesium and alkalinity to the solution. Indeed, preferential silicification of carbonate shells in this environment is widespread. Silicification should favor magnesium-carbonate shells rather than magnesium-free carbonate shells, as observed in exps. 3, 4, 9, and 10, Table 2 (Figures 2a, b and 4a, b) in which the initial sites of opal-CT precipitation were the foraminifera surfaces of the slightly more magnesium-rich *Globigerinoides ruber* and *sacculifer* (Savin and Douglas 1973).

In the pelagic environment of high-magnesium concentrations and in areas of average heatflow, the reaction opal-A → opal-CT, as mentioned, is faster in carbonate sediments than in clay-rich sediments. Based on the experimental results, in carbonate sediments the formation of nuclei with magnesium and OH^- and subsequently of opal-CT lepispheres would reduce the alkalinity and thus promote the dissolution of $CaCO_3$, which enables the reaction to continue. As a result, replacement cherts form. As the process continues, in carbonate-rich pelagic sediments, a number of radiolaria and diatom tests are being coated by opal-CT, which protects them from further dissolution. As for the foraminifera and nannofossil oozes, their continuous dissolution and replacement by silica depends on the diffusion of CO_2 out of the system; otherwise, precipitation of $CaCO_3$ on the surfaces of the calcareous tests will take place. As a result, the dissolution of the siliceous tests and the formation of opal-CT will slow down, as Keene (1976) has observed in many DSDP sediments. Indeed, in one experiment a major reduction in space for water vapor and CO_2, which was available in the sealed tubes of all experiments, resulted in $CaCO_3$ crystallization on the foraminifera surfaces while only small amounts of opal-CT crystallized.

In clay-rich sediments, however, clay minerals, in particular smectites, would initially compete for the available magnesium, and neoformed clays also for the available OH^-. As a result, the solution would be depleted in alkalinity, which in the absence of $CaCO_3$ could be replenished only through bicarbonate diffusion. Consequently, the rate of opal-CT formation is greatly reduced. In addition, as Iler (1973 and personal communication) suggested, adsorption of even trace amounts of $Al(OH)_4^-$ ions, supplied by clay dissolution, has a marked effect on amorphous silica solubility above pH 7. This effect may also be responsible for the observed reduced transformation rate of opal-A to opal-CT.

It would be erroneous to conclude that in outcrops of siliceous ooze formations a correlation between the presence of carbonate layers and the extent of opal-A diagenesis will always be observed. In this case, it will be necessary to consider the hydrology of the system as well.

In deep-sea sediments, opal-CT is widespread in both carbonate and clay-rich sediments, but in clay-rich sediments the degree of disorder of the opal-CT crystals is generally higher than in carbonate sediments of the same geologic age (Keene 1976). In addition, the quartz to opal-CT ratio is generally much higher in younger carbonate sediments (Middle Eocene) than in older clay-rich sediments (Cretaceous). Assuming that most quartz in cherts, even in carbonate sediments, had an opal-CT precursor, we suggest that the observed rate increase of opal-A to opal-CT transformation in carbonate versus clay-rich sediments also bears on the rate of opal-CT to quartz transformation, as follows: The degree of disorder of a silica phase depends on the silica concentration in solution, as implied in Figure 1. In our experiments with $CaCO_3$, silica concentrations were lower than in the experiments without $CaCO_3$ and/or with montmorillonite (for example, exps. 3 and 9 versus 1, 5, 7, and 11, Table 2; exp. 18 versus 23, Table 3). Consequently, opal-CT, which crystallizes in experiments with montmorillonite, and thus higher silica values in solution should have a higher degree of disorder than opal-CT that crystallizes from experiments with $CaCO_3$, and thus lower silica values in solution, as Keene (1976) observed in siliceous sediments of the Pacific. Carr and Fyfe (1958) and Murata and Larson (1975) further emphasized that progressive ordering of opal-CT is necessary prior to its conversion to quartz. If we accept their suggestion, accordingly, prior to the conversion to quartz the less disordered opal-CT in carbonate sediments has to undergo fewer steps of progressive ordering than the more disordered opal-CT in clay-rich sediments.

In addition, a highly disordered phase has a higher energy structure than the less disordered polymorph. Relaxation is achieved by the adsorption of impurities, which may retard its transformation to the ordered, lower-energy polymorph (G. A. Somorjai, Berkeley, California, personal communication). The higher degree of disorder of opal-CT in clay-rich versus carbonate sediments may in analogy also be responsible for the observed preservation of opal-CT (porcelanite) in pelagic clay-rich sediments of even earliest Cretaceous age.

In the Cretaceous chalk of the Wealden District, Great Britian, which was uplifted right after deposition and was exposed to fresh-water input for at least 65 m.y., P. A. Scholle (U.S.G.S., Reston, Virginia, personal communication) made a similar observation: Rocks higher in insoluble residue, mainly montmorillonite, contain less chert (quartz) nodules but are richer in opal-CT lepispheres; chert nodules occur only above the upper part of the middle chalk that contains up to 5 percent montmorillonite. Below, where the chalk contains more than 15 percent and up to 40 percent montmorillonite, silica, although present in significant amounts, occurs only as disseminated opal-CT lepispheres.

The release of magnesium during the transformation of opal-CT to quartz might be responsible for a small fraction of palygorskite and dolomite, often associated with younger and older cherts, respectively.

A comparison between the oxygen-isotope values of the quartz cores and opal-CT rims of deeply buried chert nodules might help to determine whether the quartz of these chert nodules had an opal-CT precursor. Knauth and Epstein (1975) analyzed only one shallow chert nodule (their sample 356, Table 1) and observed a 0.2–0.5 per mil difference in $\delta^{18}O$ between the opal-CT and quartz.

In conclusion, temperature strongly affects the rate of diagenesis of siliceous oozes, as suggested by Heath and Moberly (1971), Murata and Nakata (1974), and Murata and Larson (1975), and shown experimentally, for example, by Ernst and Calvert (1969). Therefore, chert should prevail in older and deeper sediments, and porcelanite in younger and shallower sediments. But as shown in our experiments and observed on land by Bramlette (1946) and P. A. Scholle (U.S.G.S., Reston, Virginia, personal communication) and in DSDP sediments by Winterer et al. (1971), Hays et al. (1972), Lancelot (1973), and Keene (1975, 1976), geochemical factors often reverse the expected relation between burial depth and the diagenetic silica polymorph.

Newly formed sea floor is generally covered with carbonate sediments. As the crust subsides below the Carbonate Compensation Depth (Sclater et al. 1971), the carbonates are covered with clay-rich sediments. The occurrence of chert in these older pelagic carbonates does not imply that age and burial depth alone were responsible for the transformation of siliceous oozes to chert. These cherts might have formed very early in the diagenetic history of these sediments as a result of geochemical factors, as described above. The time required for the transformation of siliceous oozes to porcelanites and cherts, in various host sediments, is as yet not known.

6. CONCLUSIONS

1. Temperature and time strongly affect the rate of diagenesis in siliceous oozes to porcelanites and cherts. However, based on observations on land and in DSDP sediments and on experimental work, the expected relation between burial depth and the diagenetic silica phase is often reversed by geochemical factors such as the solution chemistry and/or the mineralogy and chemistry of the host sediments.

2. Based on our experiments and observations in DSDP sediments in the pelagic environment, the rate of siliceous oozes transformation to porcelanite is greatly enhanced in carbonate sediments relative to pure siliceous oozes, and even more relative to clay-rich sediments. This observed diagenetic rate increase in carbonate sediments also has a bearing on the rate of opal-CT to quartz transformation.

3. Nuclei with $Mg : OH^- \approx 1:2$, which attract silanol groups, act as sites of opal-CT nucleation and subsequent growth of opal-CT lepispheres.

4. Reactive clay minerals retard the transformation of opal-A to opal-CT.

5. In deep-sea cherts, most quartz probably had an opal-CT precursor. But the possibility of primary quartz crystallization without an opal-CT precursor cannot be excluded, in particular during late diagenesis.

ACKNOWLEDGMENTS

We are grateful to Drs. Y. K. Bentor, Hebrew University; S. E. Calvert, Institute of Oceanography, Wormley, England; G. R. Heath, Graduate School of Oceanography, Rhode Island; D. C. Hurd, Hawaii Institute of Geophysics; R. K. Iler, Wilmington, Delaware; and R. Siever, Harvard University, for reading the manuscript and for their many constructive comments. We also thank Drs. D. O. Seevers, W. R. Riedel, and B. E. Volcani for numerous valuable discussions during this research. We thank Dr. S. W. Wise, Jr., who kindly provided the diatoms, ELTANIN core 47-15, and Dr. B. E. Volcani, who kindly provided the cultured diatoms; Chevron Oil Research Co., La Habra and Richmond, California, for the ESCA and BET analyses, and in particular Dr. J. Q. Adams for the ESCA analyses; Dr. E. Vincent for her advice in calcareous micropaleontology; Mr. R. T. Laborde and Ms. E. L. Flentye of the Analytical Facility of Scripps Institution of Oceanography; and Ms. E. Bromley and Mr. G. Galleisky for their assistance in the chemical laboratory.

This research was supported by a grant from Chevron Oil Research Co., La Habra, California, and by NSF grants OCE76–02128 (M.K.), DES 72–01410 (JMG.), and by an ONR contract No. USN N00014–75–C–0152 (J.B.K.).

REFERENCES

Alexander, G. B. (1957) The effect of particle size on the solubility of amorphous silica in water. *Jour. Phys. Chemistry* **61:** 1563–1564.

Alexander, G. B.; Heston, W. M.; and Iler, R. K. (1954) The solubility of amorphous silica in water. *Jour. Phys. Chemistry* **58:** 453–455.

Barton, D. C. (1918) Notes on the Mississippian chert of St. Louis area. *Jour. Geology* **26:** 361–374.

Berger, W. H.; and von Rad, U. (1972) Cretaceous and Cenozoic sediments from the Atlantic Ocean. In *Initial Reports of the Deep Sea Drilling Project,* Vol. XIV (eds. D. E. Hayes, A. C. Pimm et al.), pp. 787–954. U.S. Government Printing Office, Washington, D.C.

Betterman, P., and Liebau, F. (1975) The transformation of amorphous silica to crystalline silica under hydrothermal conditions. *Contr. Mineralogy and Petrology* **53:** 25–36.

Bramlette, M. N. (1946) The Monterey Formation of California and the origin of its siliceous rocks. *U.S. Geol. Survey Prof. Paper 212,* 55 pp.

Calvert, S. E. (1966) Accumulation of diatomaceous silica in the sediments of the Gulf of California. *Geol. Soc. America Bull.* **77:** 569–596.

———. (1971) Composition and origin of North Atlantic deep sea cherts. *Contr. Mineralogy and Petrology* **33:** 273–288.

Campbell, A. S.; and Fyfe, W. S. (1960) Hydroxyl ion catalysis of the hydrothermal crystallization of amorphous silica: A possible high temperature pH indicator. *Am. Mineralogist* **45:** 464–468.

Carr, R. M.; and Fyfe, W. S. (1958) Some observations on the crystallization of amorphous silica. *Am. Mineralogist* **43:** 908–916.

Corwin, J. F.; Herzog, A. H.; Owen, G. E.; Yalman, R. G.; and Swinnerton, A. C. (1953) The mechanism of the hydrothermal transformation of silica glass to quartz under isothermal conditions. *Am. Chemical Soc. Jour.* **75:** 3933–3934.

Cox, G. H.; Dean, R. S.; and Gottschalk, V. H. (1916) Studies on the origin of Missouri cherts and zinc ores. *Missouri Univ. School of Mines and Metallurgy Bull.,* Tech. Ser. 3, no. 2, 34 pp.

Davis, E. F. (1918) The radiolarian cherts of the Franciscan Group. *California Univ. Pubs., Bull. Dept. Geology* **11:** 235–432.

Duedall, I. W.; Dayal, R.; and Willey, J. D. (1976) The partial molal volume of silicic acid in 0.725m NaCl. *Geochim. et Cosmochim. Acta* **40:** 1185–1189.

Ernst, W. G.; and Blatt, H. (1964) Experimental study of quartz overgrowths and synthetic quartzites. *Jour. Geology* **72:** 461–470.

Ernst, W. G.; and Calvert, S. E. (1969) An experimental study of the recrystallization of porcelanite and its bearing on the origin of some bedded cherts. *Am. Jour. Sci.* **267-A:** 114–133.

Flörke, O. W. (1955) Zur Frage des "Hoch"-Cristobalit in Opalen, Bentoniten und Gläsern: *Neues Jahrb. Mineralogie Monatsh.* **10:** 217–223.

Flörke, O. W.; Jones, J. B.; and Segnit, E. R. (1975) Opal-CT crystals. *Neues Jahrb. Mineralogie Monatsh.* **8:** 369–377.

Fournier, R. O. (1960) Solubility of quartz in water in the temperature interval from 25°C to 300°C. *Geol. Soc. America Bull.* **71:** 1867–1868.

———. (1973) Silica in thermal waters: Laboratory and field investigations. *Interntl. Symp. Hydrochem. Biochem. Tokyo, 1970 Proc. 1,* 122–139. J. W. Clarke Co., Washington, D.C.

Fournier, R. O.; and Rowe, J. J. (1962) The solubility of cristobalite along the three-phase curve, gas plus liquid plus cristobalite. *Am. Mineralogist* **47:** 897–902.

Fyfe, W. S.; and McKay, D. S. (1962) Hydroxyl ion catalysis of the crystallization of amorphous silica at 330°C and some observations on the hydrolysis of albite solutions. *Am. Mineralogist* **47:** 83–89.

Gieskes, J. M. (1974) Interstitial water studies, Leg 25 Deep Sea Drilling Project. In *Initial Reports of the Deep Sea Drilling Project,* Vol. XXV (eds. F. S. W. Simpson, R. Schich et al.), pp. 361–394. U.S. Government Printing Office, Washington, D.C.

———. (1975) Chemistry of interstitial waters of marine sediments. *Annual Rev. Earth and Planetary Sci.* **3:** 433–453.

Harder, H. (1965) Experimente zur "Ausfällung" der Kieselsäure. *Geochim. et Cosmochim. Acta* **29:** 429–442.

Harder, H.; and Flehmig, W. (1970) Quartzsynthese bei tiefen Temperaturen. *Geochim. et Cosmochim. Acta* **34:** 295–305.

Hays, J. D.; Cook, H. F.; Jenkyns, D. G.; Cook, F. M.; Fullen, J. T.; Goll, R. M.; Milow, E. D.; and Orr, W. N. (1972) *Initial Reports of the Deep Sea Drilling Project,* Vol. IX, 1205 pp. U.S. Government Printing Office, Washington D.C.

Heath, G. R. (1974) Dissolved silica and deep-sea sediments. *Soc. Econ. Paleontologists and Mineralogists Spec. Pub. No. 20:* 77–93.

Heath, G. R.; and Moberly, R., Jr. (1971) Cherts from the western Pacific, Leg 7 Deep Sea Drilling Project. In *Initial Reports of the Deep Sea Drilling Project,* Vol. VII (eds. E. L. Winterer et al.), pp. 991–1007. U.S. Government Printing Office, Washington, D.C.

Hurd, D. C. (1972) Factors affecting solution rate of biogenic opal in seawater. *Earth Planet. Sci. Letters* **15:** 411–417.

———. (1973) Interaction of biogenic opal, sediment and seawater in the Central Equatorial Pacific. *Geochim. et Cosmochim. Acta* **37:** 2257–2282.

Hurd, D. C.; and Theyer, F. (1975) Changes in the physical and chemical properties of biogenic silica from the Central Equatorial Pacific. I. Solubility, specific surface area, and solution rate constants of acid-cleaned samples. *Adv. Chem. Series no. 147:* 211–230.

Iler, R. K. (1955) *The colloid chemistry of silica and silicates.* Cornell Univ. Press, Ithaca, N.Y., 324 pp.

———. (1973) Effect of adsorbed alumina on the solubility of amorphous silica in water. *Jour. Coll. Interf. Sci.* **43:** 399–408.

Jones, J. B.; and Segnit, E. R. (1971) The nature of opal. I. Nomenclature and constituent phases. *Geol. Soc. Australia Jour.* **18:** 56–68.

Jones, J. B.; and Segnit, E. R. (1975) Nomenclature and the structure of natural disordered (opaline) silica. *Contr. Mineralogy and Petrology* **51:** 231–234.

Jones, M. M.; and Pytkowicz, R. M. (1973) Solubility of silica in seawater at high pressures. *Soc. Royal Sci. (Liège) Bull.* **42:** 118–120.

Kastner, M.; and Keene, J. B. (1975) Diagenesis of pelagic siliceous oozes. *Internat. Sedimentological Congr., 9th, Nice,* **7:** 89–98.

Kates, M.; and Volcani, B. E. (1966) Lipid components of diatoms. *Biochim. et Biophys. Acta* **116:** 264–278.

Kato, K.; and Kitano, Y. (1968) Solubility and dissolution rate of amorphous silica in distilled and sea water at 20°C. *Oceanog. Soc. Japan Jour.* **24:** 147–152.

Keene, J. B. (1975) Cherts and porcelanites from the North Pacific, DSDP Leg 32. In *Initial Reports of the Deep Sea Drilling Project,* Vol. XXXII (eds. R. L. Larson, R. Moberly et al.), pp. 429–507. U.S. Government Printing Office, Washington, D.C.

————. (1976) Distribution, mineralogy, and petrography of biogenic and authigenic silica in the Pacific Basin. Ph.D. Thesis, University of California, San Diego, 264 pp.

Kennedy, G. C. (1950) A portion of the system silica-water. *Econ. Geology* **45**: 629–653.

Kerr, P. F.; Hamilton, P. K.; Pill, R. J.; Wheeler, G. V.; Lewis, D. R.; Borkhardt, W.; Reno, D.; Taylor, G. L.; Mielenz, R. C.; King, M. E.; and Schieltz, N. C. (1950) Analytical data on reference clay materials. *Am. Petroleum Inst. Proc.* **49, 7:** 53.

Knauth, L. P.; and Epstein, S. (1975) Hydrogen and oxygen isotope ratios in silica from the JOIDES Deep Sea Drilling Project. *Earth Planet. Sci. Letters* **25**: 1–10.

Krauskopf, K. B. (1956) Dissolution and precipitation of silica at low temperatures. *Geochim. et Cosmochim. Acta* **10**: 1–26.

Lancelot, Y. (1973) Chert and silica diagenesis in sediments from the Central Pacific. In *Initial Reports of the Deep Sea Drilling Project,* Vol. XVII (eds. E. L. Winterer, J. I. Ewing et al.), pp. 377–405. U.S. Government Printing Office, Washington, D.C.

Lewin, J. C. (1961) The dissolution of silica from diatom walls. *Geochim. et Cosmochim. Acta* **21**: 182–198.

Lier, J. A. van (1959) The solubility of quartz. *Utrecht, Kemink en Zoon,* 54 pp.

Lovering, T. G.; and Patten, L. E. (1962) The effect of CO_2 at low temperature and pressure on solutions supersaturated with silica in the presence of limestone and dolomite. *Geochim. et Cosmochim. Acta* **26**: 787–796.

Lyman, J.; and Fleming, R. H. (1940) Composition of sea water. *Jour. Marine Research* **3**: 134–146.

Mackenzie, F. T.; and Gees, R. (1971) Quartz: Synthesis at Earth-surface conditions. *Science* **173**: 533–534.

Millot, G. (1964) *Géologie des argiles.* Paris, Masson et Cie, 499 pp.

Mizutani, S. (1966) Transformation of silica under hydrothermal conditions. *Nagoya Univ. Jour. Earth Sci.* **14**: 56–88.

Moore, T. C., Jr. (1969) Radiolaria: Change in skeletal weight and resistance to solution. *Geol. Soc. America Bull.* **80**: 2103–2107.

Morey, G. W.; Fournier, R. O.; and Rowe, J. J. (1962) The solubility of quartz in water in the temperature interval from 25°C to 300°C. *Geochim. et Cosmochim. Acta* **26**: 1029–1043.

Morey, G. W.; Fournier, R. O.; and Rowe, J. J. (1964) The solubility of amorphous silica at 25°C. *Jour. Geophys. Research* **69**: 1995–2002.

Murata, K. J.; and Nakata, J. K. (1974) Cristobalitic stage in the diagenesis of diatomaceous shale. *Science* **184**: 567–568.

Murata, K. J.; and Larson, R. R. (1975) Diagenesis of Miocene siliceous shales, Temblor Range, California. *U.S. Geol. Survey Jour. Research* **3**: 553–566.

Oehler, J. H. (1973) Tridymite-like crystals in cristobalitic "cherts." *Nature Phys. Sci.* **241**: 64–65.

Rieck, G. D.; and Stevels, J. M. (1951) The influence of some metal ions on the devitrification of glasses. *Soc. Glass Technology Jour.* **35**: 284–288.

Savin, S. M.; and Douglas, R. G. (1973) Stable isotope and magnesium geochemistry of Recent planktonic foraminifera from the South Pacific. *Geol. Soc. America Bull.* **84**: 2327–2342.

Schink, D. R.; Fanning, K. A.; and Pilson, M. E. Q. (1974) Dissolved silica in the upper pore waters of the Atlantic Ocean floor. *Jour. Geophys. Research* **79**: 2243–2250.

Scholle, D. W.; and Creager, J. S. (1973) Geologic synthesis of Leg 19, Deep Sea Drilling Project results: Far North Pacific, Aleutian Ridge, and Bering Sea. In *Initial Reports of the Deep Sea Drilling Project,* Vol. XIX (eds. J. S. Creager, D. W. Scholle et al.), pp. 897–913. U.S. Government Printing Office, Washington, D.C.

Sclater, J. G.; Anderson, R. N.; and Bell, M. L. (1971) Elevation of ridges and evolution of the central Eastern Pacific. *Jour. Geophys. Research* **76**: 7888–7915.

Siever, R. (1962) Silica solubility, 0°–200°C, and the diagenesis of siliceous sediments. *Jour. Geology* **70**: 127–150.

Stein, C. L.; and Kirkpatrick, R. J. (1976) Experimental porcelanite recrystalization kinetics: A nucleation and growth model. *Jour. Sed. Petrology* **46:** 430–435.

Stumm, W.; and Morgan, J. J. (1970) *Aquatic chemistry, an introduction emphasizing chemical equilibria in natural waters.* Wiley-Interscience, 583 pp.

Taliaferro, N. L. (1934) Contraction phenomena in cherts. *Geol. Soc. America Bull.* **45:** 189–231.

Tarr, W. A. (1917) Origin of the chert in the Burlington limestone. *Am. Jour. Sci.* **44:** 409–452.

van Tuyl, F. M. (1918) the origin of chert. *Am. Jour. Sci.* **45:** 449–456.

Weaver, F. M.; and Wise, S. W. (1972) Ultramorphology of deep-sea cristobalitic chert. *Nature Phys. Sci.* **237:** 56–57.

Weaver, F. M.; and Wise, S. W. (1974) Opaline sediments of the southeastern coastal plain and Horizon A: Biogenic origin. *Science* **184:** 899–901.

Wilson, M. J.; Russell, J. D.; and Tait, J. M. (1974) A new interpretation of the structure of disordered α-cristobalite. *Contr. Mineralogy and Petrology* **47:** 1–6.

Winterer, E. L.; Riedel, W. R.; Bronnimann, P.; Gealy, E. L.; Heath, G. R.; Kroenke, L.; Martini, E.; Moberly, R.; Resig, J.; and Worsley, T. (1971) *Initial Reports of the Deep Sea Drilling Project,* Vol. VII. U.S. Government Printing Office, Washington, D.C., 1757 pp.

Wise, S. W.; Buie, B. F.; and Weaver, F. M. (1972) Chemically precipitated sedimentary cristobalite and the origin of chert. *Eclogae Geol. Helvetiae* **65:** 157–163.

Wise, S. W.; and Kelts, K. R. (1971) Submarine lithification of middle Tertiary chalks in the South Atlantic Ocean basin. *Internat. Sedimentological Congr., 8th, Heidelberg,* 110.

Wise, S. W.; and Kelts, K. R. (1972) Inferred diagenetic history of a weakly silicified deep-sea chalk. *Gulf Coast Assoc. Geol. Socs. Trans.* **22:** 177–203.

Wise, S. W.; and Weaver, F. M. (1973) Origin of cristobalite-rich Tertiary sediments in the Atlantic and Gulf coastal plain. *Gulf Coast Assoc. Geol. Socs. Trans.* **23:** 305–323.

Wise, S. W.; and Weaver, F. M. (1974) Chertification of oceanic sediments. *Internat. Assoc. Sedimentologists' Spec. Pub. 1:* 301–326.

Wollast, R. (1974) The silica problem. In *The Sea,* Vol. 5 (ed. E. D. Goldberg), pp. 359–425. Interscience, New York.

Zemmels, I.; and Cook, H. E. (1973) X-ray mineralogy of sediments from the Central Pacific Ocean. In *Initial Reports of the Deep Sea Drilling Project,* Vol. XVII (eds. E. L. Winterer, J. I. Ewing et al.), pp. 517–559. U.S. Government Printing Office, Washington, D.C.

Part VI

SILICA CYCLE IN THE OCEANS

Editor's Comments
on Papers 17, 18, and 19

17   HEATH
Dissolved Silica and Deep-Sea Sediments

18   LECLAIRE
Hypothèse sur l'origine des silicifications dans les grands bassins océaniques. Le rôle des climats hydrolisants

19   SCHRADER
Fecal Pellets: Role in Sedimentation of Pelagic Diatoms

Papers 17, 18, and 19 deal with the silica cycle in the oceans. Though not strictly part of silica diagenesis, the implications of the silica cycle are important in understanding the distribution in space and time of siliceous sediments and their diagenetic products. In turn, the silica cycle is related to paleotectonic, paleo-oceanographic, and paleoclimatic aspects. Papers 17 and 18 deal with most of these aspects, Paper 19 illustrates a specific and important part of the silica cycle.

The basic question posed by Heath (Paper 17) is whether present-day chemical conditions in the oceans can explain the deposition of ancient nondetrital siliceous sediments. The author discusses and evaluates the various parts of the silica cycle. He divides these into three groups. The first group is the sources of dissolved silica: river influx, submarine volcanism, submarine weathering of sediments and igneous rocks, dissolution of siliceous tests on the sea floor, the upward migration of silica dissolved in interstitial waters through diffusion and compaction. None of these have left a record in the geologic history. The second group is the "sinks": regions of accumulation, silica-adsorption by fine detritus. Only part of these are preserved in the geologic record. The third group comprises the transformation processes: biologic fixation, oxidative regeneration of siliceous tests, nonoxidative dissolution of siliceous tests in the water column. None of these leave a record.

The author tries to quantify these factors to establish a model for the silica cycle. However, he admits that this model can be only a first attempt because there are still many gaps in our knowledge. He bases his model on the concept of a biologically maintained steady-state. The implications of this concept are that changes in silica supply to the oceans are regulated by biologic productivity. At present the oceans are undersaturated with respect to silica. The author doubts whether the oceans ever reached saturation in the past.

Other important aspects Heath discusses in his paper are the association of siliceous sediments and the products of volcanism (compare with Papers 10 to 12) and the theory of "reverse weathering" (the uptake of silica by clay-neoformation).

Comparing Heath's quantitative estimates with Leclaire's (Paper 18) exemplifies the uncertainties still underlying this approach. For instance, the estimated amount of dissolved silica supplied annually to the oceans by the alteration of the igneous oceanic crust is 0.8×10^{14} g in Heath's paper, and 7×10^{14} g in Leclaire's paper. However, Leclaire realizes that his figure, which he obtained from Hart (1973), is based on an alteration of the oceanic crust to a depth of 2 km and thus may well be excessive. Heath's figure is based on alteration to a depth of only 100 m.

Leclaire is concerned mainly with the abundance of chert in Cretaceous and Eocene sediments (compare with Figure 1 and Table 1 in the Introduction to this volume), not only chert deposited in the oceans but also that deposited in epicontinental seas. By analyzing the present-day silica cycle, he tries to find an explanation for this abundance. He suggests that the strong undersaturation with respect to silica of the present-day oceans could possibly be explained by a gradual depletion of silica since the Eocene. The author discusses various possible explanations for the fluctuations in the supply of dissolved silica. According to Leclaire, a change in climatic conditions is the most probable explanation. During Cretaceous-Eocene times, laterite belts, formed under warm and humid climatic conditions, extended to high latitudes (see also Frakes and Kemp 1973). And it is well known from present-day conditions that run-off from laterite belts supplies large quantities of silica to the oceans. The laterite belts retreated to their present-day tropical positions from the Oligocene onwards.

Leclaire discusses another interesting point. The mineral assemblage palygorskite, clinoptilolite, and cristobalite, normally interpreted as the alteration product of volcanic ash, is also characteristic for sediments derived from lateritic terrains. This assemblage seems to be common in Eocene sediments. The author suggests that a reinterpretation of the DSDP data could provide supporting evidence for the climatic control of fluctuations in silica supply.

It is becoming more and more recognized that fecal pellets play an important role in deep-sea sedimentation (see Papers 2, 13, 17, and 19). Fecal pellets have a far greater settling velocity than their individual components. For many species, this may well be the only way their tests can reach the sea floor before dissolving to ensure their preservation in the geologic record. Schrader (Paper 19) describes the role of fecal pellets in the sedimentation of diatom frustules. Fecal pellets from planktonic herbivores have a size range between 50 and 250 mm and consequently can reach settling velocities between 40 and 400 m per day. A single frustule could have a settling velocity of only 10 m per day. Faster settling results in lower dissolution. Moreover, Schrader observed that fecal pellets of certain copepods have a membrane around the diatoms, protecting them further from dissolution. He concludes that the fecal-pellet process enhances the deposition of siliceous sediments. It can also influence the geographic distribution of siliceous sediments, as faster settling prevents the transportation over large distances by ocean currents. And finally, it may account for the undersaturation of silica in sea water.

REFERENCE

Frakes, L. A.; and Kemp, E. M. (1973) Palaeogene continental positions and evolution of climate. In Tarling, D. H., and Runcorn, S. K. (eds.): *Implications of continental drift to the earth sciences*. Academic Press, London. Vol. I: 539–559.

17

Reprinted from pp. 77–93 of *Studies in Paleo-Oceanography*, Soc. Econ. Paleontologists and Mineralogists Spec. Publ. No. 20, W. W. Hay, ed., 1974, 218 pp.

DISSOLVED SILICA AND DEEP-SEA SEDIMENTS

G. ROSS HEATH
Oregon State University, Corvallis

ABSTRACT

Most of the silica dissolved in sea water comes from silica-rich interstitial waters of marine sediments and from rivers carrying the products of subaerial weathering. Silica-secreting microplankton, diatoms and radiolarians, extract enough opal each year to strip the oceans of dissolved silica in about 250 years. Because most of the microscopic tests dissolve rapidly in the water column and at the sea floor, however, only about 4% survive long enough to be buried, and only about 2% avoid post-depositional dissolution and enter the geologic record. Opal-rich sediments are deposited beneath biologically productive surface regions where nutrient-rich deep waters upwell to the photic zone, rather than in areas where the silica enters the sea. The opal-rich deposits ultimately undergo diagenetic transformation to chert. The occurrence of volcanic material with siliceous deposits in the geologic record does not reflect a direct cause-effect relation, but rather the association of volcanism with the same tectonic processes that modify oceanographic and depositional environments so as to induce high productivity of silica-secreting organisms and preservation of the tests in protected basins. At present, 85–90% of the opaline silica incorporated in marine sediments is deposited in near-shore environments where most of it is masked by terrigenous debris. Deposition of silica by inorganic precipitation appears virtually impossible since the late Mesozoic and unlikely since the Precambrian.

INTRODUCTION

At first glance, the sedimentary geochemistries of silica and calcium carbonate have much in common. Both are important constituents of the dissolved load of rivers, both are segregated into relatively pure masses by sedimentary processes, and both are utilized in the construction of tests by important groups of marine microorganisms.

However, the surface water of the open oceans, from which the foraminifers and calcareous nannoplankton extract their tests, is usually saturated and often several times supersaturated with calcite. In contrast, the same seawater may be a thousand times undersaturated with respect to the opaline silica from which the important silica-secreting organisms, the diatoms and radiolarians, construct their tests. The basic geochemical questions are, then, what factors control the concentration and distribution of dissolved silica in the world oceans, why are the silica and carbonate systems so chemically different, and is it possible to interpret ancient non-detrital silica deposits in the light of oceanographic conditions prevailing today?

DISTRIBUTION OF DISSOLVED SILICA IN THE OCEANS

Dissolved silica in seawater has been measured for about a hundred years. However, only since the development of the colorimetric molybdate method (Isaeva, 1958; Riley, 1965) has it been practical to collect the mass of data necessary to map the distribution of silica in the oceans. Even today, the quality of determina-

tions is highly variable. Under optimum conditions (careful analyst, determinations made immediately after collection of samples, and carefully controlled laboratory conditions) the accuracy can be of the order of ±0.03 ppm (concentrations range from below detection level, <0.01 ppm, to about 11 ppm SiO_2).[1] In general, however, comparison of supposedly identical deep-sea profiles determined by different vessels suggests that errors of 10–20 percent are common and 50 percent not unusual. In near-surface waters, where the observed range of concentrations frequently exceeds 0.3 to 3 ppm SiO_2, such errors are not particularly important. In deeper waters, however, where much smaller variations must be used to evaluate vertical and horizontal advections of silica, these errors are serious and largely explain the limited value of silica as an oceanographic tracer relative to other parameters such as temperature, salinity and oxygen concentration.

Figure 1 shows four recent vertical profiles of dissolved silica in the Atlantic and Pacific Oceans. The low surface values result from biological extraction of silica. Increasing concentrations at depth reflect dissolution of falling

[1] The literature on dissolved silica in natural waters is complicated by the use of several units of concentration. Most geochemical papers including this one, use parts per million SiO_2 by weight (ppm). Biological and most oceanographic papers use microgram atoms per liter (μ gm at/l) and rarely micrograms Si per liter. The conversions are:

1 ppm SiO_2 = 16.6 μ gm at/l = 467 μ gm Si/l

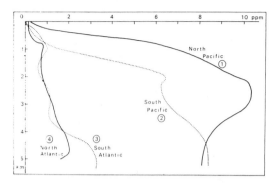

Fig. 1.—Vertical profiles of dissolved silica in the Pacific and Atlantic Oceans. (1) Zetes Stn. 13, 38°05′N, 155°04′W (Scripps Institution of Oceanography, 1970); (2) Eltanin Stn. 28–33, 43°12.6′S, 166°47′W (Scripps Institution of Oceanography, 1969); (3) Circe Stn. 245, 7°8.9′S, 21°21.1′W (Edmond and Anderson, 1971); (4) Baffin Island St. BI-0566–15, 40°51′N, 42°55′W (Grant, 1968).

tests as well as advection of dissolved silica, a subject which will be discussed in more detail in a later section. The striking difference in the concentrations of silica in the two major oceans (fig. 1) reflects the importance of oceanographic phenomena in the marine geochemistry of silica. Berger (1970) has explained this particular difference by pointing out that the Atlantic is "lagoonal" (it looses deep water in exchange for silica-poor surface water from the other oceans) whereas the Pacific is "estuarine" (it gains deep water from other oceans and

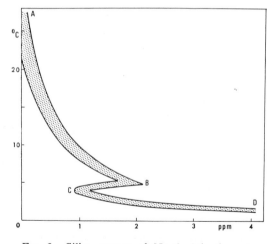

Fig. 2.—Silica content of North Atlantic waters between northeast South America and the Mid-Atlantic Ridge (after Metcalf, 1969). A-B Surface and intermediate water; C North Atlantic Deep Water; D Antarctic Bottom Water.

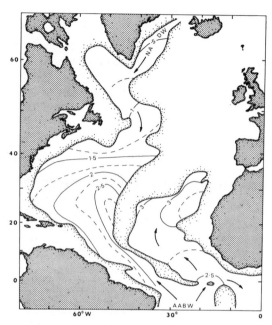

Fig. 3.—Silica content (ppm) and inferred movement of deep Atlantic water along the 2°C potential temperature surface (after Metcalf, 1969). Stippled areas lie above the 2°C surface. NADW North Atlantic Deep Water; AABW Antarctic Bottom Water.

looses surface water that has largely been stripped of silica by opal-secreting plankton). In general, the "oldest" deep waters, that is, those farthest from their last contact with the photic zone, contain the highest silica concentrations. In the Pacific Ocean, in which all the bottom waters come from the south, there is a south to north increase in silica along isopleths of thermosteric anomaly (roughly, paths of constant potential energy). In the Atlantic, deep water originates both in the arctic and antarctic, and the pattern of silica distribution is more complex. Figure 2, based on about 1500 determinations between northeastern South America and the Mid-Atlantic Ridge (Metcalf, 1969) shows strikingly the influence of hydrography on silica concentration. The maximum at about 5°C is due to Antarctic Intermediate Water that has moved north from the antarctic convergence at about 50°S, the minimum at about 4°C lies in the North Atlantic Deep Water that has moved south from the east coast of Greenland, and the silica-rich coldest and deepest water is Antarctic Bottom Water that originated in the Ross Sea area of Antarctica.

Clearly, any explanation of the marine geochemistry of silica must allow for the dynamic nature of the oceans. Figure 3, based on a criti-

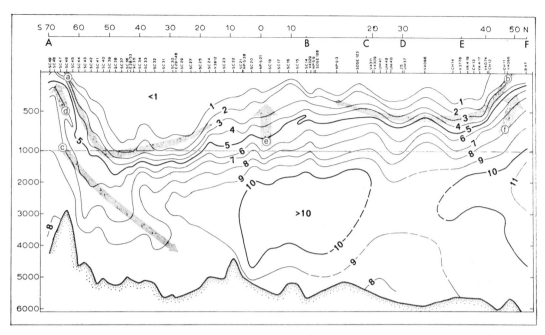

Fig. 4.—Vertical north-south profile of the distribution of dissolved silica (ppm) in the Pacific Ocean (location of profile and stations used shown on Fig. 5). Vertical scale in m, note scale change at 1000m. Station abbreviations: B, Boreas; CH, Chinook; E, Eltanin; NP, Norpac; SC, Southern Cross; SDSE, Swedish Deep Sea Expedition; UM, Ursa Major; V, Vitiaz; VK, Voeikov; Z, Zetes. Original data adjusted for consistency with Eltanin data, as far as possible, but there is probably some mismatch north of B. Stippled arrows show principal water movements determined from other oceanographic parameters (see, for example, Reid, 1965). (a) and (b) Antarctic and Arctic Intermediate Water; (c) Antarctic Bottom Water; (d) upwelling at the Antarctic divergence; (e) upwelling at the equatorial divergence; (f) upwelling south of the Aleutian arc.

cal compilation of deep North Atlantic data by Metcalf (1969) shows the movement of water along the 2°C potential temperature[2] surface, and the corresponding variations in dissolved silica. Such a pattern results from the influx of silica-depleted water from both north and south. At the present time, of the major oceans only the Atlantic displays this pattern.

The distribution of dissolved silica in the Pacific is distinctly different. A longitudinal profile such as figure 4 (location in fig. 5) shows the generally south to north increase in concentration which occurs at any level, but is particularly clear in the deep waters. Overall, the silica pattern mirrors water movements deduced from other criteria (see, for example, Reid, 1965). Upwelling in the antarctic (d), equatorial (e) and arctic (f) areas is marked by rises in the isopleths of silica concentration.

[2] The potential temperature is the temperature the water would have if brought to the surface without loss or gain of heat. Because of adiabatic compression, the *in situ* temperature is about 0.5°C higher in this case.

The surface central water masses, with their deep, stable thermoclines are low in silica due to biologic extraction, particularly in the South Pacific where the "oldest" surface water is drifting south to rejoin the circum-Antarctic mixing system. Perhaps the most convincing evidence in figure 4 for the influence of advection on the distribution of dissolved silica in deep water is the pattern along arrow (c). This arrow, which marks the salinity maximum of the core of the Antarctic Bottom Water (Reid, 1965), coincides with a lobe of relatively silica-poor water that is being carried towards the equator.

POSSIBLE CONTROLS OF THE CONCENTRATION
OF DISSOLVED SILICA

Solubility of Solid Silica Species

Figure 6 summarizes the solubility of various silica-bearing phases in seawater. As pointed out by numerous workers, and summarized by Krauskopf (1959), natural seawater at atmospheric pressure is never at equilibrium with amorphous or opaline silica. The effect of pressure on the

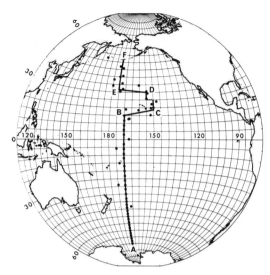

FIG. 5.—Location of stations used to construct Fig. 4, and position of the N-S profile.

Siever (1968b) has suggested that "reverse weathering" is largely restricted to the deeper portions of thick sediment sections along continental margins, where relatively high temperatures and pressures promote intense diagenesis. Finally, the distribution patterns of figures 1 and 4 are readily related to gross oceanic circulation and regions of biological productivity, but make little sense if clay-water interactions control the concentration of silica. The Atlantic Ocean, for example, receives more suspended clay-sized sediment per unit area from the surrounding continents than the other major oceans. However, its concentration of dissolved silica (fig. 1) is markedly less than the experimental values of figure 6 or the values observed in interstitial water from clayey Atlantic sediments (Bischoff and Ku, 1970, 1971). It seems that the clay minerals react too slowly and release too little silica to mask or even modify the distribution pattern resulting from

equilibrium solubility deep in the oceans is slight —at 5000 m, the solubility is only about 10% greater than its surface value at the same temperature (Jones and Pytkowicz, 1973).

Although Eocene opal and cristobalitic deep-sea cherts are less soluble than fresh amorphous silica, they should still readily dissolve in natural seawater. Of the solid silica phases, only quartz is at saturation. The slow equilibration of quartz with water and the complex vertical and horizontal variations of dissolved silica concentrations within the oceans suggest that this phase exerts little control over the silica cycle.

Solubility of Silicates

Within the past decade, Sillén (1961), Siever (1968a) and Mackenzie, Garrels and co-workers (Garrels, 1965; Mackenzie and Garrels, 1965, 1966a, b; Mackenzie and others, 1967) have emphasized the importance of reactions involving clay minerals in the weathering of silicates and in the control of the pH of seawater over geologic time. They propose that "reverse weathering," involving the uptake of silica by clay minerals in seawater, is an important geochemical process tending to maintain the concentration of dissolved silica at its present low value. The short-term importance of this process has been seriously questioned on several grounds. Pytkowicz (1967) has convincingly argued that over periods corresponding to the circulation times of the oceans (hundreds to one or two thousand years), pH is controlled by the carbonate system. In the same vein,

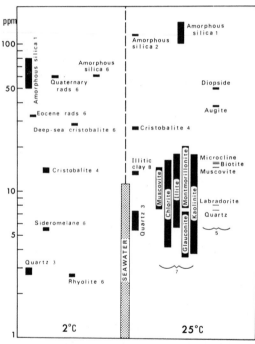

FIG. 6.—Solubility of common silica-bearing minerals at 2° and 25°C, compared to the range of dissolved silica concentrations observed in the oceans (stippled bar). Sources of data: 1—Krauskopf (1959); 2—Morey and others (1964); 3—Morey and others (1962); 4—Fournier and Rowe (1962); 5—Keller and others (1963); 6—Jones and Pytkowicz (1973), Jones (pers. comm.); 7—Mackenzie and Garrels (1965); Mackenzie and others (1967), Jones (pers. comm.); 8—Fanning and Schink (1969). References 2 to 5 refer to fresh water, but are probably also applicable to sea water at the scale plotted.

biological activity (next section) and oceanic circulation.

Biological Activity

If one compares a map of modern deep-sea siliceous sediments (Lisitzyn, 1967, 1972) with almost any measure of biological productivity in surface waters (including such diverse parameters as chlorophyll concentration, zooplankton biomass, or fish catches), the similarity is striking. Clearly, conditions that encourage the synthesis of protoplasm also favor the fixation of opaline silica. As we have seen from figure 4, high concentrations of near-surface silica (as well as nutrients such as phosphate and nitrate) occur in regions of upwelling. These regions lie along continental margins (where their impact on sedimentation is usually masked by terrigenous deposition), and along subarctic, equatorial, and subantarctic latitudinal belts. The latitudinal belts are underlain by deposits rich in calcitic and opaline tests. The accumulation of siliceous oozes requires not only high surface productivity, but also removal of calcite by solution. Thus, such oozes characteristically are found below the calcite compensation depth (a few hundred meters near Antarctica to about five kilometers in the equatorial Pacific).

Since siliceous microplankton can be found in the surface waters of all the oceans, the complete absence of opaline tests from large areas of the sea floor indicates that much of the biologically fixed silica re-dissolves rather than enters the sedimentary record.

THE SILICA CYCLE IN THE OCEANS

Figure 7 shows a number of sources, sinks and transformation processes that might be expected to influence the marine geochemistry of silica. The sources, river influx (a), submarine volcanism (b), submarine weathering (halmyrolysis) including dissolution of siliceous tests prior to burial (c), and upward migration of silica from interstitial waters (d), leave no mark on the geologic record and must be inferred from actualistic data and gross silica budgets of the past. The sinks, regions of accumulation of siliceous tests (h) and of fine detritus that has adsorbed some dissolved silica (i), are the portions of the system than can be preserved and from which past silica cycles must be deduced. Finally, transformations and cycling within the silica system (e, f, g) can be studied only in today's oceans. Our ability to reconstruct plausible ancient systems, therefore, depends exclusively on our understanding of the dynamics and stability of the present system.

In the following sections, an attempt is made

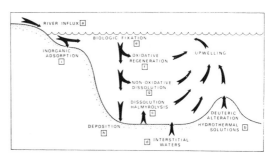

Fig. 7.—Components of the cycle of dissolved silica in the oceans. Letters refer to subdivisions discussed in the text.

to evaluate quantitatively factors (a) through (i) of figure 7. As will become obvious, there are still critical gaps in our knowledge, but a useable model does appear attainable in the not-too-distant future.

I. Sources of Dissolved Silica

River influx.—According to Livingstone (1963) river water, on the average, contains 120 ppm dissolved solids. This is equivalent to an annual supply of 3.91×10^{15} gm to the oceans. Dissolved silica forms 13.1 ppm, which is equivalent to an influx of 4.27×10^{14} gm/yr. More recent data (Gibbs, 1967) suggest that this value may be a few percent high, but such an error will not influence the crude calculations in the following sections.

Submarine volcanic sources.—Because oceanic volcanism is basic in character, it has never been considered a likely source of much silica in today's oceans. However, the association of bedded cherts with large masses of volcanic material (such as spilites) in the geologic record has been taken as evidence of a genetic relationship. Possible factors responsible for this association are considered in a later section. At the present time, primary quartz is known in sediments from the crest of the East Pacific Rise (Peterson and Goldberg, 1962) and from Henderson seamount (J. Dymond, personal communication), suggesting that some silica is supplied to the ocean both at lithospheric plate divergences (mid-ocean ridges) and from the hot spots beneath oceanic seamounts. Such silica could be released in a volatile aqueous fraction left after crystallization of the basalt, or could be freed by deuteric alteration of hot lava by seawater entering fractures resulting from thermal contraction. Both mechanisms are unsupported by experimental or field evidence. Although the work of Corliss (1970) who compared the geochemistry of quenched margins

and slowly cooled interiors of deep-sea tho-
leiites, strongly supports the deuteric release to
seawater of a number of transition metals, any
suggestion that silica is released by the same
mechanism is pure speculation in the absence of
analytical data.

Clearly, there is not yet enough quantitative
information on the subject to support a rigorous
model. However, the possible importance of volcanic silica in our system can be estimated from
the data we do have. The area of new sea floor
created at mid-ocean ridges is about 2 km²/yr
(the figure is about the same regardless of
whether a long term average based on a 200
million year lifespan for exposed sea floor, or a
short term average calculated from the average
spreading rate at active ridges is used). If the
new crust is accessible to alteration to a depth
of 100 m and has a specific gravity of 2.7 the
volume of basalt subject to deuteric alteration
each year is:

$$2 \times 10^{10} \times 10^4 \times 2.7 = 5.4 \times 10^{14} \text{ gm.}$$

Loss of 0.01 to 1 percent silica from this mass
would add 5.4×10^{10} to 5.4×10^{12} gm/yr to
the oceans—about 1/100 to 1 percent of the an-
nual river influx. Even allowing for substantial
errors in the figures used in the calculations
and adding in the effects of eruptions away
from spreading ridges, it is clear that submarine
volcanism cannot be a major source of dissolved
silica in today's oceans.

Low temperature reactions at the sea floor.—
Processes tending to release silica at the sea
floor can be grouped into three main classes:
dissolution of opaline silica tests; low temperature alteration (halmyrolysis) of oceanic basalt; and halmyrolysis of detrital silicate particles.

At the present time, dissolution of opaline
tests at the sea floor cannot be distinguished
from non-oxidative dissolution within the water
column (g of fig. 7). Such a distinction requires further experimental work on the rates
of settling and dissolution of opal in the ocean,
and a better understanding of the near-bottom
circulation. For our model, the two dissolution
components will be considered together under
(g).

In a recent paper, Hart (1970) has studied
the chemical changes accompanying low-temperature alteration of oceanic tholeiitic basalts
over periods of millions of years. For silica, he
suggests an average annual loss of 7.4×10^{-9}
gm/cm³. If we assume that such alteration proceeds to a depth of 100 m and supplies silica to
the ocean only in areas of thin or no sediments
(no more than one third of the deep ocean

basins or 1.1×10^{18} cm²), the annual release is:

$$7.4 \times 10^{-9} \times 1.1 \times 10^{18} \times 10^4$$
$$= 8.1 \times 10^{13} \text{ gm/yr}$$

Thus, although not the major source of silica in
the ocean, halmyrolysis of basalts appears ca-
pable of yielding about 20 percent of the amount
annually supplied by rivers.

Low-temperature alteration of detrital material does not appear to be a major source of
dissolved silica. Keller and others (1963)
crushed a number of common silicate minerals
and determined the concentration of silica (and
other elements) in the supernatant liquids after
equilibration (table I). If the values are reduced to allow for a bottom-water temperature
of 2°C rather than the room temperature (about
20°C) of the experiments, it appears that only
the pyroxenes of the common detrital minerals
could supply much soluble silica. The persistence of pyroxenes in Tertiary deep-sea deposits
(Heath, 1969) suggests that little alteration resulting from loss of silica prior to burial actually occurs. It should, perhaps, be emphasized
here that the effects of pressure or the use of
seawater rather than distilled water on the values cited, and the nature of the reactions involved are unknown.

Schutz and Turekian (1965) proposed that
glacigene rock flour from Antarctica supplied
as much dissolved silica to the oceans as all the
rivers of the world. Their estimate was based
on the difference between very poorly known
rates of supply of rock flour and of deposition
of glaciomarine sediments in the circum-Antarctic region. While the value of 5×10^{15} gm/
yr for glacio-marine sedimentation around Antarctica has not been revised (despite the availability of the core data from R/V ELTANIN),
a recent paper by Warnke (1970) suggests that
intense glacial erosion of Antarctica occurred

TABLE 1.—SILICA RELEASED TO SOLUTION BY MINERALS
WET GROUND IN DISTILLED WATER AT ROOM TEMPERA-
TURE (10 GMS SOLID TO 100 ML WATER)
(AFTER KELLER AND OTHERS, 1963)

Mineral	Ppm
Quartz	7.5
Labradorite	8.1
Hornblende	10.7
Olivine	12.8
Muscovite	14.8
Biotite	15.6
Microcline	15.8
Enstatite	35.3
Augite	37.9
Diopside	49.8

only during the early development of the icecap, and that modern glaciers are agents of protection rather than erosion. Thus, Schutz and Turekian's value of 14×10^{14} gm/yr for glacial rock flour derived from Antarctica may well be too large, and consequently, their mass imbalance and "missing silica" may not exist. Further evidence against the ability of glacial rock flour to add large quantities of silica to seawater comes from Muir Inlet, a branch of Glacier Bay, Alaska. The inlet, which is more than 300 m deep, is fed by outflow from five glaciers and is visibly turbid. The water column is stably stratified during the summer, with its salinity increasing from 2.6 percent at the surface to 3.1 percent at depth. Thus, below sill depth (60 m) the deep water does not suffer biologic loss of silica over a period of several months. In the fall, the concentration of glacially derived silica in the inlet should be at its highest value, yet a vertical profile of seven samples collected below sill depth by R/V YAQUINA late in August 1970 contained only 1.6 to 2.0 ppm dissolved silica.

Release of interstitial water from deep-sea sediments.—Over the past decade, an enormous number of silica determinations have been made on interstitial waters from deep-sea sediments (see, for example, Bischoff and Ku, 1970, 1971; Bruyevich, 1966; Fanning and Schink, 1969; Harriss and Pilkey, 1966; Kaplan and Presley, 1970; Manheim and others, 1970; Manheim and Sayles, 1971; Presley and Kaplan, 1970, 1971, 1972; Presley and others, 1970; Sayles and others, 1970, 1972; Siever and others, 1965). Although integration of the data is complicated by non-uniform equipment and experimental techniques, it seems clear that most interstitial waters are richer in silica than the overlying seawater. The median concentration of the nine-hundred or so determinations available to the author is close to 25 ppm, with geographic variations ranging from a 19 ppm median for the Atlantic to 60 ppm for interstitial waters from the Gulf of California and Bering Sea diatomaceous oozes.

Because the waters analysed so far do not represent a systematic sampling of deep-sea sediments (the Indian and Southern Oceans are inadequately represented, for example), and the extraction pressures and temperatures, and analytical techniques vary from study to study, it would be unwise to place too much faith in the figures quoted above (see, for example, Fanning and Pilson (1971) for a discussion of the excessively high values that can result if interstitial waters are extracted above *in situ* temperatures). Nevertheless, the existence of a concentration gradient across the sediment-water interface, which must force upward diffusion of dissolved silica, as well as the physical expulsion of interstitial water during the consolidation of sediments must inevitably add dissolved silica to the oceans. The question is, how much?

For the simple situation where reservoirs of interstitial and bottom waters differ in silica content by dS gm/cm^3, the flux of silica across the interface (Fs) is given by $Fs = -k_s \, dS/dz$. Here, k_s is an effective diffusion coefficient for dissolved silica which we will allow to cover molecular and eddy diffusion and minor advection (minor, because sedimentation rates in the pelagic realm are slow relative to diffusion rates), and dz is the thickness of the layer separating the two reservoirs. Schink (1968) reports a value of 4 cm for dz, which does not conflict with other data although the vertical spacing of most interstitial water samples has been too large to permit a good estimate of this parameter. The value of k_s in the sediments of interest is unknown, but a range of $(1.5$ to $3) \times 10^{-6}$ cm^2/sec appears conservative (see Bender (1971) and Wollast and Garrels (1971) for a discussion of this parameter). Similarly, a value of 10 ppm $(10^{-5}$ gm/cm^3) appears conservative for dS.

If we use these estimates, the flux of dissolved silica from deep-sea sediments to the ocean is:

$$(1.5 \text{ to } 3) \times 10^{-6} \times 3.2 \times 10^{18} \times 10^{-5} \times \tfrac{1}{4} \times 3.16 \times 10^{7}$$
$$= (3.8 \text{ to } 7.6) \times 10^{14} \text{ gms/yr}$$

or as much as twice the annual influx from rivers. Although the numbers used in the calculation of silica entering the oceans from below are far from adequately determined, it appears very unlikely that new data can reduce the range of $(3.8$ to $7.6) \times 10^{14}$ gm/yr to the point where its influence on the silica cycle can be ignored. The author suspects that the value may well be substantially increased as our understanding of deep-sea geochemistry improves. For example, the calculation made here ignores any dissolution within the diffusion zone, a simplification clearly at variance with observations of the degradation of siliceous tests in the upper few centimeters of cores from areas of slow sedimentation such as the red clay zones of the North and South Pacific.

II. Transformation of Silica within the Ocean

Biological fixation.—Several taxa of marine organisms extract silica from seawater and use it to construct opaline tests. Two of the taxa, the diatoms (Bacillariophyta) and silicoflagellates (Chrysophyta) are primary producers whereas radiolaria and sponges, the remaining important users, occupy higher trophic levels and, consequently, display more complex distribution patterns. Diatoms are the dominant silica users in the ocean, and extract at least an order of magnitude more opal per year than the radiolaria. Vinberg (in Lisitzyn and others, 1967) estimates that diatoms are responsible for 70 percent of the phytoplankton productivity in the oceans. Recently, however, experiments by Watt (1971), using rates of uptake of radiocarbon, suggest that the role of nannophytoplankton (dinoflagellates, etc.) may have been underestimated because of failure to recognize that their rapid metabolic rates compensate for their small biomass in most samples. Further experimental work, particularly at a number of latitudes in open ocean areas of high productivity, is needed.

The net primary productivity of the world's oceans is probably in the range 1.5 to 2 $\times$ 10^{16} gm carbon/yr (Ryther, 1959; Vinberg, in Lisitzyn and others, 1967). Numerous analyses of plankton by Russian workers (Lisitzyn and others, 1967) show a mean silica : organic carbon ratio of 2.3 : 1 in diatoms. If the analytical ratio reflects relative rates of fixation of the two components, then for the case where diatom productivity is 50 to 70 percent of total primary productivity, the mass of silica extracted per year is:

$$\frac{(1.5 \text{ to } 2) \times 10^{16}}{1} \times \frac{2.3}{1} \times \frac{(50 \text{ to } 70)}{100}$$

$$= (1.7 \text{ to } 3.2) \times 10^{16} \text{ gm/yr}$$

or 40 to 75 times the annual river influx. This range, which lies between the earlier estimates of 0.77 $\times$ 10^{16} gm/yr of Harriss (1966) and 8 to 16.1 $\times$ 10^{16} gm/yr of Lisitzyn and others (1967), must remain tentative until the relative rates of fixation of carbon and silica are better understood. Nevertheless, there can be little doubt that the rate of transformation of silica within the ocean greatly exceeds the rate of influx or removal (assuming a steady state in the system). Figure 8, modified from Lisitzyn and others (1967), shows the rate of biologic uptake of silica in the surface waters of the oceans. This figure emphasizes once again the prime role played by oceanographic factors (in this case, upwelling) in controlling the behavior of silica in the marine environment.

Oxidative regeneration.—Most of the opaline frustules constructed by diatoms during their growth phase are extremely fragile and rapidly disintegrate once their protective organic covering disappears. Gilbert and Allen (1943) recorded decreases of up to 400 times in the concentrations of diatoms at 200 m relative to overlying surface values in the Gulf of California (fig. 9). Chumakov (in Lisitzyn and others, 1967) and Kozlova (1961) report similar decreases with depth in near-surface samples from the Southern Ocean. Because fragile species are much more susceptible to destruction than species with robust frustules (Calvert, 1966), the mass fraction of biogenic silica redissolved in the upper few hundred meters of the water column is less than the cell counts would suggest. Nevertheless, it appears that at least 90 percent of the opaline tests formed in the ocean will never be preserved in the geologic record, regardless of how ideal the depositional conditions.

This rapid post-mortem dissolution of fragile siliceous tests is here referred to as oxidative regeneration because it seems to be controlled primarily by the oxidation of protoplasm enclosing the tests. Figure 10 (after Berger, 1970) shows that the increase of dissolved silica with depth in the ocean is partly correlated with dissolved phosphate (a direct product of the oxidation of organic matter), and partly independent of this nutrient. The correlation is largely restricted to the upper kilometer of the water column. Dissolution of silica at depth appears independent of other reactions in the ocean.

The quantitative importance of oxidative regeneration of silica cannot be directly estimated at present because of inadequate knowledge of the "age" of waters directly below the photic zone. Without this time parameter, the concentration information from vertical profiles, such as figure 10, cannot be converted to rate information. From our estimates of the other components of the silica cycle, and assuming no net loss or gain of dissolved silica (i.e. steady state conditions), it appears that (1.2 to 2.8) $\times$ 10^{16} gm/yr of silica redissolve during oxidative regeneration.

Non-oxidative dissolution.—As mentioned in the introductory section of this paper, the silica content of deep ocean waters rises steadily with time as long as the waters remain below the photic zone. The source of this silica is largely the dissolution of opaline tests falling through the water column or resting on the sea floor. The deep waters are everywhere undersaturated relative to opal, so that dissolution must be rate limited, rather than governed by chemical equilibrium. Experiments by Berger (1968) and Bogoyavlenskiy (1967) suggest that the rate of dissolution depends on the turbulence of the water (see also Edmond, 1970), and is unaffected by the variations in concentration of dissolved silica found in the oceans. Bogoyavlenskiy (1967) suggests that the solution rate depends only on the concentration of suspended opaline silica. His evidence for such a simple relation is meager and largely derived from laboratory experiments. Intuitively, one suspects that the surface area of the solid phase and the agitation of the water (necessary to maintain maximum concentration gradients against silica particles) are also key factors. There is no evidence that either the rate of dissolution or concentration of dissolved silica in the oceans is influenced by observed variations in pH.

In his study of the influence of deep-sea circulation on the distribution of biogenic sediments, Berger (1970) used phosphate-silica plots like figure 10 to estimate the non-oxidative silica content of deep waters of the various oceans. These data, in combination with estimated "ages" or residence times of the deep-

waters allowed him to calculate a rate of about 4.3×10^{15} gm/yr for non-oxidative addition of silica to the oceans. Berger's "ages" are based on C^{14} dating of dissolved carbonate extracted from deep water samples. Craig (1969) has pointed out that such ages ignore vertical advection of the C^{14} in particulate matter, and must be considered questionable. In the absence of better estimates, however, we will use Berger's rates and simply emphasize that this is one area where new data are needed.

Berger (1970) considered that his 4.3×10^{15} gm/yr represented non-oxidative dissolution of biogenic silica in the water column or at the sea floor. In our model it is clear that this figure includes, in addition, halmyrolysis of non-biogenic silicates, escape of interstitial dissolved silica, and influx of silica of volcanic origin (b, c, d, and g of fig. 7). Using the values estimated in preceding sections we can calculate that the rate of non-oxidative dissolution of biogenic silica in the water column and on the

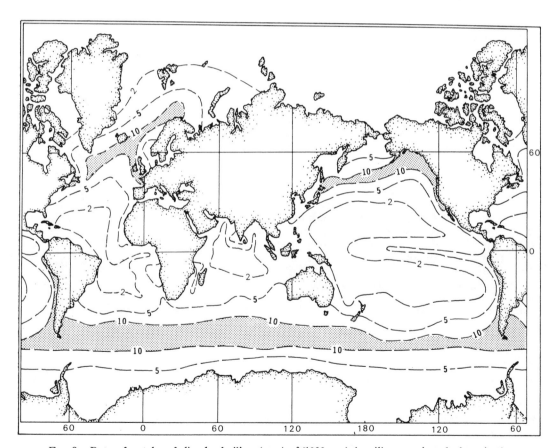

F₁ɢ. 8.—Rate of uptake of dissolved silica (gm/cm²/1000 yrs) by siliceous microplankton in the surface waters of the oceans. Modified after Lisitzyn and others (1967).

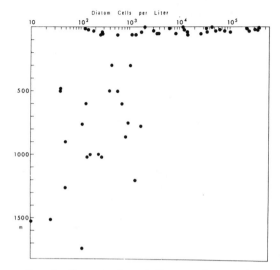

FIG. 9.—Concentration of diatom cells as a function of depth in the central Gulf of California. Data from Gilbert and Allen (1943).

sea floor prior to burial is:

Berger's value—volcanic input—vertical flux of dissolved interstitial silica—halmyrolytic input, or:

$$4.3 \times 10^{15} - 5.4 \times 10^{12} - (3.8 \text{ to } 7.6) \times 10^{14} - 8.1 \times 10^{13}$$
$$= (3.5 \text{ to } 3.8) \times 10^{15}$$

A number of independent sources of evidence suggest that fine detritus, including opal, settles through the oceans much faster than would be predicted for individual particles by Stoke's Law. The identification of fine radioactive debris at depths of several kilometers a few weeks after nuclear tests (Gross, 1967), and the virtual coincidence of boundaries between microfossil assemblages in surface waters and underlying sediments (Ruddiman, 1968) point to rapid deposition of fine particles, probably as fecal pellets of planktonic herbivores and carnivores. In view of the slow rate at which opaline silica dissolves, it appears that most of the dissolution occurs at the sea floor, rather than in the water column.

III. Sinks for Dissolved Silica

Deposition of biogenous opal.—Modern authors are almost unanimous in accepting that the input of dissolved silica to the oceans is largely balanced by the deposition of biogenous opaline tests. Only Harriss (1966), who suggests that the oceans are being stripped of silica, and Mackenzie, Garrels, and co-workers (Mackenzie and others, 1967; *see also* Burton and Liss, 1968) who favor inorganic reactions

between phyllosilicates and dissolved silica disagree with the concept of a biologically maintained steady state. Calvert (1968) has recently reviewed the alternatives and has emphatically reaffirmed the key role of siliceous microfossils in the silica system.

Direct determinations of the rate of deposition of biogenous opal in the deep ocean basins are still few and far between. In part, this lack of data is being overcome by large numbers of new determinations of deep-sea sedimentation rates by paleomagnetic-biostratigraphic and to a lesser extent radiometric dating techniques. However, the paucity of accurate opal determinations on dated cores, and the huge areas still to be studied suggest that reasonable quantitative estimates of global distribution patterns are still several years away. At this stage we can only balance input against removal to arrive at a deposition rate (excluding dissolution prior to burial) of:

River influx+volcanic influx+halmyrolysis+flux of interstitial silica−inorganic adsorption (next section) or:

$$4.27 \times 10^{14} + 5.4 \times 10^{12} + 8.1 \times 10^{13}$$
$$+ (3.7 \text{ to } 7.6) \times 10^{14} - 4.3 \times 10^{13}$$
$$= (0.85 \text{ to } 1.2) \times 10^{15} \text{ gm/yr.}$$

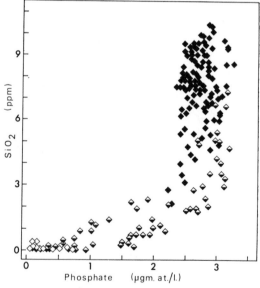

FIG. 10.—Relation of dissolved silica to dissolved phosphate in water samples from the central Pacific. Open diamonds—samples from the photic zone (less than 100 m) that are largely stripped of silica; half shaded diamonds—samples from 100–1000 m, where silica correlates with phosphate due to oxidative regeneration of biogenous matter; solid diamonds, samples below 1000 m from the region of non-oxidative dissolution. After Berger (1970).

Clearly, the "deposition rate" is strongly dependent on an arbitrary decision as to when an opal particle leaves the internal transformation portion of the silica cycle and becomes "sediment." The definition used here, that once buried a particle is effectively "deposited" seems reasonable to the author, but other definitions are possible and are considered in a later section.

Inorganic adsorption by detrital particles.—
In a classic paper Bien and others (1959) showed that part of the dissolved silica load of the Mississippi is lost by reaction with particulate matter as the river enters the Gulf of Mexico. In a comprehensive laboratory study, they showed that the process required both suspended matter and electrolytes.

Sterilization of the reactants did not inhibit the reaction. Bien and others (1959) claimed that this inorganic reaction was responsible for a major portion of the loss of soluble silica observed at the mouth of the Mississippi. However, careful examination of their data suggests that such a conclusion is not warranted. Figure 11 shows the silica loss as a function of percentage of seawater in the system (1) observed in the field, (2) observed in the laboratory, and (3) the difference between (1) and (2) representing the loss that is not observed in a sterile system. Clearly, when curves (2) and (3) sum to 100 percent (at about 85% seawater) no further silica can be lost. Process (3) is probably biogenic uptake, and removes 50 percent of the silica even if the inorganic process is independent of the concentration of dissolved silica. In fact, the inorganic uptake (process 2) probably decreases as the silica concentration drops below the laboratory values, so that process (3) actually strips most of the dissolved silica from Mississippi waters. Schink (1967) has suggested that the field results of Bien and others (1959) are erroneous because they used an excessively high value for the salinity of Gulf of Mexico waters. This criticism is not substantiated by convincing data, and is opposed by the laboratory data.

Since the Mississippi study, Liss and Spencer (1970) have reported a 10–20% abiotic loss of dissolved silica from the River Conway in Wales, whereas Stefánsson and Richards (1963) have proved the absence of a similar loss off the Columbia River. Negative findings have also been reported by Burton and others (1970), Kobayashi (1967) and Maéda (1952). The inconsistency of the various studies probably results from differences in the mineralogy of suspended matter in the rivers studied. This must remain pure conjecture, however, until a

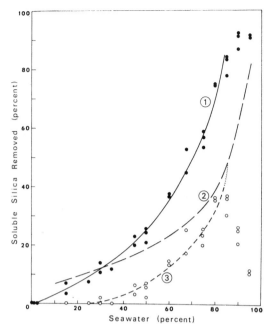

Fig. 11.—Loss of silica in excess of simple dilution from Mississippi River water entering the Gulf of Mexico, after Bien and others (1959). Solid circles and curve (1)—field data; curve (2)—laboratory data for sterile solutions; open circles and curve (3)—field data less curve (2). Curve (3) represents the minimum silica that must be removed by processes other than the inorganic adsorption described by Bien and others.

definitive comparative study is made or the actual mechanism by which the silica is removed is better understood.

For the purposes of our model, an arbitrary loss of 4.3×10^{13} gm/yr of dissolved silica is attributed to inorganic adsorption. This value, about 10 percent of the total river influx, is probably within a factor of two of the correct value, and in any case is not a dominant component of the silica cycle.

DISCUSSION

The estimated values of various components of the silica cycle are summarized in table II. For the sake of argument, one third of the non-oxidative dissolution is assumed to occur in the water (g) and two thirds at the sea floor (c2).

The figures in the last column of table II are constrained in the following ways (letters refer also to fig. 7):

1. A steady state is assumed, that is to say supply = removal, or
   $a + b + cl + d = h + i$
2. The biogenic subcycle must balance
   $e = f + g + h + c2$

TABLE 2.—APPROXIMATE MAGNITUDES OF COMPONENTS
OF THE SILICA CYCLE OF THE OCEANS IN UNITS OF
10^{13} GM/YR. LETTERS REFER TO FIGURE 7. THE
NUMBER OF SIGNIFICANT FIGURES RESULTS FROM
MATERIAL BALANCE CALCULATIONS AND NOT
FROM REAL PRECISION OF THE DATA

Components	From preceding sections	Used in discussion section
(a) River influx	42.7	42.7
(b) Deuteric/hydrothermal influx	0.004–0.54	0.54
(c) Halmyrolysis of basalt (c1)	8.1	8.1
Dissolution of tests on the sea floor (c2)	230–250	240
(d) Escape of interstitial silica	38–76	57
(e) Biologic fixation	1700–3200	2500
(f) Oxidative regeneration	1200–2800	2030
(g) Non-oxidative dissolution	120–130	125
(h) Deposition and burial of opaline tests	85–123	104
(i) Inorganic adsorption	4.3	4.3

3. Berger's (1970) value for the non-oxidative supply of silica to deep waters must be conserved:

$$b + c + d + g = 430 \times 10^{13} \text{ gm/yr.}$$

It need hardly be emphasized that these constraints only ensure that the model is internally consistent—they cannot control its relation to reality. If the final column of table II is accepted as a first approximation of the silica cycle, the values shown can be used to derive a number of secondary parameters.

RESIDENCE TIME

The residence time of a transient component in any reservoir (assuming a steady state) is given by: $\tau = A/(dA/dt)$, where A is the mass of the component in the reservoir and dA/dt is the rate of supply or removal of the component. For dissolved silica in the ocean, A is $(5.4 \text{ to } 8.1) \times 10^{18}$ gm (assuming an average concentration of 4 to 6 ppm) and dA/dt depends on the definition of "removal" or "supply." The following alternatives are of interest geochemically:

1. $\dfrac{dA}{dt}$ = biologic uptake (e)

$$\tau = \frac{(5.4 \text{ to } 8.1) \times 10^{18}}{2.5 \times 10^{16}} = 220 \text{ to } 320 \text{ years}$$

2. $\dfrac{dA}{dt} = \dfrac{\text{rate of deposition of silica on the sea floor}}{(h+i+c2)}$

$$\tau = \frac{(5.4 \text{ to } 8.1) \times 10^{18}}{(104 + 4.3 + 240) \times 10^{13}} = 1600 \text{ to } 2300 \text{ years}$$

3. $\dfrac{dA}{dt}$ = rate of burial of solid silica at the sea floor (h+i)

$$\tau = \frac{(5.4 \text{ to } 8.1) \times 10^{18}}{(85 \text{ to } 125) \times 10^{13}} = 4300 \text{ to } 9500 \text{ years}$$

4. $\dfrac{dA}{dt} = \dfrac{\text{rate of addition of silica to the geologic record}}{(h+i-d)}$

$$\tau = \frac{(5.4 \text{ to } 8.1) \times 10^{18}}{51 \times 10^{13}} = 11,000 \text{ to } 16,000 \text{ years}$$

Clearly, on a geologic time scale, the silica content of the oceans is not large enough to suppress the effects of changes in the supply or removal of silica due to variations in tectonic, climatic, or oceanographic conditions. In particular, the rate of biologic uptake of silica will react virtually instantaneously to changes in the system.

SEDIMENTATION RATES

As with the residence time, the sedimentation rate of silica depends on the definition of "sedimentation." Using options 2, 3 and 4 of the preceding section, a dry bulk density of 0.2 gm/cm³ for siliceous ooze, and deposition over 50% of the deep sea floor (1.6×10^{18} cm²), the values are:

1. For the rate of arrival of opal at the sea floor (h+i+c2):

$$\frac{3.5 \times 10^{15} \times 10^3}{1.6 \times 10^{18}} = 2.2 \text{ gm/cm}^2/1000 \text{ years}$$
$$= 10.9 \text{ cm}/1000 \text{ years}$$

2. For the rate of burial of solid silica (h+i):

$$\frac{1.1 \times 10^{15} \times 10^3}{1.6 \times 10^{18}} = 0.68 \text{ gm/cm}^2/1000 \text{ years}$$
$$= 3.4 \text{ cm}/1000 \text{ years}$$

3. For the rate of addition of solid silica to the geologic record (h+i–d):

$$\frac{5.1 \times 10^{14} \times 10^3}{1.6 \times 10^{18}} = 0.32 \text{ gm/cm}^2/1000 \text{ years}$$
$$= 1.6 \text{ cm}/1000 \text{ years}$$

The rate of deposition of diatomaceous oozes around Antarctica during the Brunhes normal magnetic epoch (0–700,000 years) locally exceeds 2.2 cm/1000 years (0.44 gm/cm²/1000 years) and has a median value of 0.74 cm/1000 years (0.15 gm/cm²/1000 years; Goodell and Watkins, 1968). Radiolarian oozes in the equatorial Pacific accumulate at about 0.4 cm/1000 years (0.08 gm/cm²/1000 years). The sedimentation rate averaged over 50% of the deep-sea floor is probably 0.2 to 0.3 cm/1000 years (0.04 to 0.06 gm/cm²/1000 years).

Clearly, the model rate, even after allowing for all the suggested dissolution possibilities, is still 5 to 8 times greater than the observed accumulation rate. This discrepancy may be ex-

336

plained in two ways: 1) values used in the model have a cumulative error of almost an order of magnitude; or 2) much of the biogenous silica is not accumulating in the deep-sea.

The first suggestion cannot be disproved using presently available data. Nevertheless, a serious discrepancy exists even if one ignores all silica sources except the relatively well-known river influx. Thus, either the second alternative applies, or there is an additional silica sink in the oceans.

Calvert (1966) has demonstrated that the Gulf of California is a global sink for dissolved silica, in that the diatomite basins in the Gulf collect about 50 gm opal/cm²/1000 years. A similar situation prevails in the Sea of Okhotsk (Bezrukov, 1955). In addition to these special environments where pure siliceous deposits accumulate, it appears that much silica is also deposited in bays, estuaries and other nearshore areas where its presence is masked by rapidly accumulating terrigenous debris. This suggestion, made by K. Turekian (personal communication, 1970) is confirmed by recent determinations in our laboratory of opal in Cascadia Basin sediments, off central Oregon. The opal is accumulating at 50 gm/cm²/1000 years now, and accumulated at more than 100 gm/cm²/1000 years during the late Pleistocene in this area of strong upwelling and high biological productivity. As a result of rapid terrigenous sedimentation, however, opal concentrations are less than 5 percent in most samples.

If the rates deduced from the model do approximate reality, it seems that 85–90 percent of the opaline silica entering the geologic record is laid down in estuaries and nearshore restricted basins where it is overlooked due to dilution by detrital particles.

TEMPORAL CHANGES IN THE SILICA CYCLE

The attempt to evaluate components of the silica cycle in the preceding sections has pinpointed numerous areas where additional observational data are needed. Because improved knowledge of the poorly known components does not require significant technical advances, we may assume that the cycle will be refined as additional manpower and resources are applied to its study. If we accept that the model presented here can be quantified to any degree we care to pay for, the question still remains as to its relevance to the geologic past.

Perhaps the most important aspect of today's silica cycle is that the deposition of opal is an oceanographically rather than a geochemically controlled phenomenon. Regardless of where dissolved silica is injected into the system, it is

deposited in identifiable masses only beneath areas of upwelling where nutrient-rich deep waters enter the photic zone. As long as siliceous micro-organisms have thrived in the oceans, a similar pattern must have prevailed. Thus, from at least the Cretaceous, diatoms have deposited silica beneath areas of high biologic productivity. From the Cambrian until the middle or late Mesozoic, radiolaria probably dominated the siliceous microplankton. Unfortunately, almost nothing is known of the environmental versus genetic control of silica-uptake by radiolarians. Thus, we cannot assert with any confidence that sudden changes in the rate of input of dissolved silica to the oceans, in the absence of changes in other constituents, would change the rate of opal secretion in a simple way.

If, however, massive amounts of silica were injected into the oceans, dissolution of siliceous tests at the sea floor and escape of interstitial silica must have been inhibited due to lower concentration gradients at the sediment-water interface. Thus, even if the productivity of the radiolaria did not vary, changes in silica input due to volcanism, emergence and submergence of land masses, or other causes, would still have led to enhanced opal deposition beneath biologically productive areas. The pattern would differ slightly from the case of enhanced extraction (as practiced by diatoms) in that silica-rich bottom waters would preserve siliceous deposits below less productive areas, as well as allowing the accumulation of thicker deposits in areas where silica would ordinarily be preserved.

This discussion inevitably leads to the perplexing association of siliceous sediments (cherts) and the products of volcanism. The arguments of the preceding paragraphs suggest that massive injections of volcanic silica (which would have to be at least two orders of magnitude greater than today's input to become a major component of the silica cycle) should lead to thicker siliceous deposits (and ultimately cherts) beneath biologically productive areas rather than near the volcanic sources. Yet a geographic relation between ancient volcanically active regions and cherts does exist.

Fortunately, Calvert's (1966) work on the siliceous deposits of the Gulf of California appears to have resolved this paradox. In the Gulf, tectonism which is associated with divergence and transform faulting along lithospheric plate boundaries has produced an oceanographic situation that is very conducive to silica-secretion by diatoms. It has also generated a series of closed basins, most of which are shielded from coarse terrigenous debris (van Andel, 1964). Finally, it is associated with late Cenozic volcanism. Clearly, the oceanographic and bathymetric features that are responsible for the diatom-rich sediments (the future cherts) are only indirectly related to the volcanism, despite the intimate association of the two rock types.

Geologically and oceanographically, there is nothing unique about the situation in the Gulf of California. It is easy to imagine that all the alpine-type chert-volcanic rock associations reflect a similar set of environmental conditions.

Finally, we must consider the possibility of inorganic precipitation of silica from a supersaturated ocean. This possibility permeates the geologic literature, yet has never been proved conclusively. Given the propensity of diatoms to extract as much silica from solution as is available (Lewin, 1962) it is hard to imagine that any part of the ocean has approached saturation since the Cretaceous, at the latest. Prior to the appearance of diatoms, the situation is much less clear. If the radiolarians have adjusted their silica uptake in the way diatoms do, so as to extract all that has been available, the concentration has probably been kept well below saturation since the Cambrian. If not, the oceans may have approached and even exceeded saturation relative to amorphous silica during the Phanerozoic.

Calcite is presently supersaturated in surface ocean waters. Because its solubility decreases with increasing temperature and decreasing pressure, calcium carbonate tends to precipitate in warm shallow areas (for example, the Bahama Banks). Silica, on the other hand, becomes less soluble with decreasing temperature and, to a slight extent with decreasing pressure. Thus, if the oceans were ever saturated with silica, precipitation should have taken place in cool, shallow areas. The Ordovician cherts of Ellesmere Island (Trettin, 1970) may reflect such an environment. Presumably, deep-sea cherts could precipitate if the effect of decrease in temperature with depth exceeded the effect of pressure increase. Since a negative temperature gradient of ½°C per kilometer cancels out the solubility increase due to rising pressure it would not be surprising to find inorganically precipitated, geologically ancient deep-sea cherts.

DIAGENESIS

Despite the enormous areas of deep-sea chert revealed by the drilling program of the "Glomar Challenger," remarkably little has been learned during the past decade of the diagenetic behav-

iour of silica in pelagic sediments. Riedel (1959) comments that pyroclastic material appears to help preserve siliceous skeletons from dissolution after burial. He suggests that this is due to liberation of silica by weathering of the volcanic material. Riedel's observations have been repeatedly confirmed by other workers, but his suggested mechanism is not supported by experimental data. Jones (pers. comm., see also fig. 6) finds that volcanic ash, regardless of composition, is much less soluble in seawater, and dissolves much more slowly than opaline silica. Conceivably, cations released by the altering ash reduce the solubility of amorphous silica (Krauskopf, 1959). Alternatively, the pyroclastic deposits may lack the clay minerals which otherwise act as a sink for dissolved interstitial silica (Mackenzie and others, 1967). Clearly, only further experimental work can confirm or refute such hypotheses.

Interstitial waters of siliceous oozes are commonly at or near saturation with respect to amorphous silica (see references cited under "Release of interstitial water from deep-sea sediments"), but diagenetic reprecipitation of silica is minimal in young, thinly covered deposits. Minor cementation of clay fillings of siliceous tests by opaline silica has been reported in near surface sediments (Heath, 1969), but true chert formation seems to require several hundred meters of overburden (and resultant elevated temperatures?) or long periods of time (tens of millions of years). The deep-sea cherts collected by the "Challenger" are described in the Initial Reports of the Deep Sea Drilling Project. In virtually all cases, it appears that biogenous opaline silica has dissolved and migrated within the sediments to form cristobalitic chert and porcellanite masses. Subsequently, the cristobalite has tended to invert to quartz, either via a solution step or by direct solid-solid transformation (Heath and Moberly, 1971). Unfortunately, despite the large number of samples recovered, the environmental factors responsible for the initiation, location and rates of the solution and inversion steps, and for the diverse textural features observed in the cherts are still virtually unknown.

CONCLUSIONS

1) The cycle of dissolved silica in the oceans is dominated by biologic extraction and oxidative dissolution of opaline tests in the upper kilometer of the water column. Each year these processes turn over an order of magnitude more silica than is supplied to or removed from the oceanic reservoir.

2) Processes supplying dissolved silica to the

oceans, in decreasing order of importance, are the escape of silica from the interstitial waters of pelagic sediments, river influx, submarine weathering of basalt, and volcanic influx. Of these, the first two supply about 90 percent of the soluble silica.

3) Deposition of opaline tests of siliceous microplankton must remove as much silica as enters the oceans each year to maintain a steady state in which the ocean is markedly undersaturated with respect to amorphous silica. The rate of dissolution of siliceous oozes relative to the turnover time of the oceans is too slow to saturate the bottom waters with silica.

4) Siliceous oozes that ultimately become chert are deposited beneath biologically productive areas, regardless of where the silica enters the oceans. The association of volcanic rocks and cherts in the geologic record apparently results from the production of suitable oceanographic and bathymetric conditions by the same tectonic processes that are responsible for the volcanism.

5) The residence time of dissolved silica in the oceans ranges from 2–300 years for biologic utilization to 11–16,000 years for incorporation of extracted silica into the geologic record. Thus, from a geologic viewpoint, changes in the rate of supply of silica to the oceans will be reflected virtually instantaneously in the rates of uptake and deposition. Such changes should not greatly affect the distribution patterns of siliceous deposits.

6) The diatom and radiolarian oozes of pelagic regions are the most conspicuous sinks for dissolved silica. However, as much as 85–90 percent of the siliceous remains may be deposited in nearshore or estuarine environments where they are masked by terrigenous debris.

7) Because of the dominant influence of biologic processes on the marine silica cycle, supersaturation and consequent inorganic precipitation of amorphous silica appears improbable since the Cambrian and virtually impossible since the Cretaceous. The final word on this subject awaits ecological studies of radiolarian uptake of silica.

ACKNOWLEDGMENTS

It is impossible to acknowledge by name all the people with whom I have discussed and argued the material presented in this review. To those not mentioned here, my sincere thanks. I am particularly grateful to T. C. Moore, Jr., and Tj. H. van Andel for penetrating comments and for reviewing the manuscript. In addition, I have been greatly assisted by discussions with W. H. Berger, J. L. Bischoff, S. E. Calvert, J. B. Corliss, J. P. Dauphin, J. R. Dymond, J. M. Edmond, J. D. Hays, M. M. Jones, A. P. Lisitzyn, H. Tappan Loeblich, F. T. Mackenzie, R. M. Pytkowicz, W. R. Riedel and K. K. Turekian. Needless to say, accountability for the ideas expressed here is mine.

REFERENCES

BENDER, M. L., 1971, Does upward diffusion supply the excess manganese in pelagic sediments?: Jour. Geophys. Research, v. 76, p. 4212–4215.

BERGER, W. H., 1968, Radiolarian skeletons: solution at depths: Science, v. 159, p. 1237–1238.

———, 1970, Biogenous deep-sea sediments: fractionation by deep-sea circulation: Geol. Soc. America Bull., v. 81, p. 1385–1401.

BEZRUKOV, P. L., 1955, Distribution and rate of sedimentation of silica silts in the Sea of Okhotsk: Akad. Nauk. SSSR Dokl., v. 103, p. 473–476.

BIEN, G. S., CONTOIS, D. E., AND THOMAS, W. H., 1959, The removal of soluble silica from fresh water entering the sea, *in* IRELAND, H. A. (ed.), Silica in sediments: Soc. Econ. Paleontologists and Mineralogists Special Pub. 7, p. 20–35.

BISCHOFF, J. L., AND KU, T. L., 1970, Pore fluids of recent marine sediments: I. Oxidizing sediments of 20°N, continental rise to Mid-Atlantic Ridge: Jour. Sed. Petrology, v. 40, p. 960–972.

———, AND ———, 1971, Pore fluids of recent marine sediments: II. Anoxic sediments of 35° to 45°N Gibraltar to Mid-Atlantic: *ibid.*, v. 41, p. 1008–1017.

BOGOYAVLENSKIY, A. N., 1967, Distribution and migration of dissolved silica in oceans: Internat. Geol. Rev., v. 9, p. 133–153.

BRUYEVICH, S. W., 1966, The Pacific Ocean. Vol. 3, Chemistry of the Pacific Ocean: Akad. Nauk. SSSR, 351 p. (Translated by Office Naval Research, Clearinghouse for Federal Sci. and Tech. Inf. Ref. AD651498).

BURTON, J. D., AND LISS, P. S., 1968, Oceanic budget of dissolved silicon: Nature, v. 220, p. 905–906.

BURTON, J. D., LISS, P. S., AND VENUGOPALAN, V. K., 1970, The behavior of dissolved silicon during estuarine mixing. I. Investigations in Southampton Water: Jour. Cons., Cons. Int. Expl. Mer, v. 33, p. 134–140.

CALVERT, S. E., 1966, Accumulation of diatomaceous silica in the sediments of the Gulf of California: Geol. Soc. America Bull., v. 77, p. 569–596.

———, 1968, Silica balance in the ocean and diagenesis: Nature, v. 219, p. 919–920.

CORLISS, J. B., 1970, Mid-ocean ridge basalts: I—The origin of sub-marine hydrothermal solutions. II—Regional diversity along the Mid-Atlantic Ridge (Ph.D. dissertation): San Diego, Univ. California, 147 p.

CRAIG, H., 1969, Abyssal carbon and radiocarbon in the Pacific: Jour. Geophys. Research, v. 74, p. 5491–5506.
EDMOND, J. M., 1970, The carbonic acid system in sea water (Ph.D. dissertation): San Diego, Univ. California, 174 p.
———, AND ANDERSON, G. C., 1971, On the structure of the North Atlantic deep water: Deep-Sea Research, v. 18, p. 127–133.
FANNING, K. A., AND PILSON, M. E. Q., 1971, Interstitial silica and pH in marine sediments: some effects of sampling procedures: Science, v. 173, p. 1228–1231.
FANNING, K. A., AND SCHINK, D. R., 1969, Interaction of marine sediments with dissolved silica: Limnology and Oceanography, v. 14, p. 59–68.
FOURNIER, R. O., AND ROWE, J. J., 1962, The solubility of cristobalite along the three-phase curve, gas plus liquid plus cristobalite: Am. Mineralogist, v. 47, p. 897–902.
GARRELS, R. M., 1965, Silica: role in the buffering of natural waters: Science, v. 148, p. 69.
GIBBS, R. J., 1967, Amazon River: environmental factors that control its dissolved and suspended load: ibid., v. 156, p. 1734–1737.
GILBERT, J. Y., AND ALLEN, W. E., 1943, The phytoplankton of the Gulf of California obtained by the "E. W. Scripps" in 1939 and 1940: Jour. Marine Research, v. 5, p. 89–110.
GOODELL, H. G., AND WATKINS, N. D., 1968, The paleomagnetic stratigraphy of the Southern Ocean: 20° west to 160° east longitude: Deep-Sea Research, v. 15, p. 89–112.
GRANT, A. B., 1968, Atlas of oceanographic sections, Davis Strait-Labrador Basin-Denmark Strait-Newfoundland Basin, 1965–1967: Atlantic Ocean Lab., Bedford Inst., Rept. AOL 68–5, 80 p.
GROSS, M. G., 1967, Sinking rates of radioactive fallout particles in the northeast Pacific Ocean, 1961–62: Nature, v. 216, p. 670–672.
HARRISS, R. C., 1966, Biological buffering of oceanic silica: ibid., v. 212, p. 275–276.
———, AND PILKEY, O. H., 1966, Interstitial waters of some deep marine carbonate sedimens: Deep-Sea Research, v. 13, p. 967–969.
HART, R., 1970, Chemical exchange between sea water and deep ocean basalts: Earth and Planet. Sci. Letters, v. 9, p. 269–279.
HEATH, G. R., 1969, Mineralogy of Cenozoic deep-sea sediments from the equatorial Pacific Ocean: Geol. Soc. America Bull., v. 80, p. 1997–2018.
HEATH, G. R., AND MOBERLY, JR., R., 1971, Cherts from the western Pacific: Leg VII, Deep Sea Drilling Project, in WINTERER, E. L., AND OTHERS, Initial reports of the Deep Sea Drilling Project: Washington, D.C., U.S. Govt. Printing Office, v. 7, p. 991–1007.
ISAEVA, A. B., 1958, On the methods of silica determination in sea water: Akad. Nauk. SSSR, Inst. Okeanol., Trudy, v. 26, p. 234–242.
JONES, M. M., AND PYTKOWICZ, R. M., 1973, Solubility of silica in seawater at high pressures: Soc. Royale des Sciences de Liege, v. 42, p. 125–127.
KAPLAN, I. R., AND PRESLEY, B. J., 1970, Interstitial water chemistry; Deep Sea Drilling Project, Leg 2, in PETERSON, M. N. A., AND OTHERS, Initial reports of the Deep Sea Drilling Project: Washington, D.C., U.S. Govt. Printing Office, v. 2, p. 373.
KELLER, W. D., AND BALGORD, W. D., AND REESMAN, A. L., 1963, Dissolved products of artificially pulverized silicate minerals and rocks. Pt. I: Jour. Sed. Petrology, v. 33, p. 191–204.
KOBAYASHI, J., 1967, Silica in fresh water and estuaries, in GOLTERMAN, H. L., AND CLYMO, R. S. (eds.), Chemical environment in the aquatic habitat: Amsterdam, Uitgevers Maatschappij, p. 41–55.
KOZLOVA, O. G., 1961, Quantitative content of diatoms in the waters of the Indian sector of Antarctica: Akad. Nauk. SSSR, Dokl., v. 138, p. 207–210.
KRAUSKOPF, K. B., 1959, The geochemistry of silica in sedimentary environments, in IRELAND, H. A. (ed.), Silica in sediments: Soc. Econ. Paleontologists and Mineralogists Special Pub. 7, p. 4–19.
LEWIN, J. C., 1962, Silicification, in LEWIN, R. A. (ed.), Physiology and Biochemistry of Algae: New York, Academic Press, p. 445–455.
LISITZYN, A. P., 1967, Basic relationships in distribution of modern siliceous sediments and their connection with climatic zonation: Internat. Geol. Rev., v. 9, p. 631–652.
———, 1972, Sedimentation in the world ocean: Soc. Econ. Paleontologists and Mineralogists Special Pub., 17, 218 p.
———, BELAYAYEV, Y. I., BOGDANOV, Y. A., AND BOGOYAVLENSKIY, A. N., 1967, Distribution relationships and forms of silicon suspended in waters of the waters of the world ocean: Internat. Geol. Rev., v. 9, p. 604–623.
LISS, P. S., AND SPENCER, C. P., 1970, Abiological processes in the removal of silicate from sea water: Geochimica et Cosmochimica Acta, v. 34, p. 1073–1088.
LIVINGSTONE, D. A., 1963, Chemical composition of rivers and lakes: U.S. Geol. Survey Prof. Paper 440-G, 64 p.
MACKENZIE, F. T., AND GARRELS, R. M., 1965, Silicates: reactivity with sea water: Science, v. 150, p. 57–58.
———, AND ———, 1966a, Chemical mass balance between rivers and oceans: Am. Jour. Sci., v. 264, p. 507–525.
———, AND ———, 1966b, Silica-bicarbonate balance in the ocean and early diagenesis: Jour. Sed. Petrology, v. 36, p. 1075–1084.
———, ———, BRICKER, O. P., AND BICKLEY, F., 1967, Silica in sea water: control by silica minerals: Science, v. 155, p. 1404–1405.
MAÉDA, H., 1952, The relation between chlorinity and silicate concentration of water observed in some estuaries: Seto Mar. Biol. Lab. Pubs., v. 2, p. 249–255.
MANHEIM, F. T., CHAN, K. M., AND SAYLES, F. L., 1970, Interstitial water studies on small core samples, Deep Sea Drilling Project, Leg. 5, in McMANUS, D. A., AND OTHERS, Initial reports of the Deep Sea Drilling Project: Washington, D.C., U.S. Govt. Printing Office, v. 5, p. 501–511.

MANHEIM, F. T., AND SAYLES, E. L., 1971, Interstitial water studies on small core samples, Deep Sea Drilling Project, Leg. 6, *in* FISCHER, A. G., AND OTHERS, Initial reports of the Deep Sea Drilling Project: *ibid.*, v. 6, p. 811–821.

METCALF, W. G., 1969, Dissolved silicate in the deep North Atlantic: Deep-Sea Research, v. 16 (suppl.), p. 139–145.

MOREY, G. W., FOURNIER, R. O., AND ROWE, J. J., 1962, The solubility of quartz in water in the temperature interval from 25° to 300°C: Geochimica et Cosmochimica Acta, v. 26, p. 1029–1043.

——, ——, AND ——, 1964, The solubility of amorphous silica at 25°C: Jour. Geophys. Research, v. 69, p. 1995–2002.

PETERSON, M. N. A., AND GOLDBERG, E. D., 1962, Feldspar distributions in South Pacific pelagic sediments: *ibid.*, v. 67, p. 3477–3492.

PRESLEY, B. J., GOLDHABER, M. B., AND KAPLAN, R. I., 1970, Interstitial water chemistry: Deep Sea Drilling Project, Leg. 5, *in* McMANUS, D. A., AND OTHERS, Initial reports of the Deep Sea Drilling Project: Washington, D.C., U.S. Govt. Printing Office, v. 5, p. 513–522.

——, AND KAPLAN, I. R., 1970, Interstitial water chemistry: Deep Sea Drilling Project, Leg. 4, *in* BADER, R. G., AND OTHERS, Initial Reports of the Deep Sea Drilling Project: *ibid.*, v. 4, p. 823–828.

——, AND ——, 1971, Interstitial water chemistry; Deep Sea Drilling Project, Leg 6, *in* FISCHER, A. G., AND OTHERS, Initial reports of the Deep Sea Drilling Project: *ibid.*, v. 6, p. 823–828.

——, AND ——, 1972, Interstitial water chemistry: Deep Sea Drilling Project, Leg 9, *in* HAYS, J. D., AND OTHERS, Initial reports of the Deep Sea Drilling Project: *ibid.*, v. 9, p. 841–844.

PYTKOWICZ, R. M., 1967, Carbonate cycle and the buffer mechanism of recent oceans: Geochimica et Cosmochimica Acta, v. 31, p. 63–73.

REID, J. L., JR., 1965, Intermediate waters of the Pacific Ocean: Baltimore, Maryland, Johns Hopkins Univ., 85 p.

RIEDEL, W. R., 1959, Siliceous organic remains in pelagic sediments, *in* IRELAND, H. A. (ed.), Silica in sediments: Soc. Econ. Paleontologists and Mineralogists Special Pub. 7, p. 80–91.

RILEY, J. P., 1965, Analytical chemistry of sea water, *in* RILEY, J. P., AND SKIRROW, G. (eds.), Chemical oceanography: London, Academic Press, v. 2, p. 295–424.

RUDDIMAN, W. F., 1968, Historical stability of the Gulf Stream meander belt: foraminiferal evidence. Deep-Sea Research, v. 15, p. 137–148.

RYTHER, J. H., 1959, Potential productivity of the sea: Science, v. 130, p. 602–608.

SAYLES, F. L., MANHEIM, F. T., AND CHAN, K. M., 1970, Interstitial water studies on small core samples, Leg 4, *in* BADER, R. G., AND OTHERS, Initial reports of the Deep Sea Drilling Project: Washington, D.C., U.S. Govt. Printing Office, v. 4, p. 401–414.

——, WATERMAN, L. S., AND MANHEIM, F. T., 1972, Interstitial water studies on small core samples, Leg 9, *in* HAYS, J. D., AND OTHERS, Initial reports of the Deep Sea Drilling Project: *ibid.*, v. 9, p. 845–855.

SCHINK, D. R., 1967, Budget for dissolved silica in the Mediterranean Sea: Geochimica et Cosmochimica Acta, v. 31, p. 897–999.

——, 1968, Observations relating to the flux of silica across the sea floor interface: Am. Geophys. Union Trans., v. 49, p. 335.

SCHUTZ, D. F., AND TUREKIAN, K. K., 1965, The investigation of the geographical and vertical distribution of several trace elements in sea water using neutron activation analysis: Geochimica et Cosmochimica Acta, v. 29, p. 259–313.

SCRIPPS INSTITUTION OF OCEANOGRAPHY, 1969, Physical and chemical data from the Scorpio Expedition, U.S.N.S. Eltanin Cruises 28 and 29: S.I.O. Ref. 69–15, 89 p.

——, 1970, Physical, chemical and biological data, Zetes Expedition, Leg 1: S.I.O. Ref. 70–5. 67 p.

SIEVER, R., 1968a, Establishment of equilibrium between clays and sea water: Earth and Planet. Sci. Letters, v. 5, p. 106–110.

——, 1968b, Sedimentological consequences of a steady-state ocean-atmosphere: Sedimentology, v. 11, p. 5–29.

——, BECK, K. C., AND BERNER, R. A., 1965, Composition of interstitial waters of modern sediments: Jour. Geology, v. 73, p. 39–73.

SILLÉN, L. G., 1961, The physical chemistry of sea water, *in* SEARS, M. (ed.), Oceanography: Am. Assoc. Adv. Sci. Pub. 67, p. 549–581.

STEFÁNSSON, U., AND RICHARDS, F. A., 1963, Processes contributing to the nutrient distributions off the Columbia River and Strait of Juan de Fuca: Limnology and Oceanography, v. 8, p. 394–410.

TRETTIN, H. P., 1970, Ordovician-Silurian flysch sedimentation in the axial trough of the Franklinian Geosyncline, northeastern Ellesmere Island, Arctic, Canada: Geol. Assoc. Canada Special Paper 7, p. 13–35.

VAN ANDEL, TJ. H., 1964, Recent marine sediments of Gulf of California, *in* VAN ANDEL, TJ. H., AND SHOR, G. G., JR. (eds.), Marine geology of the Gulf of California—A symposium: Am. Assoc. Petroleum Geologists Mem. 3, p. 216–310.

WARNKE, D. A., 1970, Glacial erosion, ice rafting, and glacial-marine sediments: Antarctica and the Southern Ocean: Am. Jour. Sci., v. 269, p. 276–294.

WATT, W. D., 1971, Measuring the primary production rates of individual phytoplankton species in natural mixed populations: Deep-Sea Research, v. 18, p. 329–339.

WOLLAST, R., AND GARRELS, R. M., 1971, Diffusion coefficient of silica in sea water: Nature, v. 229, p. 94.

18

Reprinted from *Soc. Géol. France Bull.* **16**(2):214–224 (1974)

Hypothèse sur l'origine des silicifications dans les grands bassins océaniques. Le rôle des climats hydrolisants

par Lucien LECLAIRE *

Summary†—The cruises of the drilling vessel *Glomar Challenger* (Deep Sea Drilling Project) have shown the importance and extent of silicification in the great ocean basins. This silicification especially affects the Cretaceous and Eocene sedimentary formations. They decrease considerably in magnitude from the Oligocene onwards, without marked changes in the other characteristics of the pelagic sediments.

This decrease in silicification coincides with a radical change in global climate from the Oligocene onwards. This "climatic revolution" has had as its main consequence the retreat towards the equator of the hydrolyzing climatic belts, which, since the Jurassic, had caused intensive latritic weathering of about three quarters of the total continental land mass. It seems likely that this climatic change stopped the abundant silica supply coming from these lateritic terrains, resulting in a decrease in the amount of dissolved silica in the oceans. This in turn resulted in less silicification in the oceanic sediments.

This silicification, directly depending on silica supply from weathered igneous and sedimentary continental rocks, would therefore be directly climatically controlled.

INTRODUCTION.

Cette note a pour origine une réflexion à partir de trois faits mis en évidence par l'étude du domaine océanique. Le premier est l'abondance des silicifications (cherts, porcellanite, sphérules de cristobalite, etc.) dans les dépôts du Crétacé et de l'Éocène des grands bassins océaniques. Le deuxième est la raréfaction de ces silicifications à partir de l'Oligocène. Le troisième résulte de la comparaison d'analyses d'eau de mer et d'eau de rivière : il s'avère que la silice est l'une des rares substances dont la concentration dans l'eau de mer soit nettement inférieure à sa concentration dans les eaux douces. La recherche d'une explication par l'étude des sources de la silice et des causes de fluctuations possibles des apports provenant de ces sources, nous a conduit à proposer une nouvelle hypothèse sur l'origine des silicifications intenses dans les bassins océaniques, au Crétacé et au Tertiaire inférieur. Cependant notre but n'est pas de traiter du cycle de la silice ou des mécanismes de précipitation.

SILICIFICATIONS ET AUTHIGENÈSES ASSOCIÉES DANS LES BASSINS OCÉANIQUES.

Répartition géographique et stratigraphique des silicifications.

Les données concernant cette répartition se trouvent dans les publications très détaillées (Initial Reports of the Deep Sea Drilling Project) ou plus sommaires (articles de la revues « Geotimes ») relatives aux trente-trois premières campagnes (ou Leg) du navire foreur « Glomar Challenger ». Il n'est pas question ici d'en récapituler l'ensemble mais seulement de dégager les grandes tendances[1].

Dans le Pacifique nord-occidental (Leg 6, 20, 32) les niveaux silicifiés sont particulièrement abondants dans les séries sédimentaires (craie, boues à zéolites, etc.) du Crétacé supérieur et inférieur. A titre

* Lab. de géologie, Museum nat. hist. nat., 61, rue de Buffon, 75005 Paris. Note déposée le 18 février 1974, présentée le 4 mars 1974. Manuscrit définitif reçu le 6 mars 1974.

†This summary was translated by the volume editor.

d'exemple, le forage n° 305 a traversé environ 500 m de sédiments d'âge mæstrichtien à néocomien, contenant de manière presque ininterrompue, d'innombrables niveaux silicifiés (cherts). Dans le Pacifique central et équatorial (Leg 7, 8 et 17), silex tabulaires ou en rognons, autres niveaux silicifiés (à cristobalite) sont fréquents dans les dépôts du Crétacé supérieur et encore plus abondants dans les formations d'âge éocène moyen. Très rares sont les niveaux silicifiés plus récents ; on les rencontre principalement dans le Pacifique équatorial : forages n°s 32 et 34 (Miocène moyen), 62 et 63 (Pliocène supérieur) 70, 71, 72 (Oligocène supérieur) [Lancelot, 1973].

En Atlantique oriental (Leg 14), la plupart des horizons silicifiés sont contenus dans les dépôts d'âge santonien à éocène inférieur [Berger et Von Rad, 1972]. Dans les bassins de l'Atlantique nord (Leg 1, 2 et 11), les niveaux à chert du Paléogène, notamment de l'Éocène inférieur et moyen ont été carottés à plusieurs reprises (forages n° 8, 9 12 etc.) et constituent un puissant réflecteur aisément suivi sur les enregistrements de sismique réflexion continue (horizon « A ») [Gartner, 1970]. Cet horizon silicifié s'étend sur des milliers de kilomètres. Dans le golfe du Mexique (Leg 10), des silex noirs du Cénomanien ont été carottés. En mer des Caraïbes, les niveaux silicifiés, relativement abondants, sont principalement éocènes ; on en trouve aussi dans le Crétacé supérieur. En Atlantique, seul le forage 139 a rencontré des cherts d'âge plus récent (Miocène inférieur) ; c'est une exception.

Dans l'océan Indien et le secteur australien, les niveaux silicifiés semblent moins abondants que dans les autres océans. Des cherts ont été observés en mer de Tasmanie, sur la ride de Lord Howe (forages, n° 206, 207 et 208, leg 21). Ils sont principalement d'âge éocène. Cependant quelques niveaux silicifiés sont intercalés dans une série de l'Oligocène moyen.

Sur la ride du 90e (Leg 22), des silicifications apparaissent dans les craies paléocènes et éocènes (forages n° 216 et 217).

Dans l'océan Indien nord-occidental (Leg 23), les silicifications observées sont intercalées dans des formations d'âge éocène inférieur et moyen. On peut les suivre sur les profils de sismique réflexion continue (forages n° 219, 220 et 221).

La craie de l'Éocène inférieur carottée dans le bassin de Madagascar (Leg 25) renferme une série de minces bancs de cherts au contact desquels on trouve aussi des niveaux silicifiés à sphérules de cristobalite-tridymite [Leclaire, 1974 ; Leclaire, Alcaydé et Frœhlich, 1973].

Dans l'océan Antarctique, entre le continent et l'Australie, plusieurs forages ont perforé des niveaux silicifiés d'âge oligocène et miocène infé-

rieur [Hayes, Frakes *et al.*, 1973]. L'horizon silicifié sans doute le plus récent (Pliocène) a été carotté par le navire « *Ellanin* » aux abords du plateau de Kerguelen [Weaver et Wise, 1972].

Tous les cas de silicifications ne sont pas notés dans la liste restreinte qui précède, voir par exemple Andrews, Packham *et al.* 1973, etc.

La distribution statistique des silicifications en fonction de l'âge des formations qu'elles affectent a été étudiée par Davies et Supko [1973]. Le diagramme qu'ils ont obtenu illustre de manière très significative les observations brièvement exposées ci-dessus (voir figure).

En résumé, les forages du « *Glomar Challenger* » montrent que les silicifications dans les grands bassins océaniques affectent principalement et souvent massivement, les dépôts d'âge crétacé et éocène. Un tel phénomène semble aussi avoir affecté les dépôts épicontinentaux de la plate-forme russe [Distanov *et al.* 1971]. Les silicifications plus récentes sont comparativement très rares, beaucoup moins importantes et semblent plutôt localisées aux abords de zones particulières réputées pour leur haute productivité organique : les divergences océaniques.

Répartition des authigenèses associées.

Il est maintenant bien connu que les séries sédimentaires à horizons silicifiés ont en commun plu-

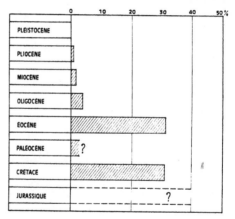

Fig. 1. — Distribution de fréquence des bancs de cherts en fonction de leur âge, dans les grands bassins océaniques [d'après Davies et Supko, 1973]. Ce diagramme a été établi essentiellement à partir des résultats de la phase I du « Deep Sea Drilling Project » L'auteur a cependant tenté de tenir compte des premiers résultats de la phase II.

sieurs caractères dont l'un des plus remarquables est sans doute un même cortège minéralogique dans lequel les rapports clinoptilolite/phillipsite, paly-gorskite/montmorillonite et feldspath potassique/plagioclases sont très élevés [2] [Lancelot, 1973]. Cependant, la présence de clinoptilolite ou de paly-gorskite n'implique pas nécessairement celle de chert.

A titre d'exemple, des études récentes dans l'océan Indien [Venkatarathnam et Biscaye, 1973] ont montré que la phillipsite se trouve souvent très abondante dans les sédiments néogènes et qua-ternaires, principalement dans les zones centrales à proximité des dorsales médio-océaniques. Par contre, la clinoptilolite est plutôt concentrée dans les matériaux déposés au Crétacé et à l'Éocène, à proximité des marges continentales. Ces résultats sont confirmés dans l'ensemble par les campagnes de forage. Par ailleurs, les boues à zéolites du Crétacé et de l'Éocène de l'Atlantique contiennent de la clinoptilolite (dominante) alors que dans des boues du Pacifique de même nature mais plus récentes, la phillipsite est dominante [Bruty, Chester et Aston, 1973]. Cependant, les observations de Heath [1969] semblent montrer que cette tendance n'est pas systématique et qu'elle peut s'inverser localement.

Dans le bassin du Mozambique (entre le plateau de Madagascar et la ride du Mozambique) s'est accumulée une curieuse série argileuse, apparamment détritique, à palygorskite dominante. Cette série contient des silts à minéraux lourds provenant de la marge africaine ou de Madagascar ; elle est riche en Radiolaires qui donnent un âge paléogène. Associée à la palygorskite, on trouve aussi en abondance de la cristobalite et de la clinoptilolite. Une association de même nature a été décelée dans les craies cénomaniennes de la ride du Mozambique. Cette séquence à palygorskite se termine par l'arrivée soudaine de sables grossiers et d'autres produits détritiques de nature diverse (Leg 25).

L'accumulation de palygorskite la plus spec-taculaire dans un bassin océanique, se trouve dans l'Atlantique, au large des côtes de l'Afrique occi-dentale, ou cinq forages du « Glomar Challenger » (n° 12, 137, 138, 140 et 141) ont traversé une séquence argileuse de 40 à 150 m d'épaisseur constituée par de la palygorskite-sépiolite souvent très pure [Pimm et Hayes, 1972]. Cette séquence argileuse d'âge paléogène (principalement éocène) se termine par une série détritique plus grossière. Ce bassin océa-nique serait gigantesque puisqu'il semble couvrir une surface d'environ 500 000 km[2].

En règle générale, et tout particulièrement en Atlantique, une série détritique à kaolinite, chlorite, illite, etc. d'âge oligocène et miocène succède à des séquences très riches en néogenèses : palygorskite, sépiolite, etc. dont l'âge est crétacé ou paléogène.

Discussion.

Il apparaît donc que dans l'ensemble, les sili-cifications dans les bassins océaniques et leurs authigenèses associées qui caractérisent, dans une certaine mesure, les dépôts du Crétacé et de l'Éocène, se raréfient considérablement à partir de l'Oligocène, sans qu'il y ait de changement net dans la nature des produits de la sédimentation pélagique. Deux explications de cette anomalie sont généralement proposées [Lancelot, 1973] : l'une suppose soit la formation très lente des cherts soit une diagenèse très tardive qui ne se déclencherait qu'au bout de plusieurs dizaines de millions d'années ; l'autre est basée sur l'hypothèse d'une modification de l'environnement océanique à l'Oligocène. Dans ce dernier cas, il faudrait admettre que certaines des conditions de sédimentation du passé soient différentes des conditions actuelles. La présence de silicifications dans les diatomites pliocènes des abords du Plateau de Kerguelen [Weaver et Wise, 1972] est un exemple qui montre que les silicifications récentes, bien qu'exceptionnelles, sont possibles. En conséquence, les auteurs s'accordent pour penser que la deuxième explication est la plus plausible.

Lancelot [1973] a, en outre, évoqué la possibilité d'un contrôle de la précipitation de la silice dans les eaux interstitielles des sédiments par les sili-cates d'alumine. En effet, selon Mackenzie et Garrels [1966], Mackenzie et al., [1967], Helgeson et Mackenzie [1970], il pourrait y avoir équilibre entre la silice dissoute et les minéraux des argiles. Ces derniers auraient la propriété de fixer la silice par exemple lorsqu'elle est en excès, empêchant de ce fait toute précipitation. Cependant, le méca-nisme invoqué, déduit d'expérience en laboratoire, n'est pas encore clairement défini et reste criticable [Perry, 1971]. De plus, il est peu vraisemblable qu'il puisse se développer suffisamment pour jouer un rôle dans la silicification des sédiments très pauvres en argiles comme certaines craies, dia-tomites ou radiolarites.

LES SOURCES DE LA SILICE DU MILIEU OCÉANIQUE.

Origine de la silice dissoute dans l'eau de mer : balance et bilan.

« Posée et résolue depuis longtemps de bien des manières, la question des sources de la silice (« de la craie ») ne comporte en fait que deux solutions générales : ou la silice est d'origine minérale ou elle est d'origine organique » Cailleux [1929]. Par origine organique Cailleux sous-entendait en pro-venance de la dissolution des squelettes d'orga-nismes siliceux. Cependant, quel que soit le rôle des organismes comme les Diatomées, les Radiolaires et les Éponges dans la concentration de la silice pouvant mener aux silex, ce ne sont en fait que des agents précipitants qui prélèvent dans l'eau de mer la silice nécessaire à l'édification de leur squelette. Le problème posé est donc finalement : quelle est l'origine de la silice dissoute dans l'eau de mer ?

Une première remarque s'impose : la saturation de l'eau de mer en silice dissoute est obtenue pour des valeurs de l'ordre de 120 à 140 ppm à 20° (80 à 90 ppm à 0°). Or, la teneur moyenne de l'eau de mer est de l'ordre de 3 ppm, alors que celle des eaux de rivières est près de cinq fois plus forte. La sous-saturation de la silice dissoute dans l'eau de mer est donc considérable ; par rapport aux eaux de rivières on est même tenté de considérer cette sous-saturation comme une anomalie. Ce fait pourrait bien être la conséquence d'un déséquilibre de la balance de la silice des océans. Plusieurs tentatives ont déjà été faites pour chiffrer cette balance [Harriss, 1966 ; Calvert, 1968 ; Gregor, 1968, etc.]. Les résultats obtenus sont peu significatifs tellement les incertitudes et les marges d'erreurs dans les estimations sont grandes. Cependant, il est peut-être utile d'avoir quelques chiffres en mémoire.

La teneur totale en silice dissoute des océans est de l'ordre de $4,3.10^{18}$ g. L'apport moyen annuel par le réseau hydrographique est de l'ordre de $4,3.10^{14}$ g. [Mackenzie et Garrels, 1966 ; Harriss, 1966, etc.]. La masse totale de silice dissoute utilisée par le plancton siliceux serait d'environ 8 à 16.10^{16} g/an [Litizsin, 1971] soit en moyenne : 12.10^{16} g/an. Selon le même auteur, un dixième à un centième seulement de cette masse de silice précipitée parviendrait au fond ; la majeure partie serait redissoute. En admettant que le centième seulement entre dans la constitution des sédiments (ce qui est sans doute une sous-estimation), on s'aperçoit que, par comparaison avec la masse des apports par les rivières (chiffre dont l'ordre de grandeur

est très vraisemblablement correct), le bilan est déficitaire d'un ordre de grandeur. Cela signifierait que la réserve de la silice des océans s'épuiserait à court termes (quelques milliers d'années) ou que le seuil létal pour le plancton siliceux (supposé être aux environs de 0,2 mg/1) serait atteint en quelques siècles, ce qui est très peu probable. La masse totale des apports doit être grandement sous-estimée. On est donc amené à supposer l'existence d'une autre source.

La découverte récente de l'expansion des fonds océaniques suggère d'emblée que l'activité volcanique des dorsales médio-océaniques où 8 km³ de basalte se mettent en place chaque année [Grégor, 1968] pourrait être cette autre source. On sait que l'altération des basaltes en milieux aqueux et notamment en milieu océanique délivre une grande quantité de silice qui passe en solution dans l'eau [Lire, par exemple, Elderfield, 1972]. Par ailleurs, Hart, [1973] a montré que l'atténuation des anomalies magnétiques constatée lorsqu'on s'éloigne suffisamment des rifts, est en relation avec une intense altération du plancher basaltique des océans. Selon ce même auteur, cette altération (dont l'un des agents les plus puissants pourrait bien être la circulation des eaux hydrothermales) se traduirait par une perte totale d'environ 7.10^{14} g/an de silice injectée dans le milieu océanique. Ce chiffre est très voisin de l'estimation de Grégor [1968]. Si ces chiffres sont corrects, la masse totale de silice sédimentée est du même ordre de grandeur que la masse totale des apports.

Le tableau rassemble quelques estimations concernant la balance de la silice dissoute dans l'eau de mer. Il apparaît qu'on ne peut tirer aucune conclusion définitive si ce n'est l'ordre de grandeur comparable. Cependant, le chiffre obtenu par l'équation

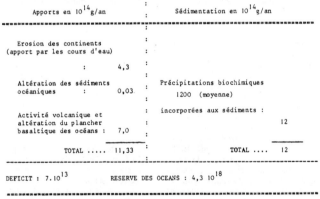

Apports en 10^{14} g/an		Sédimentation en 10^{14} g/an	
Erosion des continents (apport par les cours d'eau) :	4,3		
Altération des sédiments océaniques :	0,03	Précipitations biochimiques 1200 (moyenne) incorporées aux sédiments :	
Activité volcanique et altération du plancher basaltique des océans :	7,0		12
TOTAL	11,33	TOTAL	12
DEFICIT : 7.10^{13}		RESERVE DES OCEANS : $4,3\ 10^{18}$	

TABL. I. — Balance de la silice dissoute des océans

de Hart [1973] est basé sur l'altération d'une épaisseur de plus de 2 km du plancher basaltique, ce qui est énorme ; il est donc possible que l'apport de cette source soit surestimé. Par ailleurs, la quantité de silice précipitée par le plancton siliceux et réellement sédimentée est sans doute sous-estimée. Il est possible, en effet, qu'un cinquantième à un dixième (au lieu d'un centième) de la silice précipitée par voie biochimique entre dans la constitution des sédiments, toute chose restant égale par ailleurs. Ces remarques nous conduisent à penser que le bilan de la silice des eaux océaniques actuelles pourrait être déficitaire. Ajoutons qu'un déficit de l'ordre de 10^{13} g/an, se maintenant pendant quelques dizaines de millions d'années, suffirait à abaisser la réserve maximale possible du milieu océanique à son niveau actuel.

En résumé, la silice dissoute dans l'eau de mer a, au moins, trois sources majeures : l'altération de la surface des continents, l'altération du plancher basaltique des océans, le recyclage par redissolution de la silice précipitée par le plancton.

Origine de la silice des « accidents siliceux » et authigenèses associées dans les sédiments océaniques. Rôle de la concentration de la silice dans l'eau de mer.

Les travaux publiés au cours de la dernière décennie ont montré que la silice des silex, des sphérules de cristobalite et des authigenèses associées a, au minimum, deux origines qui sont : la dissolution des squelettes du plancton siliceux et des spicules d'Eponges enfermés dans les sédiments et la dévitrification des hyaloclastites interstratifiées ou dispersées dans les séries sédimentaires pélagiques [Gibson et Towe 1971, Calvert, 1971 ; Berger et Von Rad, 1972 ; Lancelot, 1973, etc.]. Cependant, on connaît des cas où d'épaisses couches de hyaloclastites ne portent pas trace de silicification. Par ailleurs, tous les dépôts anciens d'âge crétacé ou paléogène, à Diatomées et Radiolaires ne sont pas systématiquement silicifiés. Enfin, des sédiments riches en hyaloclastites ou en Radiolaires et Diatomées sont abondamment déposés depuis l'Éocène. Dans ces sédiments, la dévitrification des verres volcaniques a eu lieu ou est en cours de même que la dissolution des squelettes siliceux. En outre, par rapport à l'eau de mer, les eaux interstitielles profondes en général et celles des sédiments post-éocènes en particulier sont, dans bien des cas, considérablement enrichies en silice dissoute (10 à 50 ppm) [Sayles et Manheim, 1971 ; Sayles *et al.*, 1972] ; pourtant, à partir de l'Oligocène, les silicifications n'apparaissent qu'exceptionnellement. Seule la clinoptilotite semble pouvoir se développer en relative abondance, localement [Heath, 1969].

Discussion.

Ces faits laissent à penser que l'enrichissement relatif des eaux interstitielles en silice provenant soit de la dévitrification des verres volcaniques soit de la dissolution des squelettes siliceux est insuffisant pour entraîner une précipitation massive de silice dans les sédiments néogènes et quaternaires. Si l'on se réfère aux teneurs actuelles en silice dissoute des eaux de fond et des eaux interstitielles à proximité de l'interface eau/sédiment, la raison semble évidente. Les eaux de fond et les eaux interstitielles proches sont, au départ, trop pauvres en silice ; elles peuvent même, théoriquement, dissoudre non seulement la silice amorphe mais aussi la cristobalite, la calcédoine et même le quartz [Wey et Siffert, 1961]. Ceci implique en définitive que la teneur en silice dissoute dans l'eau de mer, dont un certain volume constitue par la suite l'eau interstitielle (après piégeage dans les sédiments), doit jouer un rôle : trop faible, elle empêcherait toute précipitation donc toute silicification, ce serait le cas de nos jours ; trop forte, c'est-à-dire proche de la saturation ou à saturation, elle pourrait conduire à des précipitations massives directement au niveau de l'interface eau/sédiment ou plus vraisemblablement dans les premiers mètres de dépôt encore imbibés d'eau.

LES APPORTS EN SILICE DISSOUTE : CAUSES PROBABLES DE FLUCTUATIONS.

Accélérations de l'expansion du plancher basaltique et fluctuations possibles des apports d'origine océanique.

L'étude des anomalies magnétiques du plancher océanique montre que son expansion a sans doute été affectée par des phases de paroxysme caractérisées par de fortes accélérations pendant lesquelles la quantité de basalte amenée au contact de l'eau de mer a été plus élevée par unité de temps. L'une de ces phases, vraisemblablement la plus généralisée, se serait manifestée entre — 110 et — 80 millions d'années [Hayes et Pitman, 1973]. D'autres paroxysmes et changement de direction dans l'expansion ont été mis en évidence, par exemple dans l'Océan Indien au Crétacé supérieur [Laughton, Mackenzie et Sclater, 1972] et à l'Éocène [Schlich, 1974]. Le plus récent, qui a été marqué notamment par une accélération de la séparation Australie-Antarctique et par une expansion faiblement accélérée des fonds de l'Atlantique sud [Flemming et Roberts, 1973] date du Miocène. Toutefois, cette dernière phase est nettement plus faible que les autres. Selon Hart [1973], ces phases d'expansion accélérée auraient notablement influencé la composition du milieu océanique, tout particulièrement les teneurs en calcium et en silice de l'eau de mer. Cependant, si des épisodes de silicification intense dans les bassins océaniques semblent bien coïncider avec les deux premières phases, il n'en va pas de même pour la troisième. Vankatarathnam et Biscaye, [1973] interprètent le fait que la zéolite la plus riche en silice (la clinoptilotite) — dont l'abondance caractérise les dépôts anté-oligocènes — soit localisée dans des zones proches des continents, en invoquant un volcanisme sous-marin donnant des produits moins sous-saturés pendant la période anté-oligocène que par la suite. Les produits de ce volcanisme sous-marin des dorsales médio-océaniques auraient été plus siliceux en raison de la proximité des continents. On pourrait donc penser que ce phénomène puisse être à l'origine d'un enrichissement supplémentaire en silice de l'eau de mer. Mais, il ne faut pas oublier que si les dépôts riches en clinoptilolite sont plus proches des continents que les sédiments pauvres plus récents, c'est à cause du mécanisme même de l'expansion des fonds océaniques et que point n'est besoin de faire appel à un volcanisme plus siliceux pour expliquer la relation qui semble exister entre la proximité des continents et la richesse des sédiments en clinoptilolite.

Par conséquent, rien dans l'analyse qui précède ne semble indiquer que les causes possibles de fluctuations des apports d'origine océanique aient fortement influencé le dépôt de la silice. Le fait que la dernière phase d'expansion accélérée n'ait pas été accompagnée de silicifications massives dans les bassins océaniques semblerait plutôt prouver le contraire. En outre, pendant les périodes de « repos » relatif c'est-à-dire pendant les ralentissements ou les arrêts de l'expansion, les silicifications n'ont pas été interrompues de façon marquante. Pendant les phases de paroxysme, les apports accrus de silice n'ont sans doute fait qu'amplifier les effets d'un autre phénomène.

La latérisation cause de la fluctuation des apports d'origine continentale.

De nos jours les fleuves qui, au total, apportent le plus de silice dissoute dans le milieu océanique, sont l'Amazone et le Congo. L'Amazone déverse en mer chaque année environ $0,6.10^{14}$ g de silice soit 14 % environ du total des apports (Débit moyen : 200 000 m³/sec ; teneur moyenne en silice : 10,6 mg/1.). Il est très vraisemblable que la somme des apports de l'Amazone et du Congo représente de 25 à 30 % du total des apports annuels. Le fait que la contribution de ces fleuves puisse être aussi élevée n'est pas dû au hasard. Leurs bassins versants se trouvent sous climat équatorial relativement chaud et humide. C'est sous ce type de climat que l'hydrolyse des silicates est la plus intense et la plus rapide. Les calculs isovolumétriques [Millot

et Bonifas, 1955] ont permis de chiffrer les quantités de matière libérée, matière dont le rôle dans la sédimentation a été amplement démontré [Millot, 1964 ; Ehrart, 1967]. Cette hydrolyse est particulièrement forte sous le couvert de la grande forêt équatoriale. Elle a pour résultat l'édification de sols latéritiques et une libération de silice, de chaux et de magnésie dont la majeure partie est évacuée dans les océans. De nos jours, seulement 10 à 15 % de la surface des continents sont soumis à l'action latéritisante de ce type de climat. En a-t-il toujours été de même ?

Les traces de l'action de climats latéritisants sont essentiellement de deux sortes : les paléosols latéritiques (rares) et le faciès sidérolithique. Selon Millot [1964], ce faciès est constitué notamment par des couches rouges ferrugineuses et aussi par des sables quartzeux blancs, des accumulations de kaolinites, des calcaires lacustres et des dalles silicifiées. La reprise du manteau d'altération peut s'accompagner d'une nouvelle fuite de silice et conduire aux bauxites. Parallèlement, dans les bassins de sédimentation, le dépôt de la série chimique basique avec argiles néoformées (montmorillonite, palygorskites), les silex, phosphates et séries carbonatées sont le fruit des latéritisations du continent.

G. Millot [1964] a rassemblé de nombreuses données relatives à l'âge et l'extension des latéritisations au Crétacé et au Tertiaire. D'autres informations sur ce sujet sont résumées dans l'ouvrage de Furon [1972]. Le lecteur comprendra qu'il n'est pas possible d'entrer ici dans le détail. On se limitera à quelques généralités avec quelques exemples plus précis à l'appui.

En Europe, le faciès sidérolithique s'est développé sur toutes les parties émergées, principalement au Crétacé inférieur. C'est le faciès « wealdien » du bassin de Paris, dépôt kaolinique et ferrugineux qui traduit le démantèlement de la couverture latéritique des massifs émergés comme l'Ardenne, le Massif Central et les Vosges. L'hydrolyse des massifs émergés s'est poursuivie au Crétacé supérieur. Les bassins épicontinentaux comme le bassin de Paris sont marqués par le dépôt de grandes séries crayeuses, notamment du Cénomanien au Sénonien, séries très riches en silice sous forme de silex en table, en rognons ou sous forme de sphérules de cristobalite-tridymite [Leclaire, Alcaydé et Frœhlich, 1973] avec de la clinoptilolite associée [Estéoule *et al.*, 1971]. Le faciès sidérolithique se manifeste avec une ampleur encore plus grande à l'Éocène, après les grandes transgressions. Le Massif Armoricain, notamment dans sa partie S et SE est alors atteint de la « maladie du Tertiaire » [Milon, 1930]. L'Éocène du N de l'Aquitaine est aussi un bon exemple.

En Afrique, c'est peut-être au Crétacé que les faciès continentaux sidérolithiques ont l'extension la plus grande. Citons par exemple, le « wealdien » de Tanzanie qui contient des bois silicifiés comme le « continental intercalaire » du Sahara ; les « woodbeds » (300 m d'épaisseur) à Dinosauriens qui renferment une flore analogue à celle du « wealdien » du Sussex. Ce faciès se développe aussi au Gabon, du Turonien au Sénonien. Tout le paysage émergé à l'Éocène a aussi été profondément altéré. Les séries sédimentaires des bassins tertiaires de l'Afrique occidentale dérivent des latéritisations. Citons les bassins du Rio de Oro, de Mauritanie et du Sénégal où une sédimentation chimique à calcaires, cherts et argilites est remplacée à l'Oligocène par une sédimentation détritique à sable, kaolinite et micas. A Sangalcam (Sénégal), un forage a traversé une série argileuse de 500 m d'épaisseur, série constituée à 100 % par de l'attapulgite (palygorskite) et de la sépiolite.

En Amérique, le faciès sidérolithique est aussi connu au Crétacé inférieur dans le SE des Etats-Unis, mais c'est surtout au Crétacé supérieur et à l'Éocène que, dans l'Amérique atlantique, le faciès sidérolithique prend toute son ampleur.

En Asie, les importants niveaux de bauxite dans l'Aptien et le Turonien-Santonien de la Sibérie extrême-orientale sont significatifs, de même que les couches rouges de Shantoung et du Szetchouen en Asie méridionale.

Enfin, le Crétacé inférieur du *Spitzberg* est marqué par l'existence de faciès sidérolithique à empreintes de Dinosaures et bancs de lignite, etc.

Discussion.

Ce bref survol de la répartition des faciès dérivés de l'altération latéritique des continents, survol volontairement abrégé et inévitablement très incomplet, suffit cependant à souligner la grande extension des latéritisations pendant le Crétacé et l'Éocène. Bien que la formation du faciès sidérolithique se soit poursuivie pendant l'Oligocène et même jusqu'au Miocène en Europe méridionale, il est certain que sous des latitudes plus élevées, il régresse ou disparaît dès l'Oligocène. Ce fait semble indiquer un net recul vers l'équateur des latéritisations à partir de l'Oligocène.

Compte tenu d'une certaine dérive latitudinale, il semble donc qu'au Crétacé et à l'Éocène, 60 à 80 % de la surface des continents étaient sous climat hydrolysant. On peut donc penser que, par rapport à l'époque actuelle, la quantité de silice déversée en mer par les rivières fut considérablement plus élevée.

Les preuves directes de l'influence des latéritisations sur la sédimentation en milieu océanique sont peut-être à rechercher dans la réinterprétation de données publiées dans les « Initial Reports » du « Deep Sea Drilling Project ». Dans le bassin du Mozambique (Leg 25) par exemple, une puissante série argileuse à palygorskite, clinoptilolite et cristobalite d'âge éocène a été interprétée comme le résultat d'une dévitrification de hyaloclastites. Bien que des traces de cendres volcaniques aient été observées, elles sont très rares. Si l'association minérale dominante est bien typique de l'altération de produits volcaniques [Hay, 1966], elle est aussi caractéristique des néogenèses qui prennent naissance dans les bassins sédimentaires recevant les produits libérés par la latéritisation [Millot, 1964]. Il subsiste donc une incertitude et il n'est pas totalement exclu que la série argileuse du bassin du Mozambique provienne toute ou partie d'une phase de latéritisation éocène de l'Afrique orientale ou de Madagascar.

De même, comment ne pas être tenté de voir dans la gigantesque accumulation de palygorskite-sépiolite paléogène au large de l'Afrique occidentale (voir précédemment), le prolongement des séries équivalentes qui se trouvent dans les bassins épicontinentaux tertiaires de l'Ouest africain : bassin du Rio de Oro, de Mauritanie, du Sénégal, etc. Rappelons par exemple la présence à Sangalcam (Sénégal) d'une puissante formation à palygorskite et sépiolite. Cependant, cette accumulation de palygorskite au large de l'Afrique a été interprétée de diverses manières [Pimm et Hayes, 1972] et notamment comme résultant de l'altération de cendres volcaniques par des eaux connées riches en magnésium [Peterson *et al.*, 1970].

L'extension et l'importance des faciès résiduels édifiés à la suite de phases de latéritisations intenses témoignent d'une manière directe de l'« hémorragie » de la silice des continents pendant le Crétacé et l'Éocène. Ce phénomène peut être à l'origine d'arrivées massives de silice dissoute dans les eaux océaniques. Dans cette hypothèse, le fait qu'à partir de l'Oligocène, les latéritisations perdent de leur ampleur, et n'affectent plus que les zones proches de l'équateur, se serait traduit par une diminution progressive du volume des apports à l'océan ; ce phénomène coïncidant avec la raréfaction des silicifications dans les grands bassins océaniques pourrait en être à l'origine.

L'EXTENSION DES CLIMATS HYDROLYSANTS AU CRÉTACÉ ET AU TERTIAIRE. LA RÉVOLUTION CLIMATIQUE DE LA FIN DE L'OLIGOCÈNE.

A la preuve indirecte que constitue l'intense latéritisation des continents au Crétacé et au Paléogène, s'ajoutent d'autres faits en faveur d'une grande extension latitudinale d'un climat de type tropical humide.

Furon [1972] a rassemblé de nombreuses données

permettant d'établir une première synthèse paléo-climatologique du Jurassique à l'époque actuelle. Retenons à titre d'exemple, l'extension jusqu'aux hautes latitudes des Dinosaures au Crétacé, et le fait que les gisements de charbon du Crétacé et de l'Éocène renferment généralement une flore typique des climats relativement chauds et humides. En Sibérie, le Sénonien est riche en restes d'une flore latifoliée ayant poussé sous climat chaud et humide. Au Santonien, les arbres à pain ont atteint la latitude de 70°. Un fait particulièrement frappant est souligné dans l'ouvrage de Furon : la grande uniformité dans la répartition et la composition de la flore et de la faune du Jurassique à l'Oligocène, qui indique une absence de zonation climatique sur la majeure partie des continents. Une zonation ne commence à apparaître qu'à l'Oligocène. Une chute importante de température se serait produite à la surface du globe dès l'Oligocène supérieur, chute accompagnant une importante dégradation climatique.

Ces conclusions sont confirmées par d'autres observations. L'étude des feuilles fossiles conservées dans l'W des États-Unis a conduit Dorf [1963], Wolfe et Hopkins [1967] à tracer des courbes paléo-climatiques. Ces courbes, bien que criticables, dans le détail [Axelrod et Bailey, 1969] mettent en évidence une importante dégradation climatique à la fin de l'Oligocène [Wolfe, 1971]. Par ailleurs, Mandra et Brigger [1972], partant de l'étude de Silicoflagellés prélevés dans deux carottes de l'Atlantique S et d'échantillons de Nouvelle-Zélande concluent qu'à l'Éocène supérieur, la zone tropicale s'étendait jusqu'au 60e degré de latitude S ; cette latitude représente aujourd'hui la limite septentrionale de l'extension des icebergs. En outre, à l'Éocène, le bras de mer étroit qui séparait l'Australie de l'Antarctique était constitué par des eaux chaudes, alors qu'une intense végétation forestière recouvrait le continent antarctique [Kemp, 1972]. Les forages du Leg 28 [Hayes et Frakes, 1973] dans la plaine abyssale de Wilkes (entre l'Australie et l'Antarctique) confirment en première analyse que les conditions climatiques régionales se sont considérablement détériorées au tout début du Miocène et sans doute dès l'Oligocène, avec la première manifestation d'une importante phase de glaciation.

Tous ces faits, bien que relativement fragmentaires, indiquent que le Paléogène terminal a été marqué par un recul considérable des climats chauds et humides vers l'équateur. Avec l'apparition d'une zonation latidutinale, c'est une véritable « révolution climatique » qui a bouleversé l'environnement terrestre à cette époque.

Discussion.

Les données de la paléoclimatologie sont donc bien en accord avec celles de la sédimentologie :

la fin du Paléogène a été marquée par une réduction considérable des surfaces continentales soumises à l'action des climats hydrolysants. Il est donc vraisemblable qu'il y eut, de ce fait, réduction des apports en silice dissoute dans le milieu marin. Quelles peuvent être les conséquences d'un tel événement ?

Une réduction des apports ne peut qu'affecter la balance de la silice des océans. Dans l'hypothèse d'une balance de la silice grossièrement équilibrée, on peut prévoir que la masse de silice sédimentée (dont dépendent les silicifications) soit directement proportionnelle aux apports. Cette hypothèse suffirait donc à expliquer les silicifications intenses du Crétacé et de l'Éocène et aussi leur raréfaction par la suite. On pourrait aussi s'attendre à un ralentissement brutal de la productivité de l'eau de mer en plancton siliceux qui aurait entraîné une raréfaction des sédiments siliceux d'origine biogène dès la fin du Paléogène ; or, un tel ralentissement ne semble pas s'être produit sauf peut-être sur la plate-forme russe [Distanov et al., 1971]. On a seulement constaté que le squelette des Radiolaires contenus dans les dépôts océaniques, massif auparavant, devenait beaucoup plus fin et délicat [Caulet, communication orale]. Si la productivité de l'eau de mer en plancton siliceux ne semble donc pas avoir été gravement et brutalement affectée, c'est peut-être en raison du « volant » que devait constituer la masse totale de silice dissoute dans les océans. Cette masse sans doute beaucoup plus importante pendant l'Éocène que de nos jours, aurait joué le rôle de « tampon ». Suivant cette nouvelle hypothèse, la productivité de l'eau de mer en plancton siliceux, sans doute lentement décroissante, se serait cependant maintenue à des taux relativement élevés ; la silice fixée par les organismes siliceux étant prélevée sur la réserve des océans. La conséquence logique est une diminution de la quantité totale de silice dissoute dans les océans, c'est-à-dire un appauvrissement de l'eau de mer. Cet appauvrissement a sans doute été aggravé depuis le Miocène en raison du développement d'intenses « upwellings » au niveau des zones circumpolaires et équatoriales, où la productivité de l'eau de mer en plancton siliceux et la sédimentation de silice par voie biochimique sont intenses. A titre d'exemple, les diatomites quaternaires au large de l'antarctique, sous la divergence antarctique, se sont déposées localement à des vitesses de l'ordre de 150 m/million d'années [Hayes, Frakes et al. 1973]. Rappelons que la divergence antarctique résulte des nouvelles conditions climatiques et du développement d'une vaste circulation circumpolaire rendue possible à la suite de la séparation Australie-Antarctique et de l'ouverture de passage de Drake entre l'Amérique du Sud et l'Antarctique.

Cette hypothèse de l'appauvrissement en silice des océans pourrait expliquer les teneurs anormalement faibles de l'eau de mer actuelle. Elle va dans le sens d'une balance déficitaire de la silice depuis l'Oligocène, évoquée précédemment. Toujours en partant de cette hypothèse, on peut se demander s'il est possible d'avoir une idée sur le degré de richesse en silice des océans anté-oligocènes. Deflandre [1936, 1966] nous a fait connaître la finesse et la diversité des restes organiques conservés dans les silex du bassin de Paris. Des formes aussi délicates que les organismes flagelliformes des Ophiobolacées, d'une longueur plusieurs fois supérieure à la cellule mère, ont été préservées de tout dommage. Des membranes cellulosiques de Dinoflagellés restent même colorables au moyen des couleurs d'aniline et du rouge de ruthénium. L'exceptionnelle conservation des micro-organismes dans certaines silicifications précambriennes a valu à ces silex l'appellation d'« ambre du Précambrien ». Cette image suggère évidemment la possibilité d'une précipitation directe, massive et quasi instantanée de la silice. Une telle précipitation, si elle est possible, ne peut avoir lieu que lorsque la teneur en silice des eaux de fond, où en tous cas des eaux interstitielles proches de l'interface eau/sédiment, atteint un certain degré de sursaturation. Cette sursaturation a pu se produire soit à cause d'apports massifs dans un très bref laps de temps soit à cause de l'adjonction à des eaux océaniques déjà très riches, d'une quantité limitée de silice provenant de la dissolution de squelettes siliceux ou de la dévitrification de cendres volcaniques.

Cette discussion laisse donc entrevoir tout d'abord, une réduction assez brutale et considérable des apports en silice provenant des continents à partir de l'Oligocène et aussi une diminution lente mais progressive des teneurs en silice dissoute de l'eau de mer. Telle pourrait être la nature du changement dans les conditions de sédimentation auquel il est fait allusion précédemment. Plus que la réduction des apports qui reste cependant la cause première, c'est vraisemblablement l'appauvrissement en silice de l'eau de mer qui serait directement à l'origine de la raréfaction des silicifications dans les bassins océaniques.

CONCLUSION.

Au terme de cet exposé général, il apparaît que de très importants transferts de silice des continents dans les océans ont eu lieu pendant le Crétacé et l'Éocène, essentiellement du fait de l'extension en latitude des latéritisations. En effet, du Jurassique à l'Oligocène, environ trois quarts de la surface des continents se sont trouvés, sous climats hydro-

lysants chauds et humides, plus ou moins permanents. Dans l'hypothèse d'une balance de la silice des océans à l'état d'équilibre, les apports massifs d'origine continentale, auxquels se sont ajoutés les produits de l'altération du plancher océanique, notamment pendant les phases d'expansion accélérée, ont dû être compensés par une précipitation (sédimentation) équivalente.

De fait, on trouve dans les grands bassins océaniques, des silicifications intenses affectant les séries lithologiques crétacées et éocènes. En raison d'un recul prononcé des climats hydrolysants vers l'équateur à partir de l'Oligocène, les apports d'origine continentale se sont ensuite considérablement réduits. Cette réduction coïncide avec la raréfaction des silicifications dans les bassins océaniques. La relation entre la nature du climat sur les continents et la silicification dans ces bassins est donc évidente.

Si la réduction des apports d'origine continentale permet à elle seule d'interpréter la raréfaction des silicifications dès l'Oligocène, elle n'explique pas le degré de forte sous-saturation de l'eau de mer actuelle. On est donc amené à penser que la réserve en silice dissoute du milieu océanique a aussi été affectée et qu'elle aurait considérablement diminué depuis l'Oligocène. Cet appauvrissement, sans doute lié, au départ, à la diminution des apports, a peut-être été accéléré depuis le Pliocène en raison du développement d'« upwellings » au niveau des divergences polaires et équatoriales, zones au-dessous desquelles la précipitation de silice par voie biochimique est intense mais localisée. Tout laisse à penser que pendant le Crétacé et l'Éocène, la réserve en silice du milieu océanique était beaucoup plus importante que depuis l'Oligocène. En d'autres termes, la teneur en silice de l'eau de mer de ces époques devait être beaucoup plus élevée qu'actuellement. Il n'est pas impossible que localement, la saturation ait été atteinte, voire même dépassée au moins au niveau des eaux interstitielles proches de l'interface eau/sédiment. Quoiqu'il en soit, le fait essentiel est que la silicification dans les bassins océaniques semble avoir été, indirectement, sous contrôle climatique.

Remerciements.

La reconnaissance de l'auteur va aux responsables du « Deep Sea Drilling Project » et à R. Schlich de l'Institut de Physique du Globe qui lui ont permis de participer au programme du J.O.I.D.E.S. Cette participation a joué un rôle déterminant dans la conception de cette note.

L'auteur remercie très vivement Monsieur le Professeur R. Laffite pour les conseils judicieux qu'il lui a prodigués et Monsieur le Professeur G. Millot qui, par ses suggestions et remarques pertinentes, a aidé à parachever ce travail.

1. Les numéros des campagnes et les numéros des volumes relatifs à ces campagnes sont identiques. Pour plus de détail, l'auteur renvoie donc le lecteur aux publications du « Deep Sea Drilling Project » (U.S. Government Printing Office), Washington D.C., de 1969 à 1973, sans préciser davantage.

2. La clinoptilolite est une zéolite sodi-potassique, riche en silice et voisine de l'heulandite. La phillipsite est une zéolite

calcique plus pauvre en silice. La montmorillonite est plus riche en alumine que les argiles magnésiennes, comme la palygorskite.

Selon Hay [1966], les feldspaths potassiques et l'albite peuvent être authigènes et provenir d'une transformation des zéolites.

Références citées

ANDREWS J. E. et PACKAM G. (1973). — Southeast Pacific Structures. Leg 30 : Deep Sea Drilling Project. *Geotimes*, vol. 18, n° 9, p. 18-27.

AXELROD D. I. et BAILEY P. H. (1969). — Paleotemperature analysis of tertiary floras. *Pal., Pal., Pal.*, vol. 6, p. 163-165.

BERGER W. H. et RAD U. VON (1972). — Cretaceous and Cenozoic sediments from the Atlantic Ocean. *In* Initial Reports of the Deep Sea Drilling Project (U.S. Government Printing Office), Washington, vol. 14, p. 787-954.

BRUTY D., CHESTER R. et RASTON S. (1973). — Trace elements in ancient atlantic deep sea sediments. *Nature Physical Science*, vol. 245, p. 73-74.

CAILLEUX L. (1929). — Les roches sédimentaires de France (roches siliceuses). *Mém. Serv. Carte géol. France*, 1 vol., 774 p., 30 pl.

CALVERT S. E. (1968). — Silica balance in the ocean and diagenesis. *Nature*, vol. 219, p. 919-920.

CALVERT S. E. (1971). — Nature of silica phases in deep sea cherts from the North Atlantic. *Nature Physical Science*, vol. 234, p. 133-134.

DAVIES T. A. et SUPKO P. R. (1973). — Oceanic sediments and their diagenesis : some examples from deep sea drilling. *Journ. of Sedimentary Petrology*, vol. 43, p. 381-391.

DEFLANDRE G. (1936). — Microfossiles des silex crétacés. *Ann. Paléont.*, t. 25, f. 4, p. 149-191.

DEFLANDRE G. (1966). — Microfossiles des silex crétacés. 2e éd., Publ. Lab. Micropal., École Pratique Hautes Études, Paris, 1 vol., 50 p., 18 pl.

DISTANOV U. G., KOPEJKIN V. A., KUZNETSOVA T. A. et SILANTLEV V. N. (1971). — Particularités de l'accumulation de silice dans les bassins marins au cours du Mésozoïque et du Cénozoïque. *Dokl. Akad. Nauk. U.R.S.S.*, vol. 201, p. 668-671.

DORF E. (1963). — The use fossil plants in paleoclimatic interpretations. *In* Problems in Paleoclimatology, E. M. Nairn edit. Interscience, London, p. 13-31.

ELDERFIELD H. (1972). — Effects of volcanism on water chemistry. Deception Island, Antarctica. *Marine Geology*, vol. 13, p. M1.

ERHART H. (1967). — La genèse des sols en tant que phénomène géologique. 2e éd., Paris, Masson et Cie édit., 177 p.

ESTÉOULE J., ESTÉOULE-CHOUX J. et LOUAIL J. (1971). — Sur la présence de clinoptilolite dans les dépôts marno-calcaires du Crétacé supérieur de l'Anjou. *C. R. Ac. Sc.*, Paris, 272, sér. D, p. 1569.

FLEMMING N. C. et ROBERTS D. G. (1973). — Tectono-eustatic changes in sea level and sea-floor spreading. *Nature*, vol. 243, p. M1.

FURON R. (1973). — Éléments de paléoclimatologie, Paris, Vuibert édit., sér. Sciences de la Terre, 1 vol., 216 p.

GARTNER S. (1970). — Sea-floor spreading, carbonate dissolution level and the nature of Horizon A. *Science*, vol. 169, p. 1077-1079.

GIBSON T. G. et TOWE K. M. (1971). — Eocene volcanism and the origin of horizon A. *Science*, vol. 172, p. 152-154.

GREGOR B. (1968). — Silica balance of the Ocean. *Nature*, vol. 219, p. 360-361.

HARRISS R. C. (1966). — Biological buffering of oceanic silica. *Nature*, vol. 212, p. 275-276.

HART R. A. (1973). — Geochemical and geophysical implications of the reaction between sea water and the oceanic crust. *Nature*, vol. 243, p. 76-78.

HAY R. L. (1966). — Zeolites and zeolitic reactions in sedimentary rocks. *Geol. Soc. Amer. Bull.*, Spec. pap. n° 85, p. 48-107.

HAYES J. D. et PITMAN W. C. (1973). — Lithospheric plate motion, sea level changes and climatic and ecological consequences. *Nature*, vol. 246, p. 18-22.

HAYES D. E., FRAKES L. A. *et al.* (1973). — Leg. 28. Deep Sea Drilling in the Southern Ocean. *Geotimes*, vol. 18, p. 19-24.

HEATH (1969). — Mineralogy of Cenozoic Deep-Sea Sediments from the equatorial Pacific Ocean. *Geol. Soc. Amer. Bull.*, vol. 80, p. 1997-2018.

HELGESON H. C. et McKENZIE F. T. (1970). — Silicate- sea water equilibria in the ocean system. *Deep Sea Res.*, vol. 17, p. 877-892.

KEMP E. M. (1972). — Reworked Palynomorphs from the West Ice Shelf area, East Antarctica, and paleoclimatological significance. *Marine Geol.*, vol. 13, p. 145-157.

LANCELOT Y. (1973). — Cherts and silica diagenesis in sediments from the Central Pacific. Initial Reports of the Deep Sea Drilling Project (U.S. Government Printing Office, Washington), vol. 17, p. 377-405.

LAUGHTON A. S., McKENZIE D. P. et SCLATER I. G. (1972). — The structure and evolution of the Indian Ocean. *In* 24th Int. Geol. Congr., Montréal, sect. 8, p. 65-73.

LECLAIRE L. (1974). — Late Cretaceous and Cenozoic Pelagic Deposits of the Central Western Indian Ocean. Paleoenvironment and Paleooceanography. *In* Initial Reports of the Deep Sea Drilling Project (U.S. Government Printing Office, Washington) vol. 25 (sous presse).

LECLAIRE L., ALCAYDÉ G. et FROEHLICH F. (1973). — La silicification des craies, rôle des sphérules de cristobalite-tridymite observées dans les craies des bassins océaniques et dans celles du Bassin de Paris. *C. R. Ac. Sc.*, Paris, 227, sér. D, p. 2121-2124.

LIZITSIN A. P. (1971). — Distribution of siliceous microfossils in suspension and in bottom sediments. *In* The Micropaleontology of Oceans, edit. B. M. Funnell et Riedel, p. 173-197.

McKENZIE F. T. et GARRELS R. M. (1966). — Silica mass balance between rivers and ocean. *Amer. Journ. Sc.*, vol. 264, p. 507-525.

McKENZIE F. T. (1967). — Silica in sea water : control by silica minerals. *Science*, vol. 155, p. 1404-1405.

MANDRA Y. T. et BRIGGER A. L. (1972). — Plate Tectonics,

paleomagnetism, tropical climate and Upper Eocene Silicoflagellates. *Antarctic Journ.*, sept.-oct., p. 191-193.

MILLOT G. (1964). — La géologie des Argiles, Paris, Masson et Cie édit., 1 vol., 499 p.

MILLOT G. et BONIFAS M. (1955). — Transformations isovolumétriques dans les phénomènes de latéritisation et de bauxitisation. *Bull. Serv. Carte Géol. Als. Lor.*, vol. 8, p. 3-10.

MILON Y. (1930). — L'extension des formations sidérolithiques éocènes dans le centre de la Bretagne. *C. R. Ac. Sc.*, Paris, 194, p. 1360-1362.

PETERSON M. N. A. *et al.* (1970). — Initial Reports of the Deep Sea Drilling Project (Government Printing Office, Washington), vol. 2, p. 413.

PERRY E. A. (1971). — Silicate-sea water equilibria in the ocean system : a discussion. *Deep Sea Res.*, vol. 18, p. 921-924.

PIMM A. C. et HAYES D. E. (1972). — General Synthesis. Initial Reports of the Deep Sea Drilling Project (U.S. Government Printing Office, Washington), vol. 14, p. 955-975.

SAYLES F. L. et MANHEIM F. T. (1971). — Interstitial water on small core samples Deep Sea Drilling Project, Leg. 7. Initial Reports of the Deep Sea Drilling Project (U.S. Government Printing Office, Washington), vol. 7, p. 871-881.

SAYLES F. L. *et al.* (1972). — Interstitial water studies on small cores samples. Leg. 9, Deep Sea Drilling Project. Initial Reports of the Deep Sea Drilling Project (U.S. Government Printing Office, Washington), vol. 9, p. 845-855.

SCHLICH R. (1974). — Initial Reports of the Deep Sea Drilling Project (U.S. Government Printing Office, Washington), vol. 25 (sous presse).

VENKATARATHNAM K. et BISCAYE P. E. (1973). — Deep sea zeolites ; variations in space and time in the sediments of the Indian Ocean. *Marine Geol.*, vol. 15, p. M 11-M 17.

WEAVER M. et WISE W. (1972). — Deep sea cristobalites cherts and authigenic mineral. *Nature Physical Science*, vol. 237, p. 56-57.

WEY P. et SIFFERT B. (1961). — Réaction de la silice monomoléculaire en solution avec les ions Al^{3+} et Mg^{2+}. *In* Genèse et synthèse des argiles. Coll. Int. C.N.R.S., vol. 105, p. 11-23.

WOLFE J. A. et HOPKINS D. M. (1967). — Climatic changes recorded by Tertiary land floras in northwestern North America. *In* Tertiary correlations and climatic changes in the Pacific, Hatai, K. edit., Symp. Pac. Sc. Congr., Tokyo, vol. 25, p. 67-76.

WOLFE J. A. (1971). — Tertiary climatic fluctuations and methods of analysis of tertiary floras. *Pal., Pal., Pal.*, vol. 9, p. 27-57.

19

Reprinted from *Science* **174**:55–57 (Oct. 1, 1971)

FECAL PELLETS: ROLE IN SEDIMENTATION
OF PELAGIC DIATOMS

Hans-Joachim Schrader

*Geologisches-Paläontologisches Institut und Museum der
Universität Kiel*

Abstract. *Membrane-enclosed fecal pellets of planktonic herbivores were sampled at several depths in the Baltic Sea (459 meters deep) and off Portugal (4000 meters deep) by means of a Simonsen multinet. Pellets contained mainly empty shells of planktonic diatoms and silicoflagellates. Two kinds of fecal pellets were found, those with the remains of one species (for example,* Thalassiosira baltica) *and those with the remains of several species (for example,* Chaetoceros, Achnanthes, *and* Thalassiosira). *Siliceous skeletons were protected from dissolution during settling by a membrane around the pellet.*

Pelagic diatoms are a major factor in the biology, chemistry, and sedimentation of the oceans (*1*). They are the most important primary producers at the base of the food chain (*2*) and

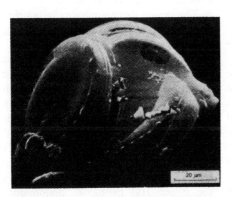

Fig. 1. Fecal pellet of *Calanus finmarchicus* Gunnerus with the centric diatom *T. baltica* enclosed within. The damage to the fecal pellet membrane occurred during coating with metal in a vacuum (scanning electron micrograph).

may be responsible for the general undersaturation of seawater with silica (*3*), because the rate at which they deposit silica exceeds the influx of dissolved silica. The excess supply of silica demanded by diatom frustules necessitates re-solution of sedimented diatom frustules to maintain the geochemical balance of silica in the water column (*4*). Evidence for dissolution in the upper centimeters of sediment has been presented (*5–7*). In addition to their important role in the modern ocean, diatoms are a diversified group of shelled plankton potentially useful in the reconstruction of ancient ocean conditions (*8*).

In discussions of the biochemical and geological aspects of diatom sedimentation, the role of fecal pellets, although occasionally mentioned (*9–11*), until recently has not received the attention it deserves (*12–14*). Membranes surrounding the fecal pellets of planktonic herbivores are common (*15*). This report gives additional evidence that certain copepods, mainly calanoid ones, form a membrane encasing voided diatom shells. The resulting pellets sink many times faster than a single shell, thus rapidly transporting silica from surface water to deeper waters. The rate of sinking for a single *Thalassiosira baltica* (Grun.) Ostenfeld frustule (50 μm in diameter) is approximately 10 m/day (*14*); the rate of sinking for a fecal pellet (100 μm in diameter), how, ever, is about 100 m/day (*16*). In addition, the membrane (Fig. 1) protects the enclosed diatoms from dissolution during their descent to the ocean floor.

For study of diatom sedimentation, samples were taken at several depths during a cruise in the Baltic Sea (*Planet* cruise 5, 1969) and an expedition off Portugal (Rossbreiten Expedition, *Meteor* cruise 19, 1970). In the Baltic the Simonsen phytoplankton multinet (*17*) (mesh size 41 μm) was used to sample at 0 to 70, 70 to 140, 140 to 210, 210 to 280, 280 to 350, and 380 to 420 meters deep in vertical hauls. About 20 m³ were filtered by the closing net towed along about a 70-m transect for each subsample. Samples were also collected with Nansen bottles from various depths (2, 40, 80, 120, 160, 200, 240, 280, and 320 m). One-half liter of the water sample was filtered through a membrane filter. The filters were then dried, and one tenth of each filter was embedded in mounting medium (Aroclor 4465) for slide preparations for examination by light microscopy. All diatoms, feces, and dinoflagellates were counted. Off Portugal, 12 samples were collected from various depths down to 4000 m in one vertical haul with the phytoplankton multinet (*18*), and about 80 m³ were filtered by the multinet along a 300-m vertical transect. Microscopical strewn slides were made from plankton sam-

ples, and all particles (feces, diatoms, dinoflagellates, copepods, and so forth) were counted. The results given here are mainly from the Landsort Basin in the Baltic (depth 459 m).

The centric diatom *T. baltica* was extremely abundant in the Baltic surface waters (25,000 cells per liter) at the time of the expedition (May 1969). Living cells of this species have a characteristic structure of fine, small circular pores with radially arranged teeth projecting from the rim into the center of the aperture (Fig. 2, top). In empty shells caught at depth there is a tendency for a reduction in the number of teeth and for pore enlargement. These morphological changes were interpreted as effects of dissolution. The relative state of preservation of these diatom frustules was measured by calculating (i) the pore density as the percentage of pore space in the total valve area, and (ii) the tooth index, that is, the number of teeth per pore times their length. Both indices were measured in fractions of micrometers. Measurements were made from photo enlargements (× 25,000) of 30 shells of *T. baltica* picked from six multinet subsamples [GIK (Geologisches Institut Kiel) No. 10,064-1; depths as in Fig. 2] and photographed with a scanning electron microscope (× 10,000).

The results demonstrate (Fig. 3) that both indices measure dissolution. Shells generally are significantly more deteriorated in deeper water than they are in shallow water. The aberrant value for the population at 280 to 350 m (*D* in Fig. 3) reflects a mixture of completely preserved (Fig. 2, top) and strongly attacked frustules (Fig. 2, bottom) found there. I suggest that the preserved shells were transported to this depth within fecal pellets. Feces were abundant in the upper 100 m (2500 diatom-containing pellets per liter). Pellets were of two kinds, those containing but one species (*T. baltica*, or chains of *Achnan-*

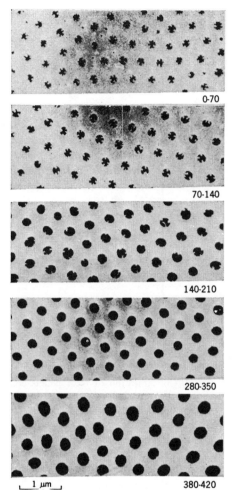

Fig. 2. Fine structure of valves of *T. baltica* frustules collected at various depths (in meters). In the top panel (collected from a transect at 0 to 70 m) the structure of a living cell is given. Other shells (collected from transects at depths of 70 to 140, 140 to 210, 280 to 350, and 380 to 420 m) represent different stages of dissolution during settling (scanning electron micrograph).

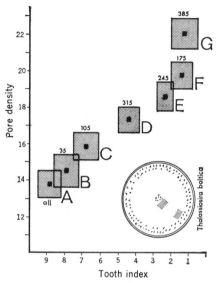

Fig. 3. Dissolution diagram of *T. baltica*. Index letters represent subsamples; the numbers over the squares give medium depths in meters. Index *A* represents parameters of living shells and those gained from intact pellets independent of depth of sampling. Indices *B, C, E, F,* and *G* represent parameters of settling frustules. The aberrant values for index *D* reflect mixture of completely preserved and strongly corroded shells. The picture of *T. baltica* on the right shows dotted areas where measurements were carried out on each specimen. See text for an explanation of method of calculation and units.

thes taeniata Grunow.) and those containing a variety of species (*T. baltica, A. taeniata,* pieces of broken *Chaetoceros* species, Ebriaceae). Fresh pellets are completely covered by the membrane (Fig. 1), and the state of preservation of the shells contained is very similar to that of living diatoms.

Thus skeletons of *T. baltica* are not damaged during passage through the gut of the copepod. *Calanus finmarchicus* Gunnerus was at that time the only predominant herbivore in the samples. The fecal pellets shown in the figures were found in guts of many *C. finmarchicus,* so it is probable that most round, ball-shaped feces containing diatoms were produced by this herbivore. As long as the fecal membrane lasts, little exchange of water can take place between the interior and the exterior of the pellet, and the diatom shells are safe from dissolution. Most fecal pellets apparently disintegrated at about

300 m, but a sizable fraction may reach the sea floor intact (Fig. 4). Moore (*19*) reports that zooplankton fecal pellets were only detected in nearshore sediments of the English coast to a depth of 166 m. In the Baltic diatomaceous feces were found on the surface of the sea floor, down to 459 m deep.

The transport of diatoms within fecal membranes potentially enriches the sediment with species that are preferentially eaten by certain herbivores, and whose shells escape harm during digestion. Selective feeding is reported for several herbivores (*20*), and differences in resistance to disintegration (*10, 21*) may characterize the different diatom species ingested. Thus, whereas *T. baltica* shells remained completed unharmed, members of the genus *Chaetoceros*, for example, were crushed to such a degree before being packed into fecal pellets that identification was very difficult. In contrast to previous findings (*11, 12*), *Calanus finmarchicus* does feed upon *Chaetoceros* species sometimes, as is obvious from its gut contents.

Phytoplankton particles smaller than 30 μm have not been observed to be actively filtered by calanoid copepods. (Some pelagic diatom species, *Fragilariopsis, Nitzschia*, some pelagic species of *Thalassiosira*, and *Coscinodiscus* are this size.) Thus a bias may be introduced against the sedimentation of very small diatom frustules by feces.

These predation processes may be of considerable importance in the formation of deep-sea sediments. Off Portugal, west of Porto, zooplankton feces which contain masses of diatom shells were found down to 4000 m deep (approximately two pellets per cubic meter at depths from 3300 to 3600 m). These pellets were packed with diatom species (*Skeletonema, Rhizosolenia alata, Eucampia*) that would never have settled singly without being completely dissolved before reaching this depth (*21,*

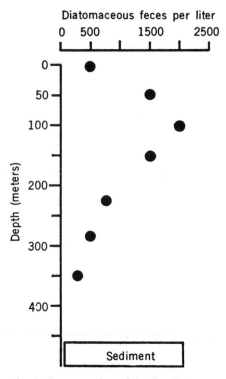

Fig. 4. Concentration of fecal pellets containing diatom shells in waters of the Landsort Basin in the Baltic.

22). The same is true for silicoflagellates, the skeletons of which were present in large numbers in some pellets; this is somewhat surprising in view of the scarcity of silicoflagellates in the upper water at the time (*23*).

The fact that diatoms, the basic staple of oceanic herbivores, are packed into fecal pellets, most of which have membranes (*15*), has the following suggested implications. By virtue of their greater size (50 to 250 μm in diameter), pellets rapidly leave the photic zone, sinking between 40 to 400 m/day (*14, 16*) and thus depleting upper waters of silica (*3*) and other nutrients. Shorter settling times provide for a better chance of preservation and, incidentally, for less drift of shells away from their place of origin, thus accounting for the fact that surface water masses can be reflected in sediments below. Preservation is further enhanced in diatoms that are eaten by copepods that pass the skeletons unharmed and encase them in a fecal membrane. Thus, much of the difference between living diatom assemblages and sediment assemblages may come from biological interactions, not simply from differential chemical dissolution favoring robust shells. Sedimentation by pellets allows formation of annual diatomaceous varves (*19, 24*), since without the accelerated sinking (*14*) diatoms would take so long to reach the ocean floor that seasonal changes would not be recorded. Sedimentation of pellets may largely provide delivery of an excess supply of silica to the ocean floor, since most silica would otherwise dissolve within the water column. This excess extraction drives the water toward undersaturation until solution rates on the ocean floor provide the necessary balance for equilibrium of the silica reservoir in the oceans.

References and Notes

1. J. H. Ryther, *Science* **166**, 72 (1969); A. P. Lisitzin, *Scientific Exploration of the South Pacific*, W. S. Wooster, Ed. (National Academy of Sciences, Washington, D.C., 1970), pp. 89–132.
2. N. J. Hendey, *An Introductory Account of the Smaller Algae of British Coastal Waters* (Her Majesty's Stationery Office, London, 1964), pt. 5.
3. S. E. Calvert, *Nature* **219**, 919 (1968).
4. W. H. Berger, *Geol. Soc. Am. Bull.* **81**, 1385 (1970).
5. K. A. Fanning and D. R. Schink, *Limnol. Oceanogr.* **14**, 59 (1969).
6. H.-J. Schrader, *Proceedings of the Second Planktonic Conference of Rome 1970*, A. Farinacci, Ed. (Edizioni Tecnoscienza, Rome, 1971), vol. 2, pp. 1149–1155.
7. ——, *"Meteor" Forschungsergeb. Reihe C*, in press.
8. T. Kanaya and J. Koizumi, *Sci. Rep. Res. Inst. Tohoku Univ. Ser. 2* **37**, 89 (1966).
9. T. J. Smayda, *Int.-Am. Trop. Tuna Comm. Bull.* **9**, 465 (1965); M. Ye. Vinogradov, *Am. Geophys. Union* **136/141**, 39 (1961) (translated from Russian); C. Apstein, *Int. Rev. Ges. Hydrobiol. Hydrogr.* **3**, 17 (1910); R. W. Kolbe, *Reports of the Swedish Deep-Sea Expedition 1947–1948*, H. Pettersson, Ed. (Göteborgs Kungliga Vetenskaps-och Vitterhets-Samahälle, Goteborg, 1955), vol. 7, pp. 151–184; *ibid.* (1957), vol, 9, p. 3.
10. J. C. Lewin, *Geochim. Cosmochim. Acta* **21**, 182 (1961).
11. D. H. Cushing, *Fish Invest. Min. Agr. Fish. Food Ser. II* **18**, 1 (1955); *ibid.* **22**, 1 (1959).
12. M. M. Mullin, *Limnol. Oceanogr.* **8**, 239 (1963).
13. ——, in *Marine Science*, H. Barnes, Ed. (Allen and Unwin, London, 1966), pp. 545–554.
14. T. J. Smayda, in *Oceanography and Marine Biology*, H. Barnes, Ed. (Allen and Unwin, London, 1970), vol. 8, pp. 353–414.
15. G. R. Forster, *J. Mar. Biol. Assoc. U.K.* **32**, 315 (1953).
16. T. J. Smayda, *Limnol. Oceanogr.* **14**, 621 (1969).
17. R. Simonsen, *"Meteor" Forschungsergeb. Reihe D* **1**, 85 (1967).
18. GIK No. 10,786–1 (0 to 300, 300 to 600, 600 to 900, 900 to 1200, 1200 to 1500, 1500 to 1800, 1800 to 2100, 2100 to 2400, 2400 to 2700, 2700 to 3000, 3000 to 3300, 3300 to 3600 m deep; Simonsen phytoplankton multinet, 41-μm mesh size).
19. H. B. Moore, *J. Mar. Biol Assoc. U.K.* **17**, 325 (1931).
20. H. W. Harvey, *ibid.* **22**, 97 (1937); E. R. Brooks and M. M. Mullin, *Limnol. Oceanogr.* **12**, 657 (1967); M. R. Reeve, *J. Exp. Biol.* **40**, 215 (1963).
21. H.-J. Schrader, *Proceedings of the Second Planktonic Conference of Rome 1970*, A. Farinacci, Ed. (Edizioni Tecnoscienza, Rome, 1971), vol. 2, pp. 1139–1147.
22. O. G. Kozlova and V. V. Mukhina, *Int. Geol. Rev.* **9**, 1322 (1966).
23. Full documentation will be given in (*7*).
24. S. E. Calvert, in *Marine Geology of the Gulf of California*, T. H. van Andel and G. G. Shor, Eds. (American Association of Petroleum Geologists, Tulsa, Okla., 1964), pp. 311–330.
25. This investigation was supported by the Deutsche Forschungsgemeinschaft. I thank Prof. E. Seibold for help and encouragement during this investigation, and W. H. Berger for fruitful discussions on silica dissolution and for criticizing the manuscript.

Part VII

DIAGENESIS AND OIL-
RESERVOIR CHARACTERISTICS

Editor's Comments
on Paper 20

20   DUNN

Excerpts from *North Sea Basinal Area, Europe—An Important Oil and Gas Province*

Diagenesis studies of deep-sea sediments have given much attention to changes in porosity. Porosity measurements are standard practice on board the DSDP vessel *Glomar Challenger*. SEM photographs have revealed the reasons for the decreasing porosity with increasing diagenesis. Though high porosity does not necessarily mean high permeability, within a specific rock type, a decrease in porosity due to diagenetic processes certainly means a decrease in permeability. Consequently, the study of the influence of diagenesis on porosity is of great economic importance in evaluating the characteristics of oil-reservoir rocks.

The final paper in this volume (Paper 20) has been chosen as an example of the application of modern diagenesis studies in providing an explanation for differences in reservoir-rock characteristics. In his paper on the North Sea oil and gas province, Dunn describes, among others, the Ekofisk Field. The primary hydrocarbon reservoir of this field is a Danian pelagic chalk, in which there are porous and tight zones. SEM photographs revealed that secondary calcite precipitation had reduced the porosity in the tight zones.

As only part of Dunn's paper discusses these diagenetic aspects, only the relevant pages, including those with SEM photographs, are being reproduced.

20

Reprinted from pp. 69, 70, 87–90, 91, and 96 of *Norges Geol. Undersøkelse* **316**:69–97 (1975)

North Sea Basinal Area, Europe
— an Important Oil and Gas Province*

W. W. DUNN

Dunn, W. W. 1975: North Sea basinal area, Europe – an important oil and gas province. *Norges geol. Unders. 316*, 69–97.

The North Sea covers the offshore part of a major sedimentary basin which extends from Norway, Scotland, and Denmark across northern Germany and the Netherlands into eastern England. Information gained from exploration efforts over the last 10 years shows that the North Sea covers several smaller sedimentary and structural basins of different geologic ages, but for descriptive purposes these can be divided into southern and northern areas. The rocks range in age from Paleozoic to Tertiary and consist of sandstone, shale, carbonates and evaporites. The most important reservoir rocks are the Lower Permian sandstones of the Rotliegendes Formation, the Upper Permian dolomites of the Zechstein Formation, the Triassic sandstone of the Bunter Formation, the Jurassic sandstones, the Maestrichtian–Danian chalk, and the Paleocene and Eocene sandstones. Significant shows of hydrocarbons have been found in 10 formations. The main source rocks are Carboniferous coal measures, Mesozoic shale and carbonates, and Tertiary shale and carbonates. The significant traps are folds and fault blocks associated with salt movement and basement faulting.

Exploration activity received its initial impetus in 1959 from the discovery of a major gas field, Schlochteren, onshore in northern Netherlands. In the early 1960s the passing of legislation favorable for the acquisition of exploration acreage offshore added further stimulus to the exploration pace. The majority of this activity was concentrated initially in the southern area, and resulted in the discovery of the first offshore commercial gas field at West Sole in 1965. This discovery was followed rapidly by other gas discoveries in the United Kingdom and the Netherlands culminating in the Leman Bank field, a major gas field by world standards. Interest and activity lagged, however, in the northern area despite reported small oil and gas discoveries in Denmark, and the discovery in 1968 of the Cod gas-condensate field in Norway. In late 1969, oil production was established at the Ekofisk field in Norway. With this discovery and subsequent confirmation as a major field, exploratory interest has shifted to the north.

W. W. Dunn, Philips Petroleum Company, Bartlesville, Oklahoma 74004, U.S.A.

[Editor's Note: Material has been omitted at this point.]

70   W. W. DUNN

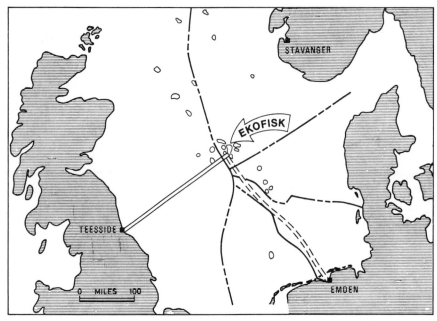

Fig. 1. Location of the Ekofisk field, North Sea, showing some of the principal oil fields. The 34" oil pipeline to Teesside and the 36" gas pipeline to Emden are also indicated.

[*Editor's Note:* Material has been omitted at this point.]

The Ekofisk and neighboring fields are located in the deepest part of the Tertiary Basin in an area containing over 1,300 feet of Danian and Paleocene sediments, approximately half of which consist of the Danian Chalk section, and the other half of Upper Paleocene clastics. The lower part of the Danian commonly contains reworked Upper Cretaceous fossils which appear also at the top of the Upper Paleocene. It can thus be assumed that at the beginning of the Danian, the Upper Cretaceous Chalk was subjected to submarine erosion wherever it may have been structurally elevated. Isopachous mapping of the Upper Cretaceous suggests the existence of elevated areas which may have been shallow relative to sea level and which were the erosional source areas for part of the Danian section. The bulk of the section, however, was deposited in deep water and consists principally of coccolith remains and lime-muds. The Danian Chalk is the primary hydrocarbon reservoir in the Ekofisk area.

The transition from Danian to Upper Paleocene sediments is usually marl, but in places may be shales and sand. The overlying Upper Paleocene section is predominantly clastic. Shale and silt are characteristic on the margins of the basin and in areas of isopach thins. Sand becomes an important component in areas of thick section, for example in the Cod Field area. Thin marl and lime-stone beds constitute a minor part. The Upper Paleocene section, like the Danian, is a deep water deposit. The sands in the axial part of the Northern Tertiary Basin and in the Cod Field are turbidites — a feature which is

362

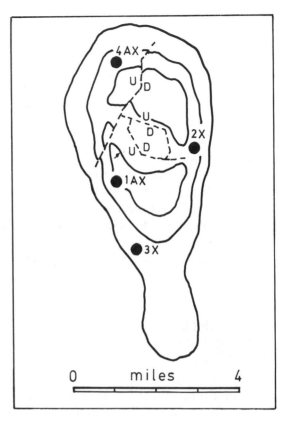

Fig. 18. Ekofisk Field, structural form-lines and positions of the 4 original wells.

4AX

U D

U
  D

U D

2X

1AX

3X

0   miles   4

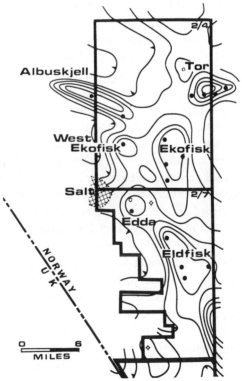

2/4

Albuskjell

Tor

West
Ekofisk

Ekofisk

Salt

2/7

Edda

Eldfisk

NORWAY
U K

0   6
MILES

Fig. 19. Greater Ekofisk, structural form-lines to top of Danian.

consistent with the deep water origin of the total section. Sand sources may have been Mesozoic and older sediments eroded from the Mid-North Sea High. Other sources and other environments of deposition are, of course, possible and may be found outside the immediate confines of the Northern Tertiary Basin. The probable distribution of the sands is north-westward toward the Forties Field.

The structural configuration on top of the Danian in the Ekofisk Field is indicated in Fig. 18, and the line contours of the Greater Ekofisk area are drawn in Fig. 19. The main Ekofisk structure is north-south oriented and 7.5 miles long by 4.5 miles wide. An interesting aspect is that the productive limits do not appear to be entirely structurally controlled. Test information indicates:

1. that in the Ekofisk Field we are dealing with one reservoir only, and
2. that Ekofisk and West Ekofisk are separate reservoirs. Since the productive column exceeds the spill point, there should be continuity between Ekofisk and West Ekofisk. The probable explanation is that the reservoirs are controlled by porosity and permeability.

Fig. 20 shows a suite of logs over the productive interval of an Ekofisk well. Note the zone of reduced porosity from about 10,435' to 10,510'. This interval can be recognized in all wells drilled to date and was originally thought to separate the upper and lower productive intervals into two reservoirs. Core analysis data would tend to substantiate this contention but test data do not.

The reservoir is a chalky limestone of Danian and Upper Cretaceous age with the high porosities and low permeabilities characteristic of this type of microgranular sediment. Intensive fracturing increases the low primary permeability from less than 1 millidarcy to an average of 10–12 millidarcies. Looking at the lithological characteristics of this carbonate in more detail and in particular at the distinction between 'porous' and 'tight' zones, the fine-grained, homogeneous limestones of the two zones are surprisingly similar in appearance even though their porosities are $> 30\%$ and $< 10\%$, respectively.

It was obvious that conventional thin-section examination would not give the resolution necessary to study such a fine-grained formation, and so a high-powered scanning electron microscope, SEM, was used to investigate the difference between the high and low porosity intervals. In the Fig. 21 photomicrograph, both the porous (32.7%), right, and the tight (8.2%), left, look almost alike. They both basically have the appearance of a foraminiferal micrite, with foraminifera of deep-water facies. Figs. 22 and 23 show with increasing SEM magnification two samples, one of high porosity (32.7%, right) and the other of low porosity (8.2%, left). These Figures illustrate, I believe most spectacularly, the difference between the productive and highly porous section, and the non-productive low porosity section. The highly porous section consists practically exclusively of coccolith fragments and platelets, whereas the low porosity sample clearly shows the secondary calcite growth which reduces the porosity.

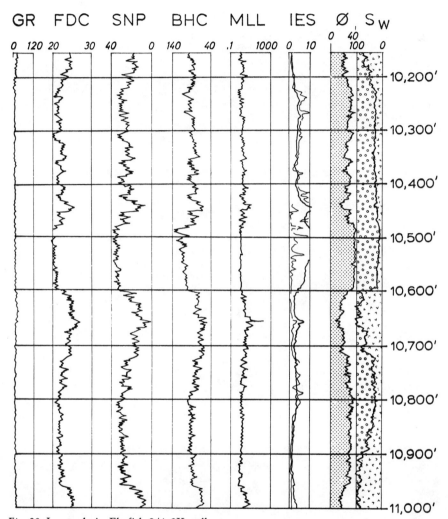

Fig. 20. Log analysis, Ekofisk 2/4–2X well.

[*Editor's Note:* Material has been omitted at this point.]

Fig. 21. SEM photomicrograph of an Ekofisk Danian limestone.

Fig. 22. SEM photomicrograph magnification of the fields indicated in Fig. 21. Danian limestone, Ekofisk.

Fig. 23. SEM photomicrograph magnification of the fields indicated in Fig. 22. Danian limestone, Ekofisk. For brief explanation, see text.

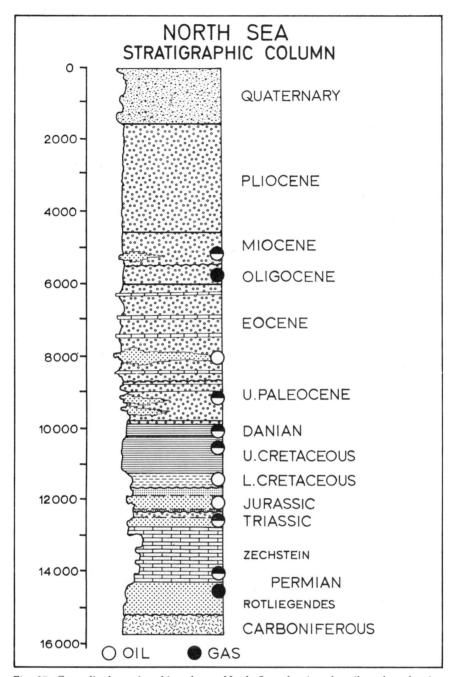

Fig. 27. Generalized stratigraphic column, North Sea, showing the oil- and gas-bearing formations. Thicknesses in feet.

AUTHOR CITATION INDEX

SUBJECT INDEX

For quick reference, see Table 2, page 7.

375

About the Editor

GERRIT J. VAN DER LINGEN is Research Sedimentologist at the Sedimentation Laboratory of the New Zealand Geological Survey, University of Canterbury, Christchurch, New Zealand. He has held this position since 1965. From 1961 to 1965 he worked for the Surinam Government Geological and Mining Service (South America).

His interest in the diagenesis of deep-sea sediments was a direct result of his participation in two cruises of the Deep Sea Drilling Project (legs 21 and 30). He was coauthor of two publications on this subject for the *Initial Report Series* of the DSDP.

Dr. van der Lingen received his geological education at Utrecht University in the Netherlands. His Ph.D. thesis, completed in 1960, was on Paleozoic sediments in the Spanish Pyrenees.

In 1971 he held a fellowship in West Germany, awarded by the Deutscher Akademischer Austauschdienst. He has also been on several lecture tours in the United States, Europe, and Asia. Internationally, he is active in several functions: Council-member of the International Association of Sedimentologists, member of the Editorial Board of the journal *Sedimentary Geology*, and New Zealand Correspondent for the Royal Geological and Mining Society of the Netherlands, and the Australasian Sedimentologists Group.